W9-AFB-981

MOONS AND PLANETS

O vast Rondure, swimming in space
Covered all over with visible power and beauty,
Alternate light and day and the teeming spiritual darkness,
Unspeakable high processions of sun and moon and countless stars above,
Below, the manifold grass and waters,
With inscrutable purpose, some hidden prophetic intention,
Now first it seems my thought begins to span thee.

Down from the gardens of Asia descending
Adam and Eve appear, then their myriad progeny after them,
Wandering, yearning, with restless explorations,
with questionings, baffled, formless, feverish,
with never-happy hearts, with that sad incessant refrain,
—"Wherefore unsatisfied soul? Whither O mocking life?"

Ah who shall soothe these feverish children?
Who justify these restless explorations?
Who speak the secret of the impassive earth?

Yet soul be sure the first intent remains, and shall be carried out,
Perhaps even now the time has arrived.
After the seas are all crossed,
After the great Captains have accomplished their work,
After the noble inventors,
Finally shall come the poet worthy that name . . . .

Walt Whitman

Wynken, Blynken, and Nod one night,
Sailed off on a silvery shoe.
Sailed on a river of misty light
Into a sea of dew.
"Where are you going and what do you wish?"
The old moon asked the three.
"We have come to fish for the herring fish
That live in this beautiful sea.
Nets of silver and gold have we,"
Said Wynken, Blynken, and Nod.

WILLIAM K. HARTMANN

MOONS AND PLANETS: An Introduction to Planetary Science

BOGDEN & QUIGLEY, INC.
PUBLISHERS

Tarrytown-on-Hudson, New York / Belmont, California

Cover design by Winston G. Potter

Text design by Science Bookcrafters, Inc.

Library of Congress Catalog Card No.: 70-170777

Standard Book No.: 0-8005-0032-6

Printed in the United States of America

1 2 3 4 5 6 7 8 9 10—76 75 74 73 72

For G
A M
Y
L
E, who helped

PREFACE

THE LAST DECADE of space exploration has revealed a new solar system. This book is for any reader—whether scientist, amateur astronomer, or student—who wants a contemporary view of that new solar system.

In the more traditional disciplines, a fixed college level dictates the mathematical content, number of references, and general coverage for an author. However, this interdisciplinary field—so new that its boundaries are scarcely defined—may be encountered by students for the first time as freshmen or first-year graduate students. This consideration has made designing a text difficult.

Probably no one has a wide-enough background to do full justice to such a broad subject. Yet the need for a unified survey remains, because collections of papers by various authors lack the cohesion necessary for anyone new to the field; most astronomy and geology textbooks devote insufficient space to planetary exploration; and finally few books bring together all the data gathered in the last few years on the moon, in space, and from ground-based observatories. Recently I was again reminded of the need for such a unified survey as I listened to meteorologists, geologists, and spectroscopists trying to agree on the interpretation of data just transmitted from the Mariner 9 spacecraft in orbit around Mars.

In response to this need, my publishers and I opted for a nonmathematical, descriptive presentation which we hope will interest students and general readers, regardless of their background. For this reason the first three chapters stress background material and definitions. The remaining chapters trace the development of the solar system chronologically. Individual sections lead the reader from an elementary survey to an account of problems now at the frontiers of research. Questions and problems, many with solutions, appear throughout the text and are set off by stars, ★. Paragraphs explaining mathematical theory are also starred to indicate that they may be skipped without destroying the continuity. The index lists all technical terms and thus serves as a glossary. Detailed references are given to encourage readers to pursue the various topics in greater detail.

My attempt to fill this gap would not have been possible without the encouragement of a number of colleagues who reviewed the chapters relating to their own fields. Although I accept responsibility for the final content, I thank the following in this regard: Edward Anders, Thomas Arny, Michael Belton, Alan Binder, Clark Chapman, Dale Cruikshank, Donald Davis, Tom Gehrels, Brian G. Marsden, Thomas B. McCord, Carl Sagan, Spencer Titley, Charles Wood, John Wood, James H. Zumberge, and especially Elizabeth Roemer and T. S. Smith, who offered valuable suggestions on the entire manuscript.

My thanks go also to Faye Larson and Micheline Wilson for invaluable help in typing; to Joyce Rehm for assistance in preparing photographic material; to Edward Quigley, Robert Fletcher, Barbara Zeiders, and Lois Lasater for their technical assistance and cooperation during publication; to many close friends for encouragement; to my family for their understanding; and especially to my wife for assistance in all stages of the effort.

Pasadena, California WILLIAM K. HARTMANN

CONTENTS

INTRODUCTION

FITTINGLY ENOUGH, this book was started on July 16, 1969. That morning in Florida, Armstrong, Aldrin, and Collins sat in their command module atop the 320-foot Saturn V. Walter Cronkite announced that there was one minute to lift-off. Lift-off came with a burst of smoke and brilliant flame. The first flight to the moon had begun. At that moment we didn't know if they would make it.

Apollo 11 achieved lunar orbit on July 20, 1969. The ship came around the limb of the moon and reestablished contact with Houston. They were "go" for the landing. Armstrong and Aldrin left Collins for their long voyage down. Newsmen Huntley and Brinkley, who had been keeping up a continuous commentary for hours, announced that there was simply nothing more they could say during these historic minutes; so they let us listen to the dialogue between Houston and the LM. "Picking up some dust. . . ." Contact. "Tranquillity Base here. The Eagle has landed." 1:18 P.M. M.S.T.

At 7:30 P.M. M.S.T. they opened the hatch. Armstrong came down the ladder. He moved faster and surer than we had expected. He stepped down onto the landing pad and as quickly off, with his left foot, onto the moon, as if there were nothing to it. "That's one small step for a man; one giant leap for mankind." 7:56 P.M. M.S.T., July 20, 1969.

WHY STUDY THE PLANETS?

That week will not be forgotten. The President called it the greatest event since creation. We witnessed an evolutionary quantum jump: man's first step across the void from the world where he first appeared to another world.

What can we hope to find there? We hope to find knowledge of how planets evolve; knowledge about what governs their crustal structure and their climates; knowledge that can help us live more successfully in our own environment. Eventually we may even find new worlds to live in or other life to communicate with. We hope to leave for our children better answers to the questions that men have asked for 10,000 years: questions about where we and our world came from; the kinds of questions that someone tried to answer millenia ago in a book called Genesis—the same book that three men read to us as they made the first manned flight around the moon on Christmas eve, 1968.

Our generation is faced with the problem of whether to turn inward upon itself and surrender to what my congressman called "a serious crisis of the spirit," in which ". . . for the first time since the early 1930's . . . there is a real questioning about whether life is really going to be better" (Udall, 1969); or whether somehow to acquire the knowledge and organization and courage to go ahead.

The exploration of space has already given us a new perspective on this problem, as shown by two passages quoted by NASA Associate Administrator, Oran Nicks (1970). The first passage was written in 1948 by British astrophysicist Fred Hoyle:

> Once a photograph of the earth, taken from the outside, is available—once the sheer isolation of the earth becomes plain—a new idea as powerful as any in history will be let loose.

The second passage, written after the exploration of space had become reality, is quoted from a letter by John Caffrey:

> I date my own reawakening of interest in man's environment to the Apollo 8 mission and to the first clear photographs of the earth from that mission. My theory is that the views of the earth from that expedition and from the subsequent Apollo flights have made many of us see the Earth as a whole, in a curious way—as a single environment in which hundreds of millions of human beings have a stake. I suspect that the greatest lasting benefit of the Apollo missions may be, if my hunch is correct, this sudden rush of inspiration to try to save this fragile environment . . . if we still can.

The exploration of space may affect man's outlook in another way. Catherine Drinker Bowen (1963) writes of the sense of optimism engendered in Europe by the exploration of the new world in the late 1500s:

> Were God's heavens then no longer immutable? And not only heaven but earth was shifting its geography. Francis Drake sailed home, having seen the limitless Pacific . . . People spoke of America as today we speak of the moon, yet far more fruitfully. [Francis Bacon spoke of] "That great wind blowing from the west . . . the breath of life which blows on us from that New Continent. . . ." Columbus, he said, had made hope reasonable.

Contemporary scientists have sounded a similar theme while reflecting upon the lunar landings. Freeman Dyson (1969), of the Institute for Advanced Study, has written:

> We are historically attuned to living in small exclusive groups, and we carry in us a stubborn disinclination to treat all men as brothers. On the other hand, we live on a shrinking and vulnerable planet which our lack of foresight is rapidly turning into a slum. Never again on this planet will there be unoccupied land, cultural isolation, freedom from bureaucracy, freedom for people to get lost and be on their own. Never again on this planet. But how about somewhere else?
>
> . . . Many of you may consider it ridiculous to think of space as a way out of our difficulties, when the existing space program . . . is being rapidly cut down, precisely because it appears to have nothing to offer to the solution of social problems . . . If one believes in space as a major factor in human affairs, one must take a very long view.

Our goal in studying planets is not just the collection of new facts and third-order theories about the universe, but the gaining of a basic understanding of how our world works. Perhaps just as important to us is an awareness that there really is a frontier there—scientifically, psychologically, and spatially.

A VIEW OF SCIENCE

Many people have a misconception that science is like a tree branching outward; in time the divisions of science branch continually finer and finer so that science becomes more and more complex, more and more specialized. Science is like this only in a limited sense: The data proliferate.

Science should really be viewed in the opposite way, as a process that synthesizes many complex observations into theories that enable one man to understand a wide range of natural phenomena. By "understand" I mean predict the right answers to problems. Take mechanics, for example. During the Renaissance, scientists made endless observations of levers, falling objects, inclined planes, rotation, orbital motion, and so on. Even the most learned men were not able to predict a wide variety of experimental results until Newton and his successors saw the interrelationships among the observations. Today, a college sophomore studying physics can "understand" (i.e., quantitatively predict) most of the phenomena Newton studied, and a senior's understanding surpasses that of Newton.

In other words, it is like working our way *down* the tree instead of up. First, we can understand only individual events—the twigs. Then we see how they all join together and we can grasp a branch of knowledge, and so on until we can understand whole portions of the tree. (Nobody has yet reached the trunk!)

This is the approach we must use in our study of planetary science. Here on earth we have long known specific examples of meteorological, geological, and geophysical effects. Now, as we look out toward the planets, we realize that we can observe other examples of these effects under different conditions in different environments. By such observations we begin to understand the laws and relationships that eventually help us to explain the nature of the solar system in general and the earth in particular.

I hope in this book to illuminate differences and similarities among planets, to enable the reader to command a wide range of facts and effects. In keeping with this approach, certain simple physical concepts will be stressed that allow us to understand, qualitatively at least, varied solar-system phenomena. Students of science should master the sections headed *Mathematical Theory,* which will provide not only a qualitative understanding, but quantitative insight into many of these phenomena. The "theory" sections may be skipped, however, without affecting the continuity.

PLANETARY SCIENCE

Planetary science, *planetology* as it is often called, may be thought of as the physics of planets. Originally, *physics* was defined as "the branch of knowledge treating the material world and its phenomena." In that sense all science is physics. The discipline now commonly called "physics" concerns quite isolated systems in rather ideal circumstances. The physicist may study friction, or turbulent motion, or the atomic nucleus, but he always tries to isolate the specific phenomenon as much as possible in order to distinguish the basic laws of the particular system.

Other sciences are forced to deal with the physics of more complicated systems in the real world, where many phenomena may operate at once and specific subjects cannot be isolated. Chemistry, for example, is the physics of matter interacting at the electronic level. Biology is the physics of life. Astronomy is the physics of stars and stellar systems. Meteorology is the physics of the atmosphere. Geophysics is the physics of the earth as a body. Geology is the physics of the earth's outer layers, in particular the crust. In this sense planetary science is the physics of planets.

To study the material realm, we must consider it as a four-dimensional system—it extends over three dimensions in space and a fourth in time. This means simply that nothing—not even a whole planet—is static; everything evolves. This is the view we shall take in this book. The planets are to be studied not as static three-dimensional spheres as they appear at this epoch in time but as evolving systems. In fact, the book is laid out in a more or less chronological pattern of topics reflecting the origin and evolution of our solar system.

The four-dimensional approach is being applied in modern studies of the earth as well as other planets. Traditional geological investigations were often limited to two-dimensional traverses over two-dimensional landscapes, the results being plotted on two-dimensional maps. Little evidence was gathered by drilling into the earth's crust (the third dimension) or from radioisotope dating techniques (the fourth). An excellent example of the new approach is the contemporary geological study of the volcano Kilauea, in Hawaii. The whole mountain has been instrumented with seismographs, tiltmeters, and other geophysical instruments whose data output is transmitted to a central observatory on the rim of the main crater. You can stand in one room and watch an array of graphs which are recording not just the two-dimensional scene a man would perceive walking on the volcano's surface, but the workings of its interior. Motions of molten lava, for example, can be traced deep below the mountain's summit as the lava works its way toward the surface.

Planetary science is a highly interdisciplinary field; most scientists now working in the field were trained in one of the older disciplines. Geologists interpret the lunar surface, meteorologists construct theoretical models of the Venus atmosphere, and astronomers observe the spectrum of Mars. The planetologist attempts to achieve whatever synthesis can be found: He considers the rate of impact of asteroidal fragments and comets upon the moon; he asks why Venus should have a carbon dioxide atmosphere while the earth (which is about the same size) has a radically different nitrogen–oxygen atmosphere; he expects some day to receive samples of Martian rocks and gas rather than rely on earth-based observations.

Thomas Kuhn points out in his essay on scientific revolutions (1962) that a new field of scientific inquiry is not established until the participants agree on a body of basic facts and theories and approaches to their work. Planetary science is such a new field, and its agreed-upon body of knowledge is only now emerging. Planetary scientists are not yet in complete agreement on the processes by which planets are formed; we cannot explain why the continents drift, and controversy continues about the cause of some of the moon's most important features. New journals, such as *Icarus, Planetary and Space Science, Earth and Planetary Science Letters,* and *The Moon* are emerging in an effort to consolidate planetary studies; in the meantime planetary science papers are still being published in such widely scattered journals as *Science, Nature, Journal of Geophysical Research, Geochimica et Cosmochimica Acta,* and *Astrophysical Journal*. Certain areas of vigorous study have come to have names of their own. *Space science* has been used to designate the study of the interplanetary medium, high atmosphere, solar–planetary interactions, and cosmic rays; *astrogeology* has been used to designate geological and stratigraphic studies of lunar and planetary surfaces and is a popular term within the United States Geological Survey; and *xenology* is used to refer to studies of xenon and its isotopes, which are important in determining the early history of meteorites and the planets.

In these subdisciplines, a great deal of work is now being published simply because the measurement techniques are agreed upon, the necessary instruments are available, and some theoretical foundation has been laid. In this book, however, we shall focus neither on the methodology nor the highly developed subdisciplines but on the fundamental problems of planetary development.

EVOLUTIONARY VIEWPOINT OF THIS BOOK

This book is arranged from a chronological or evolutionary viewpoint. After early chapters describing basic observational facts, definitions, and matters of celestial mechanics, we begin with a view of the origin of the solar system (Chapters 4 through 6). Small bodies of the solar system are then discussed, with a reminder that they give evidence on the earliest state of the planets (7 through 9). Planets themselves are then discussed (10 and 11), along with their atmospheres (12). Finally, we consider the development of life (13).

As a final introductory note, consider a sticky philosophic problem. Most of our contemporaries would not quarrel with the assertion that we live in an evolving system. Yet this is not self-evident, as the British philospher and cosmologist H. Dingle reminds us (1960):

> Nearly 100 years ago Philip Gosse, in order to reconcile the facts of geology with the Hebrew scriptures, advanced the theory that, . . . "there had been no gradual modification of the surface of the earth, or slow development of organic forms, but that when the catastrophic act of creation took place, the world presented, instantly, the structural appearance of a planet on which life has long existed." The beginning of the universe on this theory occurred some 6000 years ago. There is no question that the theory is free from self-contradiction and is consistent with all the facts of experience we have to explain; it certainly does not multiply hypotheses beyond necessity since it invokes only one; and it is evidently beyond refutation by future experience. If, then, we are to ask of our concepts nothing more than that they shall correlate our present experience economically, we must accept it in preference to any other. Nevertheless, it is doubtful if a single person does so.* It

* Here the philosopher is optimistic. Although this "theory" is hardly popular in our society, readers who have traveled in America by car will at-

would be a good discipline for those who reject it to express clearly
their reasons for such a judgment. . . .

Can we—as "enlightened" planetary scientists—express clearly our reasons for judging that evolution of the cosmos is a meaningful concept? We find in nature landscapes showing ordered strata, strata showing fossils that proceed in an orderly sequence terminating in the present, rocks containing patterns of radioactivity that give the same sequence and consistent ages; and we find in physical theory a scheme that enables us in our minds—or rather our computors—to construct models that begin with what appear to have been the ancient conditions and then evolve into the present conditions.

This internal consistency of our evolutionary viewpoint does not prove that it is the ultimate truth. However, it does give us assurance that our philosophy is pragmatic and that future discoveries will continue to refine present results. This is the strength of the scientific method: It continues to converge on a more and more detailed, consistent, satisfying, and practical view of the universe. With a profound interest and a rather religious sense of approaching an ultimate reality we pursue the evolutionary thread that leads backward to what seems—according to recent results—to have been the creation of the universe and leads forward into the future.

References

Bowen, C. D. (1963) *Francis Bacon* (Boston: Little, Brown and Company).

Dingle, H. (1960) "Philosophical Aspects of Cosmology," *Vistas in Astronomy, 1,* 162.

Dyson, F. (1969) "Human Consequences of the Exploration of Space," *Bull. Atomic Scientists, 25* (Sept.), 8.

Kuhn, T. S. (1962) *The Structure of Scientific Revolutions* (Chicago: University of Chicago Press).

Nicks, O. W., ed. (1970) *This Island Earth* (Washington, D.C.: *NASA SP-250*).

Udall, M. K. (1969) "The Environment—What You Can Do," *Congressman's Report, 9* (May 20).

---

test that dominating the American evening radio waves in vast areas of the country are fundamentalist evangelical radio programs that insist on precisely this view—that all the seemingly evolved strata, fossils, and radioisotopes were put into their complex pattern just to tempt scientists into error. We begin our planetary studies with our feet planted in an interesting sociological soil with roots extending far back in history.

BASIC INFORMATION

THE PURPOSE OF THIS CHAPTER is to provide a basic description of the planets for readers unfamiliar with the solar system.

"THE NINE PLANETS"

The solar system is commonly said to have nine planets. In order, moving out from the sun, they are Mercury, Venus, Earth, Mars, Jupiter, Saturn, Uranus, Neptune, and Pluto. A traditional mnemonic device for remembering this sequence is "Men Very Early Made Jars Stand Upright Nicely (Period)." (Surely today's students can do better than this!) Because confusion often sets in near *S*aturn, *U*ranus, and *N*eptune, remember that the *SUN* is in the system, too.

The first four planets are sometimes called the *terrestrial planets* because of their nearness to the earth. The four planets from Jupiter through Neptune are sometimes called the *giant,* or *Jovian, planets* because of their size and similarity to Jupiter.

One might think that the concept "planet" could be easily defined. After all, there are nine bodies that seem to be uniquely distinguished by size, position, orbit, and so on. But the situation is not really so clear.

Among the ancients, a "planet" was any of seven bodies that changed position from day to day among the stars. These seven were the sun, the moon, and the first five extraterrestrial planets out to Saturn. The rest had not yet been discovered.

The modern definition is that *a planet is any body excepting a comet, asteroid, meteoroid, or satellite orbiting around the sun*. But even this definition is contradicted: The planet Pluto was probably once a satellite of Neptune (see Chapter 6). If we try to define planets by the regular spacing of their nearly circular orbits, we face the fact that Pluto has a markedly noncircular orbit and can on occasion come closer than Neptune; the orbits "overlap." If we try to define planets by size, a critic could point out that the third satellite of Jupiter is larger than the planet Mercury. If we try to define planets by a genetic criterion, we face the problem that origins and differences among planets, asteroids, comets, and satellites are not entirely understood.

As will be seen in Chapter 6, there are legitimate reasons for suggesting that a planet may be defined genetically as any solar system body larger than about 350 km (220 miles) in diameter. By this uncommon definition there would be some 30 "planets" in the solar system. Sometimes, to clarify these problems, the nine familiar planets are called the *principal planets*. This problem is essentially semantic and hence not too important for our purposes. Usually we shall use the first and most common modern definition quoted above, but certain other bodies (e.g., the moon) will, for convenience, sometimes be called planets.

The uncommon definition mentioned above illustrates that a planetologist has far more than nine major bodies to study and correlate. Aside from the principal planets, there are 32 satellites, three asteroids larger than 350 km in diameter, many smaller asteroids, and innumerable comets a few kilometers in diameter. In addition, uncounted pieces of rock and metal as small as 1 meter (1 yard) across and smaller are drifting in the solar system.

BODE'S LAW

Since it is useful to be able to remember the distances of the planets from the sun, *Bode's law,* a helpful aid in this task, should be memorized.

In 1772 the German astronomer Johann Bode popularized this simple empirical rule.* Write down a sequence of 4s and add the sequence 0, 3, 6, 12, . . . , doubling each time (Table 2-1). By putting in the appropriate decimal point, you get the distances of the planets from the sun (Table 2-1). These distances are thus given in terms of the mean distance from the earth to the sun,

* The rule was apparently first discovered by another German astronomer, Johann Titius; thus it is often called the Titius–Bode law.

Table 2-1
Bode's law and planetary symbols[a]

	Mercury	Venus	Earth	Mars	Asteroids	Jupiter	Saturn	Uranus	Neptune	Pluto
	4	4	4	4	4	4	4	4		4
	0	3	6	12	24	48	96	192		384
Predicted distance	0.4	0.7	1.0	1.6	2.8	5.2	10.0	19.6		38.8
Actual distance	0.4	0.7	1.0	1.5	2.8	5.2	9.5	19.2	30.0	39.4
Symbol	☿	♀	⊕	♂	①[b]	♃	♄	♅	♆	♇
No. known satellites	0	0	1	2	0	12	10	5	2	0

[a] Giving distances in terms of astronomical units.
[b] The well-observed asteroids (more than 1750 being known) are numbered; the symbol is the encircled number.

which is defined as an *astronomical unit.* The abbreviation is A.U., and one would thus say that Jupiter is 5.2 A.U. from the sun.

Bode's law lacked any theoretical justification, but it passed its first test in 1781 when the English astronomer William Herschel discovered Uranus at 19.2 A.U. Thereupon, a search was made for the "missing planet," which was supposed to lie between Mars and Jupiter. This led to the discovery of the first asteroid, Ceres, on the first night of the new century, January 1, 1801, by the Italian astronomer Giuseppe Piazzi. Discoveries of more asteroids followed in subsequent years.

To this day, Bode's law lacks any theoretical justification, although it has proved unsettlingly accurate. It is generally supposed that the formation of the solar system must have involved some dynamical effect that partitioned the early solar system into regularly spaced zones, each dominated by a single planet.

A set of symbols for designating the planets is the one useful contribution of astrology. The symbols are listed in Table 2-1 and form convenient subscripts and datum-point symbols in theoretical discussions and graphs.

THE PLANETS CONSIDERED FROM A DISTANCE

The following is a brief descriptive survey of the planets from a remote point of view, that is, emphasizing properties that can be measured from earth or space. In Chapters 11 and 12 we shall discuss the surface and atmospheric properties of each planet. Numerical data are tabulated separately in the Appendix.

Mercury

Mercury is the smallest principal planet and the closest to the sun. It has no detectable atmosphere. Its surface has a slightly pinkish cast, and through the telescope one can see faint darkish markings reminiscent of the appearance of the moon to the naked eye. These dark regions may be lava-covered areas similar to the dark lava plains of the moon. A comparison of the moon and Mercury is given in Figure 2-1.

Because Mercury's orbit is so close to the sun, Mercury is usually difficult to see. As a result, a number of famous astronomers lived their whole lives without ever seeing the planet. It can be seen only at twilight or in the daytime sky, and even with a telescope it is hard to observe.

Prior to the 1960s it was thought that Mercury kept the same face toward the sun at all times, just as the moon keeps the same face toward the earth. In fact, astronomers had made maps of the planet showing the markings that were supposed to be fixed in position with respect to the subsolar point. To the embarrassment of these observers, it was discovered by radar techniques

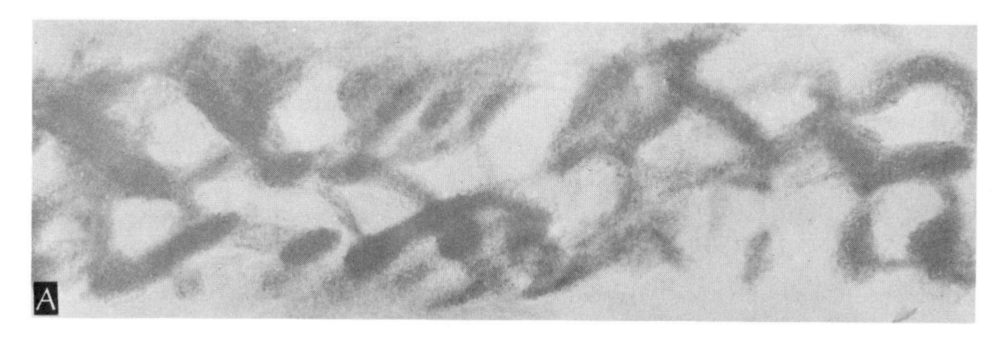

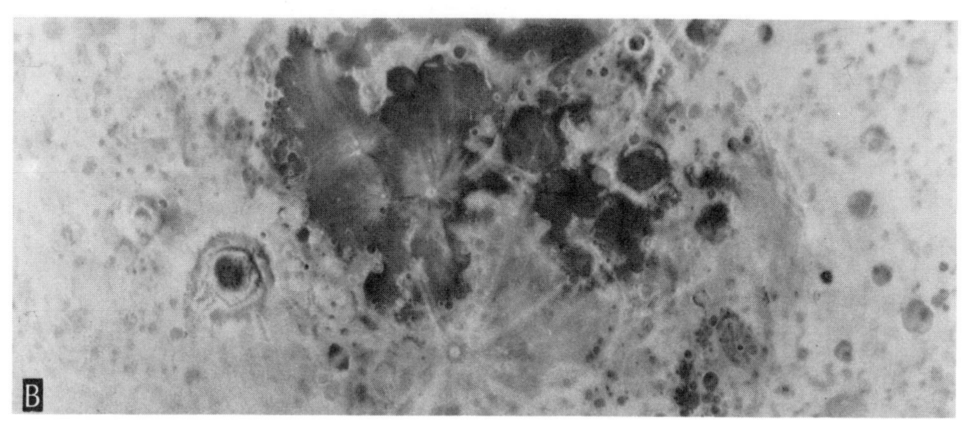

Figure 2-1

Comparison of Mercury with our moon. A: Map of Mercury, showing patchy dark markings barely visible under good conditions with large telescopes. (C. R. Chapman.) B: Map of the moon. Although the moon and Mercury are similar in many ways, the moon's dark markings are much less symmetrically distributed, being concentrated (for unknown reasons) on the side facing earth. Compare the full-moon photo, Figure 11-12. (B. Vigil, Lunar and Planetary Laboratory.) Both maps are centered on the planetary equators and extend to ±60° latitude. North is up.

(Pettengill and Dyce, 1965) that Mercury has a rotation period of about 59 days and does not keep one side to the sun. The saving grace for the visual observers was that the 59-day period causes a peculiar recurrence of configurations between the earth, Mercury, and the sun, such that on certain occasions when Mercury is favorably placed for observation, it tends to have nearly the same side toward the sun. The earlier maps of Mercury have thus been amended to take this effect into account (Cruikshank and Chapman, 1967). Figure 2-2 shows one of the best photographs presently available, barely showing the dusky surface markings.

A spacecraft mission to photograph Mercury is being planned for the mid-1970s. Such a mission might resolve the nature of the dark markings and show whether Mercury has been heavily cratered like the moon and Mars.

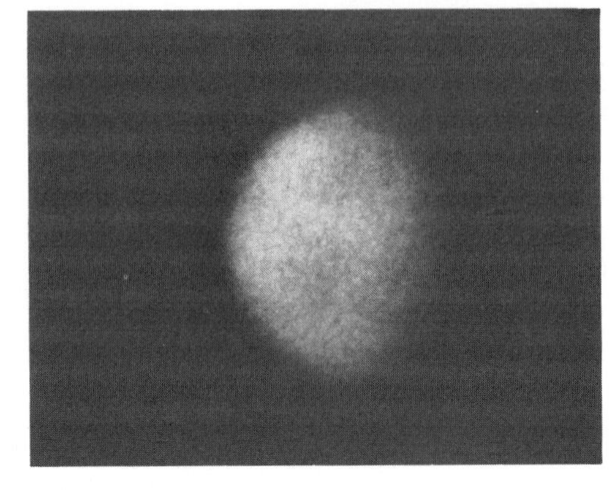

Figure 2-2
Mercury. This 1969 photo is one of the few in existence to show the planet's surface markings. (New Mexico State University Observatory.)

Venus

Venus is sometimes called earth's sister planet. It most nearly matches the earth in size and its orbit is closest to the earth's orbit. It has a very dense atmosphere composed mostly of carbon dioxide (CO_2). Opaque clouds and fog-like mists in the Venus atmosphere make the surface invisible. Through the telescope, therefore, Venus appears as a blank white or yellowish disk. Occasionally, faint nebulous cloud patterns are seen. These cloud patterns are much more prominent in ultraviolet light because some of the cloud—or atmosphere—constituents apparently absorb the ultraviolet. Photographs taken with ultraviolet filters clearly reveal the cloud patterns, which are somewhat reminiscent of the earth's cloud-circulation patterns, although the Venus clouds, as shown in Figure 2-3, tend to lack the great cyclonic whorls so characteristic of the earth's low-level clouds.

Because of the cloud cover we cannot watch the rotation of the solid surface of the planet, and until recently the period of rotation was a mystery. Repeated attempts to determine the rotation by watching motions of the cloud markings on ultraviolet photographs were frustrated by rapid changes in the clouds and their ill-defined patterns. In an interesting example of "reinforcement" of unfounded ideas in science, a rotation period of about 20 to 30 days gained popularity among one group of observers, while others favored a 24-hour period, and others a rotation that kept one side always facing the sun. In the early 1960s radar techniques were used to bounce signals off Venus, with the totally unexpected result that Venus was found to rotate in 243.1 days, not in the same direction as the earth and most other bodies, but backward, from east to west (Dyce, Pettengill, and Shapiro, 1967)! This peculiar situation may be abetted by dynamical resonance between earth and Venus (Goldreich and Peale, 1967), although the resonance is apparently not exact. There is evidence from modern ultraviolet photography that high-atmosphere cloud markings circulate in the east-to-west direction with a period of about 4 or 5 days (Smith, 1967).

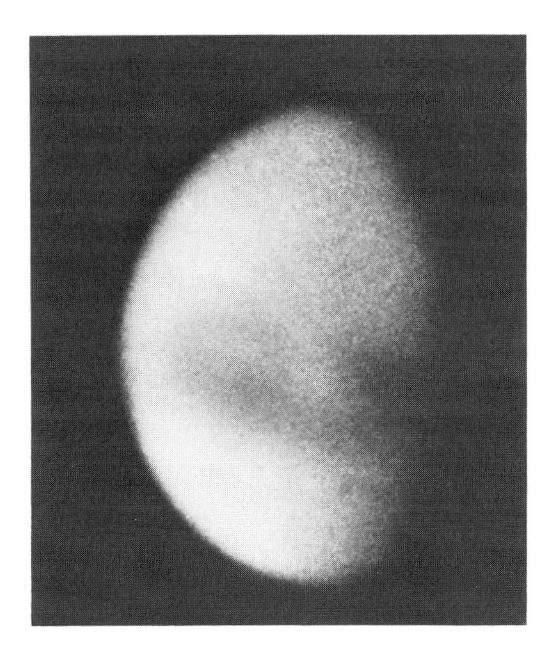

Figure 2-3
Venus, photographed in ultraviolet light, showing cloud patterns. (Lunar and Planetary Laboratory.)

Little is known about the surface of Venus. Earth-based instruments and the Soviet spacecraft that landed on the surface show that the temperature and pressure are very high, about 750°K (890°F) and 90 atmospheres, respectively. We shall discuss the cause of these high temperatures when we consider planetary atmospheres in Chapter 12. There is no liquid water and virtually no water vapor in the atmosphere, and the surface is probably dry and dusty.

Because Venus has highly reflecting clouds and approaches close to the earth it appears very bright. On occasions when it moves into a position to be our "evening star" or "morning star" (setting or rising a few hours after or before the sun), it is the third brightest object in the sky. It is then some 15 times brighter than the brightest star (Sirius) and can cast shadows.

Venus was first observed at close range on December 14, 1962, by the U.S. spacecraft Mariner II, which passed 38,854 km (21,645 miles) from the surface of the planet. The first contact with the planet was achieved October 18, 1967, when the Soviet spacecraft Venera 4 parachuted into the atmosphere and radioed back data. On December 15, 1970, Venera 7 landed on Venus.

Earth

From space, the earth is most prominently characterized by its shifting white cloud patterns against a bluish background, as shown by Figure 2-4. The atmosphere is composed mainly of nitrogen (78 percent by volume), oxygen (21 percent), and argon (1 percent). The surface is dominated by oceans of liq-

Figure 2-4
Earth, photographed by Apollo astronauts. Cloud patterns are the dominant features. (NASA.)

uid water that cover 71 percent of the surface area. The land areas are characteristically tan and greenish in color although frequently obscured by clouds.

The most important characteristic of the earth is its life. Plant and animal life is widespread, both on the land and in the oceans, although evidence of this life is not easy to detect from space. The principal threat to the life now seems to be twofold: (1) the proliferation of weapons and other technological means of accidentally modifying the entire planetary environment, and (2) famine or deterioration of the quality of life due to the population explosion. The population of the earth is roughly 3.5×10^9 (3,500,000,000), and the mean population density for the land area of the whole world is about 61 persons/mile^2, including virtually uninhabited regions. This compares to respective figures for the United States (54 persons/mile^2), India (395), and England (578), based on estimated 1966 populations. The population is expected to double in less than 40 years, even allowing for a slight decline in birth rate.

Mars

Mars is about half the size of the earth. It has a very thin atmosphere, composed mostly of carbon dioxide but with much lower air pressure at the surface than is encountered at the top of Mount Everest or by high-flying jets. A spacesuit would be needed by a man on Mars.

Mars displays a number of interesting features, which can be seen in Figure 2-5. Most prominent are the bright polar caps, which are probably com-

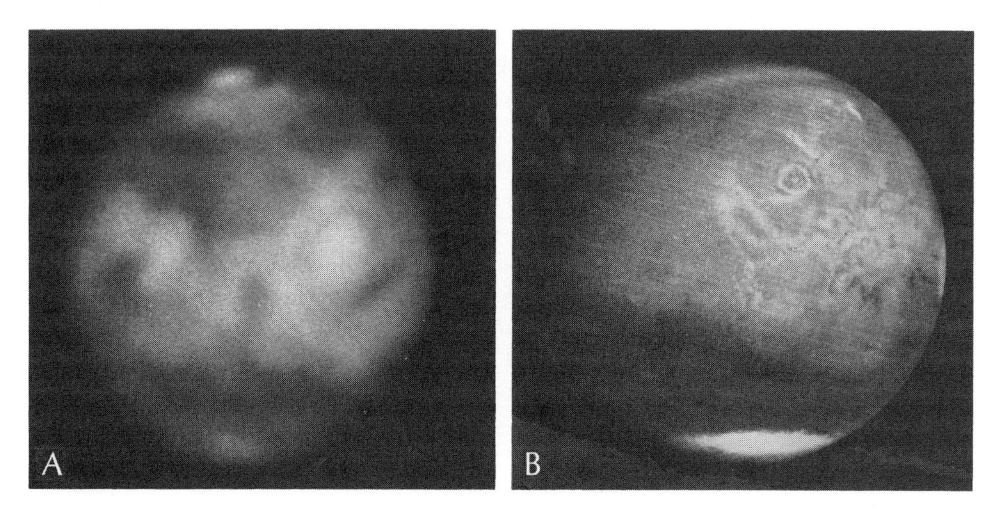

Figure 2-5

Mars. A: One of the best earth-based photos, showing a small north polar cap and associated dark "collar." (Lunar and Planetary Laboratory.) B: Spacecraft photo, showing south polar cap, haze over the north pole, and bright cloud or frost patterns too small to be seen from earth. (NASA.)

posed of thin layers of frozen carbon dioxide (CO_2), familiar to us as dry ice. The polar caps may also contain water frost (H_2O). Dark markings are superimposed on the orange "deserts" that give Mars its characteristic ruddy hue. "Desert" is probably not a bad choice of word to describe the orange areas. They are apparently composed of a powdery material that can be borne aloft in the winds of the thin atmosphere. Evidence for this is found in the yellowish clouds that occasionally form and obscure darker markings; they are thought to be dust storms. The largest observed dust storm began in September 1971 and spread until it obscured almost all of the planet's markings. Photography in November 1971 from the Mars-orbiting Mariner 9 spacecraft showed an almost completely blank disk! In addition to the yellowish clouds there are bluish-white clouds that appear to be formed by ice crystals and probably resemble thin cirrus veils on earth.

The most intriguing aspect of Mars is the possibility that primitive life exists there. One of the first observed hints of this was the changing appearance of the dark markings. During the Martian spring, as the polar cap dwindles, the nearby markings turn considerably darker. As the season wears on, the darkening extends farther from the pole toward the equator. One interpretation is that a small supply of frozen water tied up in the thin polar cap is being released and diffuses away from the pole, allowing primitive vegetation to flourish. In the Martian fall, the markings fade. Early visual observers reported the dark areas to be more green or bluish-green than the deserts, also suggesting plant life. Recent calibrated studies show that the dark areas are not green, but brown (slightly less red than the bright areas) and that the areas do not get any greener as they darken (McCord, 1969). Instead of indicating life, the changing markings may be caused simply by windblown dust deposits altered by seasonal winds.

A feature often discussed in popular literature on Mars is the "network of canals." In 1877 the Italian astronomer, Schiaparelli, gave the name "canali" to more-or-less linear, faint, dusky streaks crossing the orangish deserts. This term was poorly translated as "canals," and the American astronomer Percival Lowell popularized the "canals" in speculative writings as straight, artificial waterways constructed by Martians to irrigate their drying, dying planet. However, even before the recent Mariner photographic spacecraft flights, experienced observers had described the canals not as narrow lines but as patchy alignments of dusky features. The photographs by Mariner spacecraft confirm this; from nearby space the canals are not at all striking, simply aligned dark patches and borders between regions of different tone. The Martian canal affair came about mostly because visual observers (such as Schiaparelli and Lowell) are prone to exaggerate alignments of faint markings, apparently because of a physiological effect whereby the eye tends to interpret rough alignments as linear streaks. Figure 2-6 shows a modern map, and it can be seen that the "canals," at best, are not prominent.

Figure 2-6

Map of Mars. The nature of the Martian dark markings is unknown, with estimates ranging from ash or lava deposits to simple vegetation. North is up. Compare Figure 2-2 for similar maps of Mercury and the moon. (S. Larson, Lunar and Planetary Laboratory.)

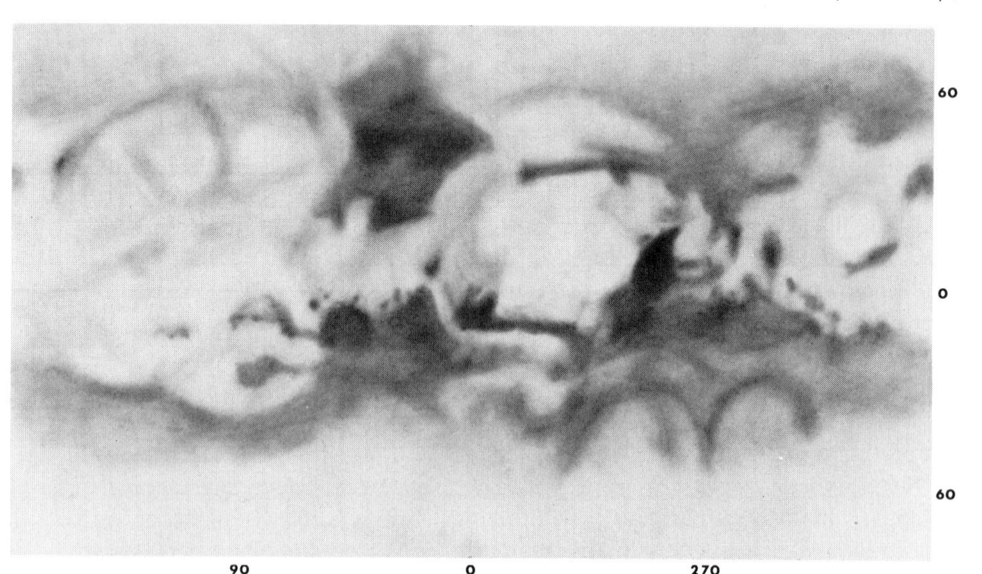

The first closeup observations of Mars took place on July 15, 1965, when the American spacecraft Mariner IV flew past and sent back photographs and measurements. Later Mariners made similar flybys and got still better data. The 1965 observations were first to show that Mars is heavily cratered, like the moon. This suggests that most planets go through an early cratering process due to meteorite impacts and some volcanic activity. On the earth, evidence of this ancient activity has been obliterated by mountain building, erosion, sediment deposits, and so on, but on Mars erosive processes were evidently less efficient and the ancient craters are still prominent. A continuing series of spacecraft studies of Mars is planned, including the Viking spacecraft, intended to soft-land on Mars. Such studies may help determine whether life exists there.

Jupiter

Jupiter is by far the largest planet, having more mass than all the other planets put together. It has more than three times the mass of the next largest planet, Saturn, and about a 20 percent greater diameter. Astronomers have noted that dynamically the solar system can be thought of as composed of two main bodies, the sun and Jupiter. However, Jupiter has only $\frac{1}{1000}$ the mass of the sun.

Jupiter has a dense atmosphere characterized by bands of clouds arranged parallel to the equator, shown in Figure 2-7. We cannot see any solid surface because of these cloud belts. The clouds are a variety of colors—oranges,

Figure 2-7

Jupiter, photographed from earth in two colors. The Great Red Spot lies near the lower right limb. Part A was made with a blue filter, causing the Red Spot to look dark. Photograph B, taken almost simultaneously, is in red light and hence shows the Red Spot as a light area. (Lunar and Planetary Laboratory.)

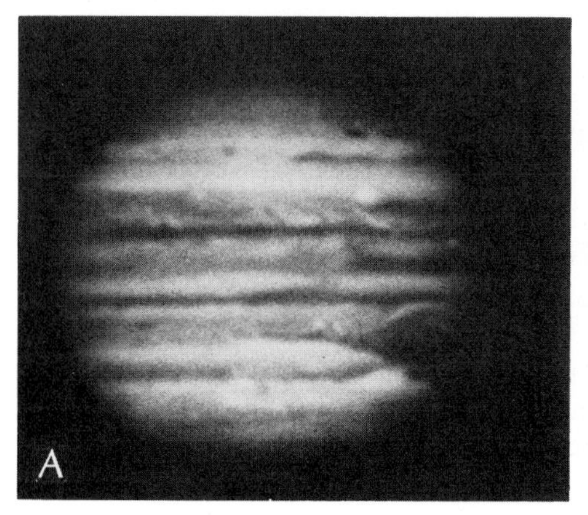

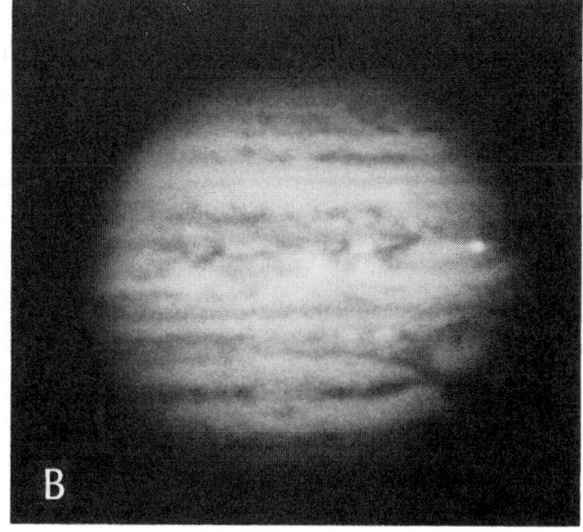

browns, tans, and reds. Various kinds of features typify these cloud belts and are known as disturbances, festoons, knots, and so on, depending on size and shape. The chemistry of the atmosphere is uncertain, but the colors must result from some particular chemical compounds. Gaseous constitutents in the atmosphere include hydrogen (H_2), ammonia (NH_3), and methane (CH_4) (Owen, 1970).

A famous feature of Jupiter is the Great Red Spot, which can be seen in Figure 2-7. This was probably first described by Jean Cassini in 1665 and was named in 1878 when it became very prominent and was rediscovered (Chapman, 1968). Recent observations by Reese and Smith (1968) show that the Red Spot is characterized by circulating currents; small clouds caught in the Red Spot spiral around it in a counterclockwise pattern reminiscent of a leaf caught in a whirlpool. It is a giant whirlpool indeed; somewhat variable in size, it can reach four times the size of the entire earth.

Jupiter and Saturn present an interesting problem with respect to rotation. Different parts of the atmosphere can move independently (with accompanying shear and turbulence) and thus it is found that the equatorial zones rotate faster than the higher latitudes. In the case of Jupiter, two different rotation rates are used to keep track of the cloud markings: $9^h50^m30^s$ for the equator (System I) and $9^h55^m41^s$ for the high latitudes (System II). The famous Red Spot drifts along with a variable period, drifting sometimes ahead of and sometimes behind other features in System II (Peek, 1958). Which of these rotation rates represents the true rotation of the solid planet? Probably none. Clearly the Red Spot cannot be related to a fixed surface feature. Radio signals from Jupiter give a slightly different period, and they may be "static" from electromagnetic phenomena of the magnetic field around Jupiter, but the search continues for a single, precisely periodic observable quantity that would mark the rotation of the solid planet.

Jupiter has 12 satellites, the largest number of any planet. There are four large ones, about the size of our moon, called the *Galilean satellites* because they were discovered by Galileo. The discovery date was January 7, 1610, and on the next night they were independently discovered by the German astronomer Marius, who named them Io, Europa, Ganymede, and Callisto (in order outward from Jupiter), after associates and paramours of the Greek god Jupiter. Another convention for naming them uses Roman numerals; the Galilean satellites are I, II, III, and IV in the order given above, and subsequent numbers are assigned in order of discovery. Faint surface markings have been suspected on the surfaces of the Galilean satellites. Figure 2-8 is a specially processed photograph of Io, indicating such markings. Number V's orbit is inside that of I. The outer seven occur in two distinct groups. Satellites VI, VII, and X move in very similar orbits; the other four move in a direction opposite to that of the rest. The orbital characterisitics of the outer satellites are quite different from the regularly spaced, circular orbits of the Galilean satellites, and at least some are suspected to be bodies captured in Jupiter's strong gravitational

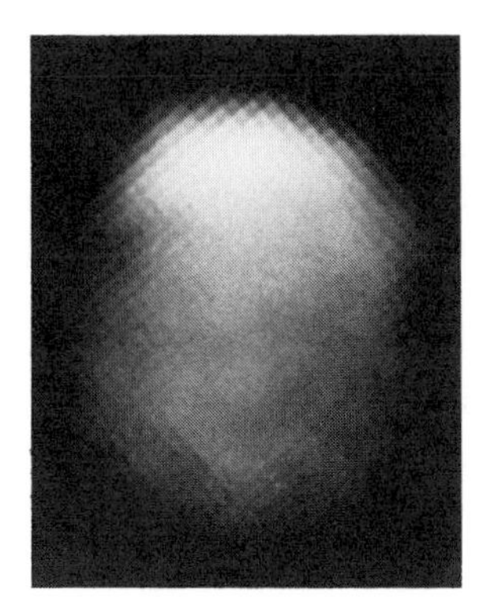

Figure 2-8

Jupiter's satellite Io, showing evidence of surface markings and a possible bright polar cap. This photograph was made by an unusual process of compositing many images, using computer removal of atmospheric blurring that degrades most astronomical photography. (Princeton University Observatory.)

field rather than "original satellites" formed as part of a single system with Jupiter. Satellite VIII has an unstable and rapidly changing orbit and was twice lost and recovered after its 1908 discovery.

It is planned to send two spacecraft, Pioneer F and G, through the asteroid belt to pass near Jupiter in the mid-1970s.

Saturn

Saturn is famous because of its rings, which are shown in Figure 2-9. When Galileo first turned his crude telescope on the planet in 1610, he could not see the rings clearly and drew Saturn as a triple planet, since the rings appeared to him as appendages, one on each side. Their true nature remained a source of

Figure 2-9

Saturn, showing cloud belts and rings. Note the shadow of the planet on the rings (upper left) and the shadow of the rings on the planet. (Lunar and Planetary Laboratory.)

controversy until the 1660s (Alexander, 1962). In 1859 the famous English physicist James Clark Maxwell argued that the rings could not be a solid plate but must be made up of innumerable particles each moving in an independent orbit around Saturn. The American astronomer James Keeler (1895) became the first to prove this observationally when he detected the varying orbital velocities of different parts of the rings; it was one of the first triumphs of spectroscopic astronomy. Although the rings are 270,000 km (170,000 miles) in diameter, they are extremely thin. Recent measures by the French astronomers Focas and Dollfus (1969) indicate that they are only about 2.8 km (1.7 miles) thick. It is thought that the rings may be debris left from a satellite that broke up (or never formed) because it was inside the Roche limit (see Chapter 3). They are apparently composed of ice or ice-coated particles. The mean particle size is unknown, proposed values ranging from microns (10^{-4} cm) to meters (Alexander, 1962, p. 440).

The rings are subdivided. The dusky outer ring is called ring A and is separated by a gap from the brighter ring B. The gap is called *Cassini's division.* On the inner side of ring B is a very faint, tenuous ring, C. In late 1969 the French observer Guerin (1970) announced the discovery of a still fainter and more tenuous innermost ring, D, which extends from the planet toward ring C but is separated from C by another gap. The new ring D has been confirmed photographically but cannot be seen visually even in a large telescope because it is too faint.

The planet itself is rather like Jupiter but smaller and with less structure in its cloudy atmosphere. Yellowish and tan cloud belts parallel the equator. Occasional bright and dark markings disturb these belts.

Saturn has the lowest mean density of all the planets, 0.71 g/cm^3, compared to 1.00 g/cm^3 for water. If a sufficiently immense ocean were provided, Saturn could float in it!

Saturn has a system of 10 satellites.* Number VI, Titan, is one of the largest satellites in the solar system, and is about as large as the planet Mercury. It was the first satellite found to have an atmosphere, a discovery made by the American astronomer Kuiper (1944). Titan is easily visible in a small telescope of a few inches aperture.

Uranus

(Pronounced U'-ran-us.) Now we come to the outer three planets, which were not known to the ancients. Uranus was discovered accidentally on March 13, 1781, by the English musician-turned-astronomer, William Herschel. Herschel was observing star fields at the time with his telescope; later studies

* The reality of satellite X, whose discovery in 1966 was reported by the French astronomer Dollfus, is questioned by some. It is said to lie near the outer edge of the rings.

showed that other accidental observations of Uranus had been made earlier but that the older observers had all mistakenly plotted the planet as a star.

Uranus resembles Jupiter—or more accurately Saturn without the rings. Its diameter is only 41 percent that of Saturn, yet is nearly four times the size of the earth. It has a pale greenish tinge, and various visual observers with large telescopes have reported bright spots, as well as dusky bands parallel to the equator, as on Jupiter and Saturn. As shown in Figure 2-10, Uranus presents so small a disk that these details are barely visible, even in large telescopes. Spectroscopic and theoretical studies indicate that hydrogen, methane, and ammonia are major constituents of the atmosphere.

Uranus has a peculiar dynamical property. Its axis of rotation, instead of lying nearly perpendicular to the plane of the solar system as is true of the other planets, lies almost in the plane of the solar system. This means that sometimes the "north" pole of Uranus will be pointed toward the sun, while half a

Figure 2-10

Image sizes of Uranus and Neptune as compared to Jupiter, all photographed from the earth. The entire disk of either of the outer planets is comparable to the size of the smallest details photographed on Jupiter, accounting for the difficulty of recording detail on Uranus or Neptune by conventional photographic techniques. Compare with Figure 2-7 for different cloud patterns on Jupiter. (New Mexico State University Observatory.)

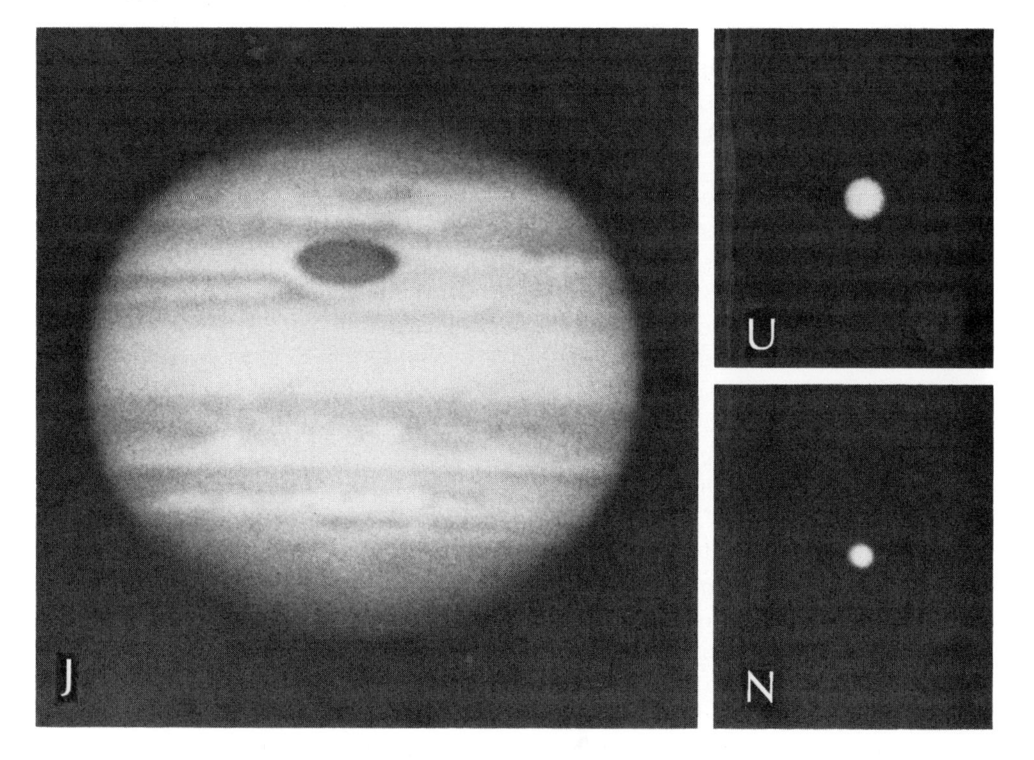

revolution later the "south" pole will be pointed toward the sun. This is more than a curiosity; it is a property that must be reasonably accounted for in any theory of the origin and evolution of planets (see Chapter 6). The rotation period—10.8 hours—was determined not by watching markings, but by measuring Doppler shifts in the spectrum (see below).

Uranus has five known satellites. The satellites share the peculiar dynamical situation; they revolve in orbits parallel to the planet's equator, nearly perpendicular to the plane of the solar system. This, too, is an important clue about planetary development: the revolution of satellite systems is tied in with the rotation of the parent planet, not with the revolution of the solar system.

Neptune

Immediately after Uranus was discovered in 1781, mathematicians and celestial mechanicians tried to determine its orbit using both the new observations by Herschel and the older observations to determine its orbit. Soon they found that no orbit would fit the observations. The theorists then tried discarding the old observations, supposing them to be in error, and computed a new orbit in 1821. Soon it was found that Uranus did not follow that orbit either. It became clear that some other planet must lie beyond Uranus, attracting Uranus by gravitational force and causing Uranus to depart from the predicted orbits.

An interesting chapter in the history of science ensued, as recounted by Lyttleton (1968). Unknown to each other, two astronomers, an Englishman and a Frenchman, set out in the 1840s to predict where the new planet should be. Since Bode's law had been abundantly verified in the discovery of Uranus and the asteroids, both men logically assumed a solar distance of nearly 38 A.U.—ironically, it turned out, since Neptune is the one serious failure of Bode's law. The error turned out not to be fatal. Adams, who had just finished his undergraduate work at Cambridge, predicted a position but had trouble getting his professors interested in searching for the planet with a telescope. Some months later, the French dynamical astronomer Le Verrier also predicted a position. A month or so after publication of Le Verrier's result, the English astronomers began a rather desultory search, in July and August 1846. On several occasions they spotted what was in fact the planet but, because they lacked enthusiasm, did not realize what they had stumbled onto. Meanwhile, Le Verrier chanced upon two young astronomers in Germany who were enthusiastic about the project. Armed with Le Verrier's predictions and new star charts, they located the new planet within half an hour of starting their search on September 23, 1846. It was confirmed the next night, and the observatory director, Encke, announced the discovery on September 25. Adams and Le Verrier are now both credited with the discovery, but at the time it became an international scandal because of the British failure to grasp their opportunity. On top of all this, Lyttleton, who has recently reviewed the calculations by Adams and Le Verrier, states that they are rather crude, and that the quick discovery of the planet involved no small amount of good luck.

Neptune is so far away that features are difficult to observe, as shown in Figure 2-10. It is thought to have cloud bands like those of Jupiter. Neptune is slightly smaller but more massive than Uranus, and is slightly more dense. Another interesting progression from Jupiter to Neptune is observed in the atmospheric constituents. Jupiter has large amounts of ammonia and smaller amounts of methane. Proceeding outward, spectra of the outer planets show diminishing amounts of ammonia but relatively stronger absorptions due to methane. The explanation is that the low temperatures at these large distances from the sun cause more of the ammonia to freeze into ammonia-ice crystals in the outermost planets. With the ammonia gas gone, the methane absorption becomes relatively more prominent.

A rotation period of 15.8 hours has been measured by means of the Doppler effect, the surface markings being insufficiently seen to reveal rotation.

Neptune has a peculiar satellite system, with one large satellite, Triton, revolving in a highly inclined orbit from east to west, contrary to the common direction for large satellites, and a small satellite in a highly elliptical, properly west-to-east orbit discovered by Kuiper in 1949. Triton is shown in Figure 2-11, an example of the long photographic exposures used to search for faint satellites.

Figure 2-11

Overexposed image of Neptune showing its satellite, Triton. The image of Neptune is surrounded by a halation ring and diffraction spikes, photo-optical artifacts caused by the long exposures necessary to record the satellite. (McDonald Observatory.)

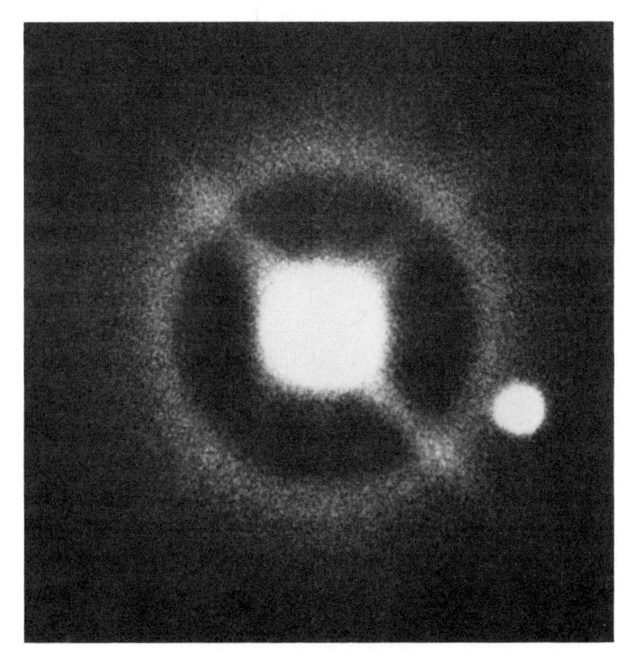

Table 2-2
Useful facts to memorize

Body		Mass (g)	Radius (cm)	Known satellites
Sun		2×10^{33}	7×10^{10}	—
Mercury				0
Venus				0
Earth		6.0×10^{27}	6.4×10^{8}	1
	Moon	7.4×10^{25}	1.7×10^{8}	
Mars				2
	Phobos			
	Deimos			
Asteroids		$\leqslant 10^{22}$	$\leqslant 2 \times 10^{7}$	
Jupiter		2×10^{30}	7×10^{9}	12
	Io			
	Europa			
	Ganymede			
	Callisto			
Saturn				10(?)
	Titan			
Uranus				5
Neptune				2
Pluto				0
Comets		10^{15}–10^{19}		

Gravitational constant	$G = 6.67 \times 10^{-8}$ dyne cm²/g
Astronomical unit	1 A.U. $= 1.49 \times 10^{13}$ cm (150 million km)
Boltzmann constant	$k = 1.38 \times 10^{-16}$ erg/deg
Stefan–Boltzmann constant	$\sigma = 5.67 \times 10^{-5}$ erg/cm² deg⁴ sec
Velocity of light	$c = 3.00 \times 10^{10}$ cm/sec
Mass of H atom	$M_H = 1.67 \times 10^{-24}$ g
Planck's constant	$h = 6.63 \times 10^{-27}$ erg sec
Mass of sun	$M_\odot = 2.00 \times 10^{33}$ g
Luminosity of sun	$L_\odot = 4 \times 10^{33}$ ergs/sec
Solar constant	$F_\odot = 1.36 \times 10^{6}$ ergs/cm² sec

Pluto

Neptune's mass and motions did not entirely explain the irregularities of Uranus's orbit; therefore a new search for another planet was instituted by Percival Lowell at the Lowell Observatory in Arizona. The new planet was detected photographically there in 1930 by an assistant, Clyde W. Tombaugh. It was given the name Pluto, and a new planetary symbol was created from the first two letters, PL, which were also Lowell's initials.

Partly because of the peculiarities in Neptune's satellite system and partly because of Pluto's eccentric and inclined orbit, which overlaps that of Neptune, Lyttleton (1936) and Kuiper (1956, 1957) suggested that Pluto may have originated as an escaped satellite of Neptune. One idea was that Pluto, originally in a Neptunian satellite orbit, passed close to Triton. Triton's orbit was altered by gravitational forces to its present unusual configuration and Pluto was ejected into an orbit around the sun. Recent studies of the dynamics of the Neptune satellites support this possibility (see Chapter 6).

Pluto is so far away that very little is known about its physical nature. Studies of its light variations indicate that it is rotating with a period of 6.39 days (Walker and Hardie, 1955). Measurements of its mass and diameter are notoriously uncertain, and until 1970 all measures gave an unacceptably high density; a number of determinations gave densities exceeding 8 g/cm^3, compared to 5.5 for the earth, 1.3 for Jupiter, and 0.7 for Saturn. Recent determinations have given lower values, but considerable uncertainty remains because of Pluto's great distance. Goldreich and Soter (1966) give an independent argument that the density is close to 2 g/cm^3, a value much more compatible with other outer planets and their satellites.

OTHER BASIC DATA

Readers who intend to pursue scientific careers should remember certain first-order data about the sun and planetary bodies. Table 2-2 contains the most useful data about the most commonly discussed bodies as well as a set of physical constants useful in quantitative studies. The *solar constant,* given in Table

Table 2-3
Classes of solar-system bodies

	Sun	Jovian planets	Terrestrial planets	Small bodies[a]
Mass (g)	$\sim 10^{33}$	$\sim 10^{30}$	$\sim 10^{27}$	$\sim 10^{21}$
Diameter (cm)	$\sim 10^{11}$ (1 million km)	$\sim 10^{10}$ (100,000 km)	$\sim 10^{9}$ (10,000 km)	$\sim 10^{6}$ (10 km)
Density (g/cm^3)	1	1	5	2
Distance from sun (A.U.)	0	>3	<3	Scattered
Atmosphere	—	Extensive	Thin or none	None
Satellites		Many	Few or none	None
Composition	Mostly hydrogen	Mostly hydrogen	Silicates, some metals	Ices and silicates, some metals

[a] Comets, asteroids.

2-2, is defined as the rate at which energy is received from the sun by a 1-cm^2 surface facing the sun at the mean distance of the earth's orbit (ergs per square centimeters per second). In other words, it is the mean *flux* of sunlight at the top of the earth's atmosphere. Helpful facts to remember are that the sun's mass is about 1000 times that of Jupiter and that the sun's radius is about 10 times that of Jupiter, which in turn is about 10 times that of the earth.

Table 2-3 compares four categories of solar-system objects by size.

MISCELLANEOUS TERMS

Figure 2-12 illustrates a number of terms common to planetary astronomy. The term *apparition* refers to an extended period of visibility of a planet, for example the 1956 apparition of Mars (when it was on our side of the sun) or an apparition of Venus (when it is within a month or two of elongation and visible in the evening or morning sky). *Elongations* occur for inner planets and can be either eastern elongations (appearing east of the sun, in the evening sky) or western elongation (west, in the morning). (Mnemonic: east, evening.)

Certain dynamical terms should be noted, for instance the prefixes "peri-" (designating the point of closest approach of the smaller body to the primary body) and "ap-" (designating the farthest point). If one wants to speak in general terms, without specifying a particular primary body, one uses the suffix "-apsis"; thus, *periapsis* and *apapsis* can be used to designate points in orbits around any planet or satellite. For specific bodies there are specific suffixes (usually Greek)—earth (perigee, apogee), sun (perihelion, aphelion), and moon (perilune, apolune) being the bodies most frequently discussed.

Certain other terms are of very general use. The *ecliptic* is the plane of the earth's orbit. *Inclination* is the angle between the plane of a planet's or satellite's orbit and the earth's orbit. The earth's orbit, or ecliptic plane, was chosen long ago as the most common standard of reference. Celestial mechanicians sometimes refer to the *invariable plane,* which is defined by the total angular momentum of the entire solar system. The distinction is small, because Jupiter's inclination is only 1°3. The planets with the greatest inclinations are Pluto (17°1) (which we noted is probably not a "real" planet in the genetic sense and thus may be excused) and Mercury (7°0). Sometimes the "inclinations" given for satellite orbits are the angles between the orbits and the planet's equatorial plane, not the plane of the solar system. Thus a listing under "inclinations" of the Uranus satellites might be given as about a degree or as 98°, and one must be careful when checking references to determine which definition is being used.

Additional definitions are illustrated in Figure 2-13 and discussed below. *Revolution* is the motion of one body around a second body. *Rotation* is the spinning motion of a body around an axis within itself. Rotation and revolu-

	Configuration (plane of solar system, viewed from north-ecliptic pole)	*Appearance in Sky*
Elongation (inner planets) Greatest elongation is illustrated.	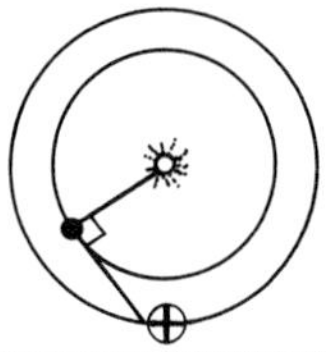	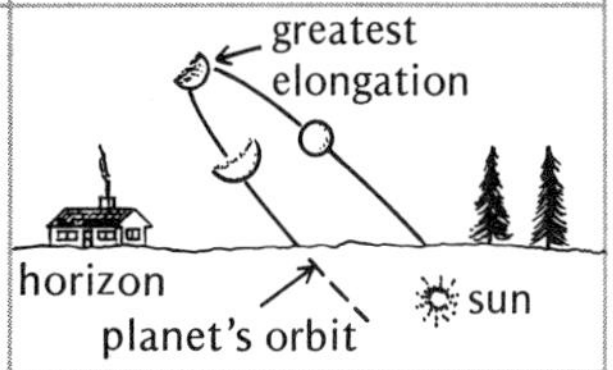 greatest elongation horizon planet's orbit sun
Opposition (outer planets)	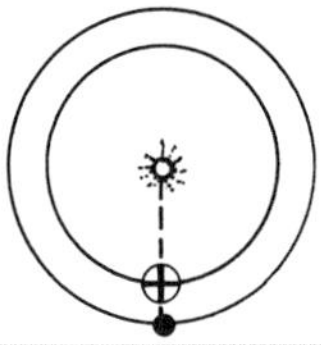	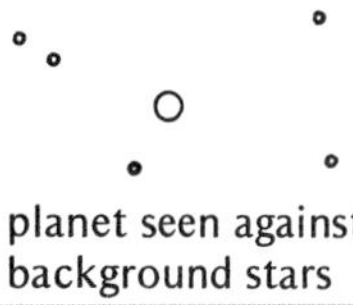 planet seen against background stars
Conjunction Two or more planets close together in the sky.	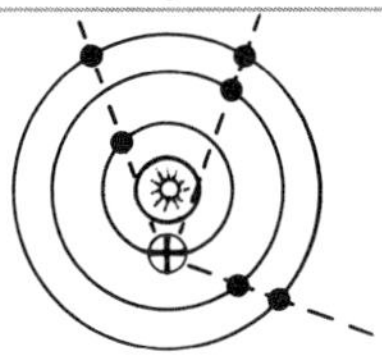	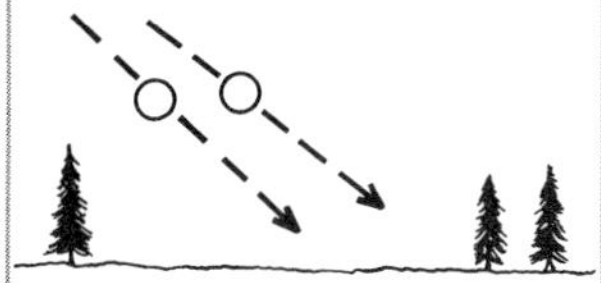
Occultation Small body passes behind larger body.		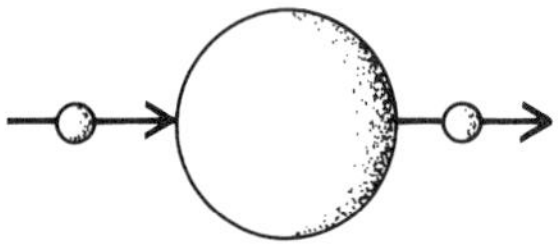
Eclipse One body passes into the shadow of another.	 shadow	
Transit Small body passes in front of larger body.		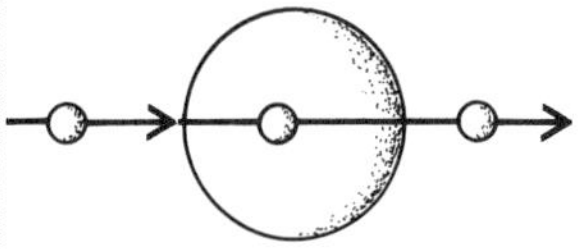
Examples: Jupiter satellite system A–A′ Satellite and its shadow B Not visible (occulted) C Not visible (eclipsed) D Just out of eclipse E In transit	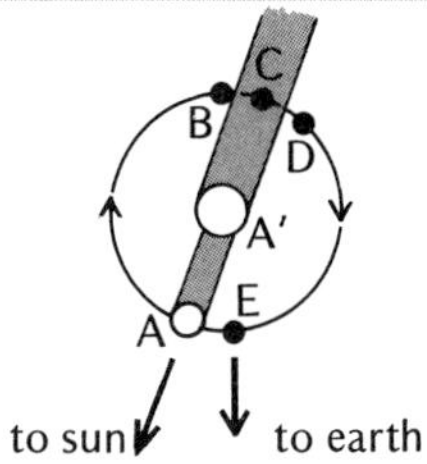 C B D A′ A E to sun to earth	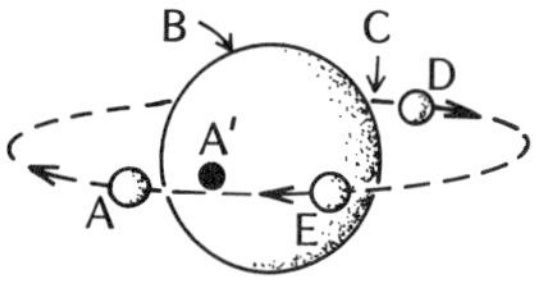 B C D A′ A E

Figure 2-12

Terms used in planetary astronomy. (See the text for discussion.)

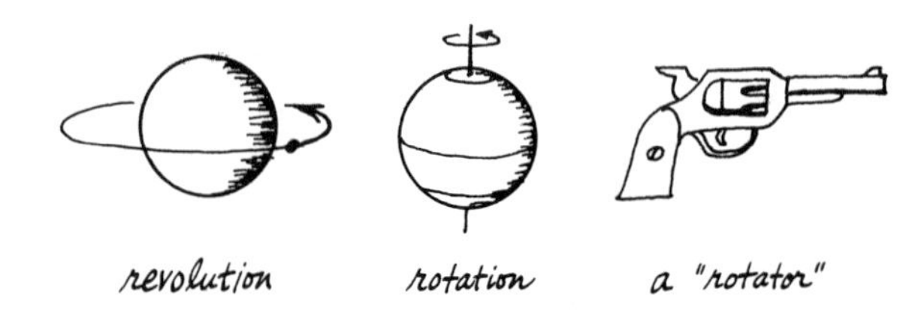

Figure 2-13
Revolution refers to motion of one body around another. *Rotation* refers to spinning around an axis within the body. A revolver, which has a rotating chamber, is misnamed.

tion are commonly confused, especially in everyday speech (to be entirely proper we should speak of rotating doors and rotators).

Prograde motion is the normal mode of revolution and rotation in the solar system, that is, west-to-east motion or counterclockwise motion when seen from the north ecliptic pole. As we have pointed out, Jupiter's outer four satellites, Saturn's moon Phoebe, and Neptune's Triton have *retrograde* orbits. Venus rotates in a retrograde sense, as does Uranus. Retrograde motions are indicated by inclinations and obliquities (see below) of greater than 90°. *Obliquity* is the angle between a planet's axis of rotation and the pole of the orbit. For the earth it has the familiar value $23\frac{1}{2}°$. Most of the other planets have similar values, except Venus (which is virtually "upside down," with a value near 180°), Uranus (with the unusual value of 98°), and Jupiter (with the small value of 3°). The fact that the largest planet is the most accurately aligned may be significant in terms of theories of planetary origin. Obliquity is sometimes incorrectly labeled "inclination" in uncritical literature. It is obliquity that is responsible for seasons on the planets, since obliquity causes one hemisphere or the other to be tipped toward the sun in a given part of a planet's orbit.

Albedo is a measure of the reflectivity of a surface, that is, the percentage of sunlight that the surface reflects. An albedo can be calculated for each color, or an average albedo can be given for all the colors of sunlight (i.e., averaged over the wavelength range of sunlight), which is the more common practice. A complication arises because of phase effects. *Phase* is defined as the angle between the sun and the observer as seen from the planet. The Bond albedo, the most commonly used definition of albedo, refers to the total percentage of sunlight reflected *in all directions*. To determine the Bond albedo, we have to observe a planet over a wide range of phase angles to determine how much light is reflected in each direction. Different surface particles, such as fine dust, rocks, or cloud droplets, have different phase functions. If the planet can be observed only over a narrow range of phases, as is the case with the giant planets, for example, the phase behavior must be estimated. For this reason the Bond albedo is broken down into two factors, p and q. Thus

$A = pq$. The factor q is known as the phase integral and can be either observed or theoretically calculated, as indicated above. The factor p, known as the *geometric albedo,* gives the percentage of light reflected at zero phase angle. Both Bond and geometric albedos can be given for specific colors or averaged over all colors. For example, the visual *Bond albedo* measures the percentage of light reflected in all directions at wavelengths to which the eye is most sensitive.

To clarify the situation by a brief example, we note that for the giant planets and their satellites, the geometric albedo can be observed directly if the planet or satellite diameter is known, but the Bond albedo must be based on an estimate of the brightness variations with phase. Albedos are indicative of surface and atmospheric properties, as will be discussed in Chapter 11. Venus, because of its nearly white clouds, has a Bond albedo of 0.76. The earth has an intermediate value of about 0.36 (averaging over clouds, oceans, and land). The moon has very low values, mostly in the range 0.06 to 0.10, depending on the region. These values are also typical of Mercury and most asteroids. One of the ways to identify materials on other planets is to compare their albedos in different colors to those of known materials on earth, for example clouds and different types of rocks, such as granite, basalt lavas, and iron-bearing colored minerals. The method is at best, however, only indicative.

The *Doppler shift* is an apparent shift of the wavelength, or color, of light from an object that is moving toward or away from the observer. The same effect occurs with sound waves and light waves. The waves reach the observer at shorter intervals when the object is approaching, and so the wavelength seems shorter. The pitch of a car's engine seems higher as it approaches and suddenly lowers as the car zooms past. Light from an approaching object has a *blue shift,* and light from a receding object has a *red shift.* By measuring the Doppler shift, astronomers can determine motions including rotation, which carries one limb of an object toward the observer while the other limb recedes. The Doppler shift is measured by comparing the observed wavelengths of certain features in the object's spectrum with the normal, unshifted positions.

Aeon is a term that has recently come into use to designate 10^9 (1,000,000,000) years. In the American system of numbers this is called 1 *billion* years, and this system has been followed in this book. However, one must beware of confusion that arises because the British use billion to refer to 1 million million (10^{12}).

TELESCOPIC APPEARANCE OF THE PLANETS

Figure 2-14 shows in rough-sketch form something of the appearance of each of the planets as seen through a moderate-sized telescope of, say, 6 to 12 inches aperture. Trained observers under good conditions can see much more detail than is indicated, and, in general, visual observation reveals more than

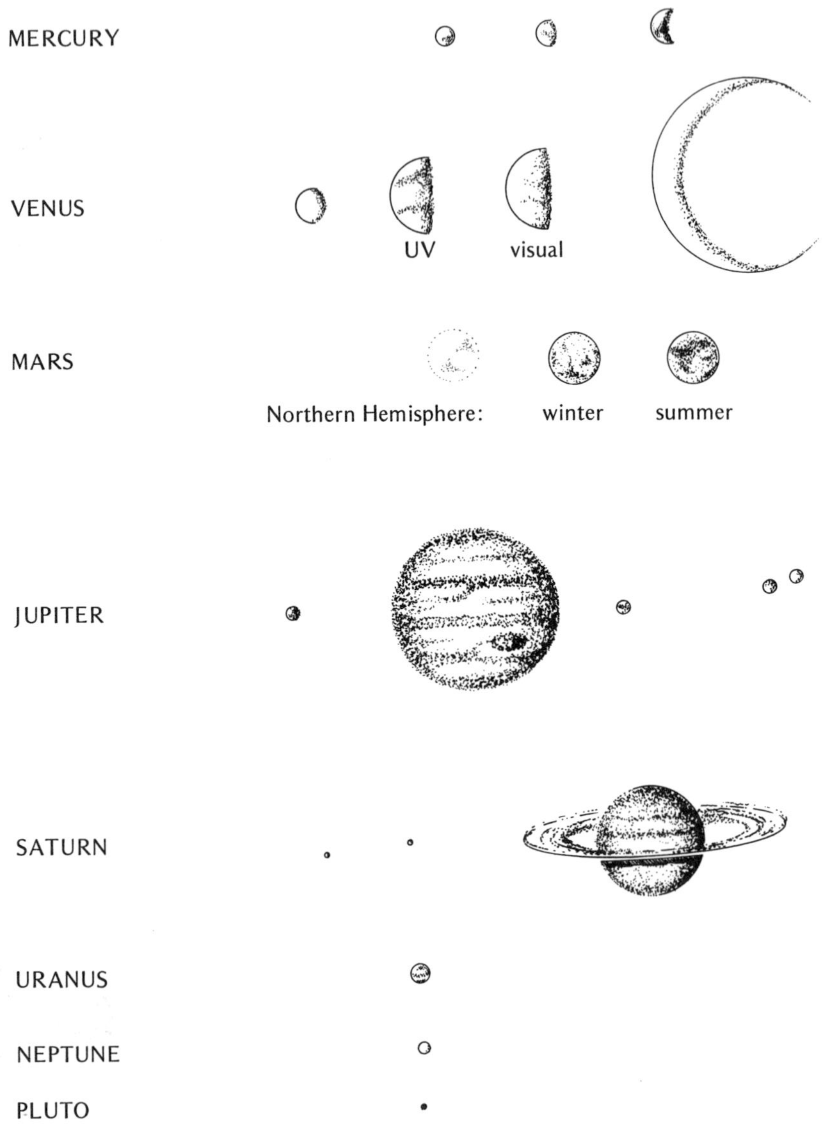

MERCURY

VENUS

UV visual

MARS

Northern Hemisphere: winter summer

JUPITER

SATURN

URANUS

NEPTUNE

PLUTO

Figure 2-14
Appearance of the planets as seen by a visually experienced observer, using a moderate-sized telescope.

can be photographed, because the observer can take advantage of moments of good conditions to which the camera does not respond quickly enough. The sizes of the sketches indicate very roughly the apparent sizes of the planets seen from the earth. Note that the inner planets display the largest apparent size when they are in crescent phase between the earth and the sun.

Some satellites of the giant planets are easily visible in small telescopes of a few inches aperture. These include the four "Galilean" satellites of Jupiter, and Saturn's satellite Titan.

References

Alexander, A. F. O'D. (1962) *The Planet Saturn* (New York: The Macmillan Company).

Chapman, C. R. (1968) "The Discovery of Jupiter's Red Spot," *Sky and Telescope, 35,* 276.

Cruikshank, D. P., and C. R. Chapman (1967) "Mercury's Rotation and Visual Observations," *Sky and Telescope, 34,* 24.

Dyce, R. B., G. H. Pettengill, and I. I. Shapiro (1967) "Radar Determinations of the Rotations of Venus and Mercury," *Astron. J., 72,* 351.

Focas, J. H., and A. Dollfus (1969) "Optical Characteristics and Thickness of Saturn's Rings Observed on the Ring Plane in 1966," *Astron. Astrophys., 2,* 251.

Goldreich, P., and S. J. Peale (1967) "Spin-Orbit Coupling in the Solar System II: The Resonant Rotation of Venus," *Astron. J., 72,* 662.

Goldreich, P., and S. Soter (1966) "Q in the Solar System," *Icarus, 5,* 375.

Guerin, P. (1970) "The New Ring of Saturn," *Sky and Telescope, 40,* 88.

Keeler, J. E. (1895) "A Spectroscopic Proof of the Meteoric Constitution of Saturn's Rings," *Astrophys. J., 1,* 416.

Kuiper, G. P. (1944) "Titan, a Satellite with an Atmosphere," *Astrophys. J., 100,* 378.

______ (1956) "The Formation of the Planets, I, II, III," *J. Roy. Astron. Soc. Can., 50,* 57, 105, 158.

______ (1957) "Further Studies on the Origin of Pluto," *Astrophys. J., 125,* 287.

Lyttleton, R. A. (1936) "On the Possible Results of an Encounter of Pluto with the Neptunian System," *Monthly Notices Roy. Astron. Soc., 97,* 108.

______ (1968) *Mysteries of the Solar System* (New York: Oxford University Press, Inc.).

McCord, T. B. (1969) "Comparison of the Reflectivity and Color of Bright and Dark Regions on the Surface of Mars," *Astrophys. J., 156,* 79.

Owen, T. (1970) "The Atmosphere of Jupiter," *Science, 167,* 1675.

Peek, B. M. (1958) *The Planet Jupiter* (New York: The Macmillan Company).

Pettengill, G. H., and R. B. Dyce (1965) "A Radar Determination of the Rotation of the Planet Mercury," *Nature, 206,* 1240.

Reese, E. J., and B. A. Smith (1968) "Evidence of Vorticity in the Great Red Spot of Jupiter," *Icarus, 9,* 474.

Smith, B. A. (1967) "Rotation of Venus, Continuing Contradictions," *Science, 158,* 114.

Walker, M. F., and R. Hardie (1955) "A Photometric Determination of the Rotational Period of Pluto," *Publ. Astron. Soc. Pacific, 67,* 224.

CELESTIAL MECHANICS

CELESTIAL MECHANICS is the science that attempts to describe and predict the motions of objects in space. Historically, celestial mechanics developed through efforts to keep track of planets, comets, and other bodies, and to regulate the world's clocks. Recently, the science has achieved new importance and refinement in the incredibly precise calculations of orbits that make spaceflight possible. Celestial mechanics is one of the most accurate sciences; the moon's position, for example, is known within a matter of meters!

This chapter discusses miscellaneous celestial mechanical topics that are needed for an understanding of the subsequent chapters.

HISTORICAL DEVELOPMENT THROUGH THE RENAISSANCE

Some of the Greeks had a fair conception of the solar system. Aristarchus of Samos (ca. 270 B.C.) is said to have advocated that the earth rotated and revolved around the sun, but his writings are now lost. Hipparchus (ca. 160–125 B.C.) was able to determine correctly the shape of the moon's orbit and he attempted to measure its distance. Although some Greeks thought the sun to be the central body of the solar system, the most popular concept was based on erroneous ideas championed by Ptolemy (ca. 73–ca. 151 A.D.). According to the *Ptolemaic system,* the earth is stationary at the center and all other bodies revolve around it. Not only the moon and planets, but also the sun, were supposed to be satellites of the earth.

In the Middle Ages the Greeks were held in esteem as the greatest of scholars, and the Ptolemaic idea of the solar system was handed down in dogmatic fashion for more than 15 centuries.

Five names dominate the overthrow of the Ptolemaic system, a revolution that shook science, theology, philosophy, and perhaps the psychology of man. They are

Nicolas Copernicus	1473–1543	Circular motion of earth around sun (1543)

Tycho Brahe	1546–1601	Observations (ca. 1600)
Johann Kepler	1571–1630	Analysis, Kepler's laws (ca. 1610)
Galileo Galilei	1564–1642	Telescopic observations supporting Kepler (ca. 1610)
Isaac Newton	1642–1727	Laws of gravity (*Principia,* 1687)

The work of these five scientists produced a steady advance in our knowledge. During the rise of the spirit of free enquiry, the Polish astronomer Copernicus questioned the Ptolemaic assumption of the earth's central position. The motions of the planets across the sky could be much more simply explained, he found, by assuming that the earth rotated on its own axis and that the earth and other planets revolved around the sun. Copernicus suspected that the planets moved in perfect circles around the sun.

Inspired by the idea of making a new star catalog, measuring the planetary motions and illuminating the controversy between the Ptolemaic and the Copernican theories, the Danish astronomer Tycho Brahe built a private observatory and started recording planetary positions night by night. There were no telescopes in those days, but Tycho (as he is called) had elaborate instruments for naked-eye observations.

Tycho hired a German assistant, Kepler, who inherited the stacks of observations. Kepler studied the data for years before discovering the empirical rules that describe how the planets move. He found that planets moved in ellipses, not circles. His three basic rules, known as *Kepler's laws,* are discussed in the next section.

Meanwhile, the Italian observer Galileo became the first to apply the new telescope (probably invented in Holland) to planetary observations. His discoveries included craters and mountains on the moon and rings around Saturn, and overturned earlier, idealized conceptions of the planets as "perfect" spheres. One of Galileo's most important discoveries was that Jupiter has a family of four large satellites, now called the *Galilean satellites,* which *revolve around Jupiter,* just as predicted by the Copernican theory. Legend tells of some defenders of the Ptolemaic theory who refused to look through Galileo's telescope. A black day in the history of formalized religion ensued when the Church also refused to accept the evidence and forced Galileo to recant. Galileo remarked with just cause,

In questions of science, the authority of a thousand is not worth the humble reasoning of a single individual.

Support for the Copernican theory began to snowball. One support was the first discovery of planetary rotation outside the earth. The French-Italian astronomer Jean Cassini (1625–1712) in 1666 described his observation of a large spot on Jupiter (probably the now-famous Red Spot), whose motions he timed as it crossed Jupiter's disk. The discovery that other planets rotated on their axes helped convince skeptics that the earth also rotated.

In the year of Galileo's death the English physicist Newton was born.* Newton, who said he made his discoveries "by always thinking about them," thought about the meaning of planetary motions. At this time, the motions of the planets were known empirically, but no one knew any theoretical reasons for Kepler's laws or any rationale for why planets move in ellipses as opposed to, say, squares. Newton reasoned from his observations and experiments that masses were endowed with gravitational fields and that this caused bodies to attract each other with a force proportional to the product of their masses but inversely proportional to the square of the distance between them. This discovery is known as *Newton's universal law of gravitation* and is discussed in his book *Principia* (1687), probably the most important scientific book ever written. Through his law, Newton was able to derive all of Kepler's laws.

Newton's accomplishment exemplifies how science, at its best, works. Science takes a lot of seemingly unrelated observations and combines them into a *theory* that one man can use to predict correct answers to questions about nature. Of Newton, Alexander Pope wrote,

> Nature and Nature's laws lay hid in night:
> God said, "Let Newton be!" and all was light.

KEPLER'S LAWS

Kepler's laws can be stated as follows:

1. Each planet moves in an ellipse with the sun at one focus.
2. The line between the sun and planet sweeps out equal areas in equal amounts of time. (Law of areas.)
3. The ratio of the cube of the semimajor axis to the square of the period is the same for each planet. (Harmonic law.)

The laws would read the same for any two bodies besides the sun and a planet; the two bodies could be a planet and satellite, the earth and a spaceship, or the sun and a comet, except that the ratio in the third law would be different for each pair of bodies, since it depends on the sum of their masses.

Each ellipse has two *foci,* as shown by Figure 3-1; ellipses can have different shapes, the circle being a particular kind of ellipse where the two foci coincide in the center. The farthest distance from the center of an ellipse to the periphery is called the *semimajor axis* (a); in a circle it is the same as the radius. A measure of departure from circularity is the *eccentricity* (e). The *period* is the amount of time required for the body to go all the way around the orbit.

* A certain science-fiction story has it that Galileo and Newton were one and the same spirit, transferred and reincarnated in 1642!

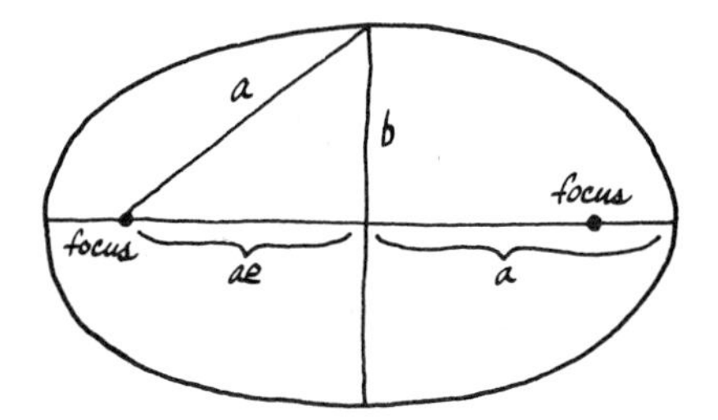

Figure 3-1
Geometric properties of an ellipse. *a*, semimajor axis; *b*, semiminor axis; *e*, eccentricity.

★ MATHEMATICAL THEORY ★

The laws can be simply stated mathematically:

1. $r = \dfrac{a(1-e^2)}{1+e\cos\theta}$ the equation for an ellipse; a = semimajor axis, e = eccentricity.
2. $\dfrac{dA}{dt} = \text{constant}$ A = area.
3. $\dfrac{a^3}{P^2} = \text{constant}$ P = period.

It is important to be able to derive Kepler's laws from Newtonian theory. This is very simple in the case of the second law, because the law of areas is a restatement of one of the basic conservation laws that are so important in physics: the law of conservation of angular momentum. Angular momentum is given by (linear momentum) × (radius), mvr. Thus conservation of angular momentum says that

$$mvr = \text{constant}$$

or

(1) $$m(r\,d\theta/dt)r = \text{constant} \qquad \text{substituting for } v.$$

But from Figure 3-2 we see that the area swept out in a time dt (being $\frac{1}{2}$ × the base × the height of the triangle) is $dA = \frac{1}{2}r\,d\theta\,r$. Thus

$$m(2dA/dt) = \text{constant} \qquad \text{substituting into (1).}$$

Since the mass of the planet, m, and 2 are constant, we now have the law of areas, $dA/dt = \text{constant}$. Another important exercise, beyond the scope of this book, is to derive Kepler's first and third laws from Newtonian physics.

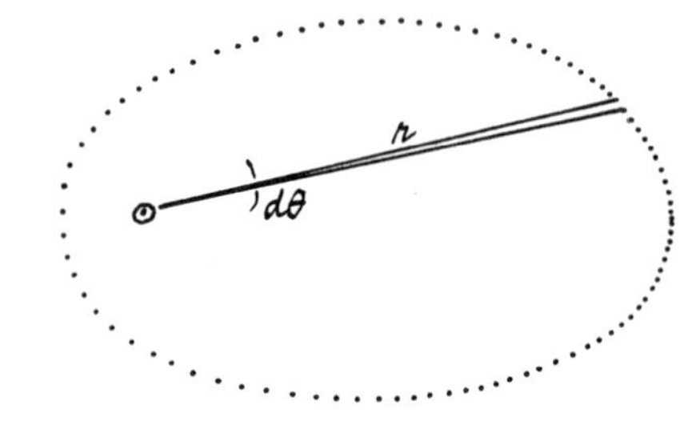

Figure 3-2
Orbital motion in an elliptical orbit. The dots are spaced to indicate equal time intervals, indicating fastest motion at perihelion and slowest at aphelion.

NEWTON'S LAWS

Newton clarified the meaning of the concepts force, motion, energy, and so on. He realized that if any body is *not* moving on a straight line, some force must be acting on it to deflect it from straight-line motion. In the case of the planets, the principal force is gravity. Newton determined that gravitational attraction between the sun and a planet must be proportional to the mass of the sun and to the mass of the planet, and he showed what mathematical form the force law has. This formulation is his *law of gravitation.*

★ MATHEMATICAL THEORY ★

The law of gravitation states that

$$F = \frac{GMm}{r^2},$$

where F = force between two bodies
M = mass of larger body
m = mass of smaller body
r = distance between the two bodies
G = a constant, called the *gravitational constant*

The equation gives the force exerted on a small body by a large body, or vice versa. The earth attracts familiar objects with a force which we call their weight. If you are given the value of the gravitational constant, the mass of the earth M, the radius of the earth, and the mass of a rocket, you could compute the rocket's weight in pounds, which equals the force necessary to lift it off the ground. This is why the force exerted by rocket engines is usually expressed as "*pounds* of thrust."

★ PROBLEM ★

Prove that a man weighs only 1/6 as much on the moon as on the earth, given that the moon's mass is 1/81 of the earth's mass and the moon's radius is 1/3.7 of the earth's.

Solution

A concise solution is as follows: In the law of gravitation, G is a constant and the mass of the man, m, stays constant, so in going from earth to moon the only things that change are the mass of the planet, M, and the distance, r, from its center to its surface. M decreases by a factor 1/81, and r^2 decreases by a factor 1/13.7 when we transfer from earth to moon. Combination of these effects shows that F decreases by 13.7/81, or 1/6.

ORBITS

A body anywhere in the solar system is under the gravitational influence of the sun. Assuming that the body is not too close to a planet, the sun will be the dominant influence. If the body is given a velocity, v, in any direction, it will start to move in some orbit influenced by the sun. Body A is Figure 3-3 is pushed to the left and starts to curve downward toward the sun. Body B is pushed away from the sun but eventually turns back. Both bodies are in orbit

Figure 3-3
A body at point P can go into a variety of different orbits around the sun, depending on its initial velocity. Two possible orbits are shown.

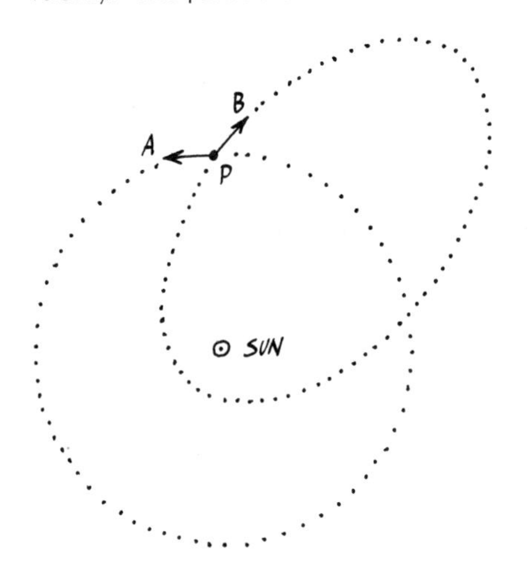

around the sun. Similarly, a football thrown by a quarterback is in orbit around the center of the earth. (In this case the orbit intersects the surface of the ground.)

Whether the object is in orbit around the sun, earth, or some other body, the principle is the same. The body that causes the dominant influence, like the sun in the case of the solar system, is called the *primary body*. Remembering Kepler's first law, we note that the primary body will be at one focus of the orbit. Detailed analysis shows that the focus lies at the center of gravity of the two bodies, which is near but not exactly coincident with the larger body. The sun is so massive, relative to the planets, that it virtually coincides with the focus.

Perturbations

Unfortunately for our efforts to keep things simple, the universe consists of more than two bodies. The earth, for example, is influenced not only by the sun but also by all the other planets and by the moon. Since the sun dominates the planets, the interplanet forces are minor and cause only small effects. The planetary orbits, therefore, are not perfect ellipses but ellipses with minor sinuosities superimposed on them. These departures from perfect elliptical orbits are called *perturbations*. Mars, for example, is said to be *perturbed* by Jupiter, by the earth, and so on.

Circular Velocity

Circular velocity is the speed of a satellite in a circular orbit about a primary body. The circular velocity decreases with increasing distance because the pull of gravity declines with distance.

It is instructive to imagine a body being inserted into orbit at various speeds. Suppose that a spacecraft has reached a point *P* above the earth's atmosphere* (see Figure 3-4). Its attitude-control system has aligned it parallel with the ground, but its position is stationary with respect to the center of the earth. If a rocket motor is not fired, the spacecraft would fall directly to the ground on path *A*. If a short burst is fired, it would fall into an elliptical orbit around the center of the earth, but the orbit, *B*, would intersect the ground like that of the football. With a higher speed the rocket could go nearly all the way around, striking the ground on the far side, on orbit *C* (*a la* ICBM). A still higher speed would put the rocket into a complete orbit, *D*, with perigee on the far side. A certain, still faster, velocity is needed to put the rocket into a precisely circular orbit, *E*. The description remains the same whatever the primary body and whatever the smaller body.

* If the spacecraft were in the atmosphere, atmospheric friction would produce a resistive force called *drag*, which would perturb the motions described here.

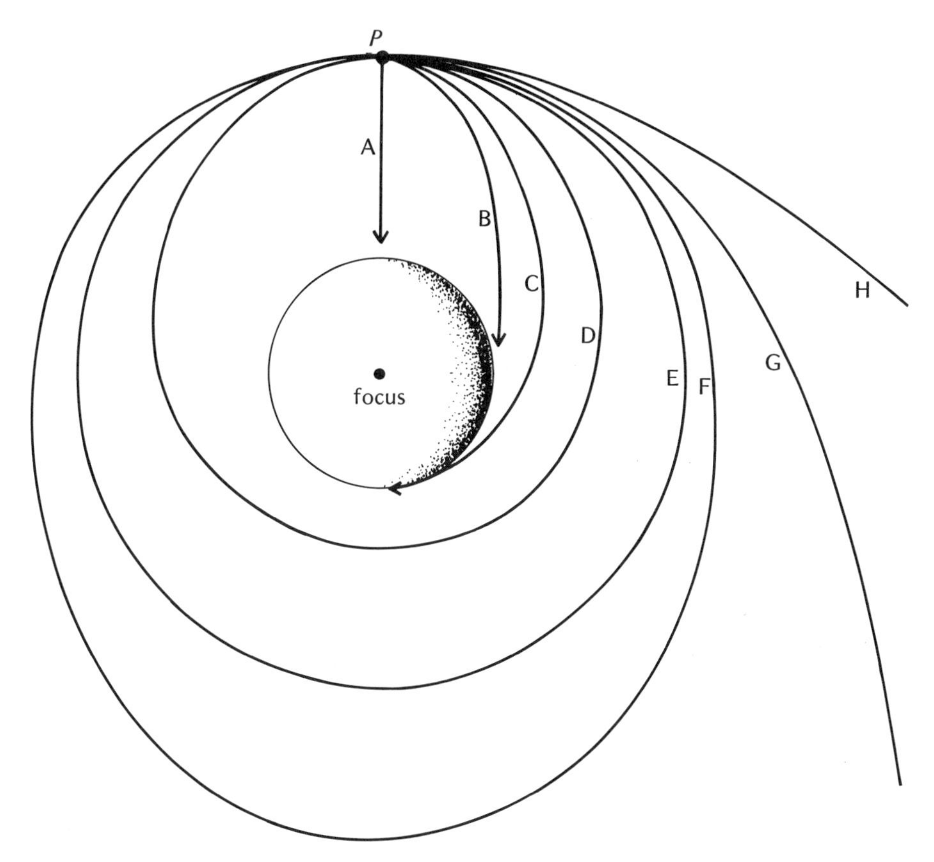

Figure 3-4
A rocket at point *P* above the earth can go into a variety of different orbits. Cases A through H correspond to firing the rocket parallel to the ground, but with different velocities (see the text).

Near-circular orbits were chosen for the first spaceflights because the rockets were barely powerful enough to climb to the top of the atmosphere, and the space capsules had to skim along at a constant height to avoid atmospheric drag.

★ MATHEMATICAL THEORY ★

It is easy to remember how to derive the circular velocity, because gravity must be just matched by centrifugal force:

$$\frac{GMm}{r^2} = \frac{mv^2}{r}.$$

Thus

$$v_{\text{circ}} = \sqrt{\frac{GM}{r}}\,.$$

Note that this gives v as a function or r and that as r increases, v decreases.

Escape Velocity

Escape velocity is the speed required to project a body completely free of a primary body. Like circular velocity, the escape velocity is less at larger distances from the primary.

If our imaginary spacecraft is Figure 3-4 is accelerated to an even higher speed than that required for circular orbit *E*, it climbs into an elliptical orbit, *F*, with apogee on the far side of the earth. At the critical speed of escape velocity, the apogee is at infinite distance and the rocket never comes back. Instead, it keeps slowing down as it moves out on path *G*; the speed approaches but never quite reaches zero. Another way of saying the same thing is that the rocket with escape velocity has a kinetic energy just equal to the earth's gravitational potential energy. The orbit, *G*, has the shape of a parabola and is called a *parabolic orbit.* At each point on *G*, the speed will just equal escape velocity (sometimes called *parabolic velocity*) at that point. If the rocket is made to move faster than escape velocity, it goes into a different orbit, *H*, with a slightly different shape from orbit *G*. Again the rocket never comes back, but in this case it has a kinetic energy *greater* than the earth's gravitational potential energy, and the rocket is always moving faster than the escape velocity at each point in its orbit. This path is called a *hyperbola* or a *hyperbolic orbit.*

Ellipses, circles, parabolas, and hyperbolas are a geometrically related family known as *conic sections,* since they can all be produced by slicing cross sections through a cone at various angles.

Another way of viewing escape velocity is to imagine a body dropped from rest at an infinite height. It falls toward the ground with increasing speed. Its speed at any point will be the escape velocity at that height.

★ MATHEMATICAL THEORY ★

It is easy to derive escape velocity, since the gravitational potential energy must just equal the kinetic energy:

$$\frac{GMm}{r} = \tfrac{1}{2}mv^2$$

$$v_{\text{esc}} = \sqrt{\frac{2GM}{r}}\,.$$

Note the important and useful fact that

$$\text{escape velocity} = \sqrt{2} \times \text{circular velocity}.$$

If you can remember one you can remember the other (escape velocity is obviously the higher).

★ QUESTION ★

Relative to the center of the sun, the earth's orbital speed is about 29 km/sec (18 miles/sec); how fast would a rocket have to leave earth's orbit in order to reach interstellar space?

Astrometry and Orbit Determination

If you could watch a spaceship returning from the moon night after night from your backyard, how could you determine what its orbit was, in three-dimensional space? The task of measuring positions is a branch of celestial mechanics called *astrometry*. In principle, only three observations of position, appropriately spaced, are required to determine an orbit, but in practice many observations are used to refine the orbit as accurately as possible. In the 1800s, when many planets and asteroids were being discovered, the laborious computations needed to derive the orbit had to be done by hand and might require months; today they are done by computer in minutes (although the computer programmer may have to work for days to get the computer routine in working order).

★ VELOCITY EQUATION
(MATHEMATICAL THEORY) ★

Suppose that an orbit has been determined for a body and you want to know how fast it is moving at a particular point, relative to the primary body. A useful equation, sometimes called the *vis visa*, or *velocity equation*, gives the velocity as a function of position (see Figure 3-5). With *a* the semimajor axis and *r* the distance to the object,

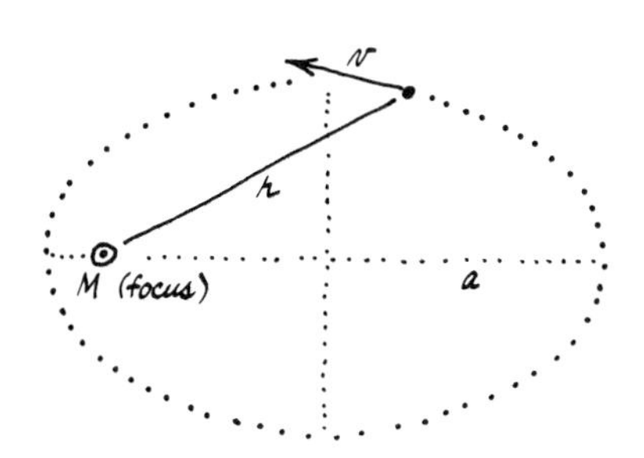

Figure 3-5
Orbital parameters used in the velocity equation.

$$v^2 = GM\left(\frac{2}{r} - \frac{1}{a}\right).$$

This equation is a much more general form of our equations for circular and escape velocity, since it allows calculation of velocities required for any desired orbit.

★ PROBLEM ★

Show that the equation reduces to those already cited for circular and escape orbits. (*Hint:* What value does the semimajor axis take in each case?)

★ PROBLEM ★

A comet in orbit around the sun has an aphelion 100 A.U. from the sun and its perihelion is at the earth's orbit. Find its velocity at perihelion relative to the sun. (*Hint:* Remember that all values must be in cgs or some other consistent system of units.)

★ PROBLEM ★

If the comet were in a direct orbit, as the earth is, and if it passed near the earth, how fast would it be moving with respect to the earth?

THE THREE-BODY PROBLEM

All the above discussions of orbits exemplify the *two-body problem.* The physical system consists of two bodies and no other influences. Given one body, we describe the orbit of another body around it. Because the planets are small and have only minor perturbative influences on each other, two-body theory gives a good description of the orbit of each planet and accounts for Kepler's laws.

The *three-body problem* is to describe the motions of three bodies when they are big enough to influence one another or at least when two of them are big enough to influence the third. An example would be a spacecraft moving in the earth–moon system, where both the earth and moon are big enough to influence the spacecraft's trajectory. There is no single general analytical solution that describes such motion, contrary to the known equations in the case of the two-body problem. Therefore, the only way in general to predict motions in a three-body system is by *numerical integration,* the process of solving the problem in small steps. A computer could be set up to start with the spacecraft in position 1, compute the forces on it from the earth and moon, let it move a small distance under the influence of these forces to position 2, and so on for hundreds or thousands of positions, depending on the accuracy needed.

The three-body problem applies not only in the earth–moon–spaceship case, but also in many other solar-system cases, such as a sun–Jupiter–comet system. Some special cases are interesting also, as revealed by the next section.

LAGRANGIAN POINTS

Certain special cases of the three-body problem have specific analytic solutions. One of these is the case of the Lagrangian points. In a system of two major bodies with nearly circular orbits (such as the earth and moon), there are five points where the gravitational forces of the two bodies plus the centrifugal force just balance, so that a third body put in one of these points would tend to remain in a fixed position with respect to the other two.

The five positions are called *Lagrangian points* after the mathematician Lagrange (1736–1813), who first discovered their locations. The Lagrangian points are shown schematically in Figure 3-6. Of the five points, only two, L_4 and L_5, are truly stable. The other three are only quasi-stable; if a small body were put at L_1, L_2, or L_3, it would stay only until an outside influence (e.g., the sun) moved it out of position, whereupon it would drift away. A body put at L_4 or L_5, if slightly disturbed, would oscillate around its original position.

A good analogy to the Lagrangian points is the circumstance of a steel ball

Figure 3-6
Location of the five Lagrangian points in an orbiting system.

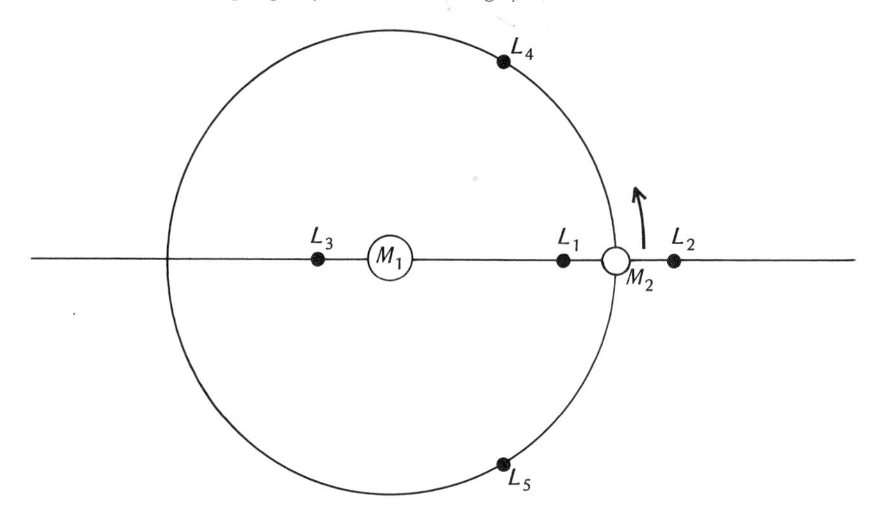

on a hummocky plastic surface. Points L_4 and L_5 are like depressions; the ball will stay in the bottom and if slightly disturbed will only oscillate around the bottom. Points L_1, L_2, and L_3 are like hilltops, or saddle-shaped regions; if the ball is carefully put there it will stay, but the slightest disturbance will send it rolling downslope. Such a surface is, in fact, a good analogy to the gravitational potential field, and in theoretical parlance the L_4 and L_5 points are often called *potential wells*.

A key idea in considering Lagrangian points is that the whole system is rotating because the two main bodies are in orbit around their common center of gravity. In the analogy above the plastic surface is like that of a phonograph record, with the biggest mass near the center and the second body halfway out. L_2 is then near the rim and L_4 and L_5 are ahead and behind the second body by angles of 60°. Objects at a Lagrangian point would seem stationary with respect to the system (to an observer on the phonograph record), but of course no part of the system is stationary with respect to the surrounding stars.

The Lagrangian points are practical realities. The sun and Jupiter form a system of two massive bodies, with the L_4 and L_5 points 60° ahead of and behind Jupiter in its orbit. Two groups of asteroids have been discovered occupying these positions. They are called the *Trojan asteroids* and are named after the Homeric heroes. Until 1906, when the first of the Trojans was discovered by the German astronomer Max Wolf, the Lagrangian points were mathematical abstractions, but now more than a dozen Trojans are cataloged and many more are known to exist. Their orbits are seriously perturbed by other planets and they may drift as far as 20° out of the L_4 and L_5 positions.

A second example of Lagrangian-point observations is shrouded in controversy. The Polish astronomer K. Kordylewski (1961) announced that he had detected faint clouds of particles in the L_4 and L_5 points of the earth–moon system, 60° from the moon. Variable results were reported from subsequent searches for these clouds, including naked-eye searches. However, sensitive photographic observations have failed to confirm their existence (Roosen, 1968; Bruman, 1969).

TIDAL EFFECTS

Tides

When two bodies are in orbit around each other, each exerts a force on the other, according to Newton's law of gravity. Consider the earth and moon as an example. The side of the moon facing the earth has a stronger force on it than the far side, because the facing side is closer. This means that the body of the moon experiences a *net differential force* acting along the earth–moon line at each moment. The moon stretches slightly along this line. The amount of stretching is limited by the elasticity of the solid-rock interior of the moon. The stretching forms approximately symmetric *tidal bulges* toward and away from the earth. These bulges, raised in the solid body of the moon, are called *body tides* (Figure 3-7).

Figure 3-7
Body tides raised in the moon by the differential gravitational attraction of the earth.

The same situation arises in the case of the earth but with two differences. First, as shown in Figure 3-8, the earth has oceans that are free to flow and form *ocean tides,* bulges *A* and *B*. Body tides form as well, but the ocean tides are

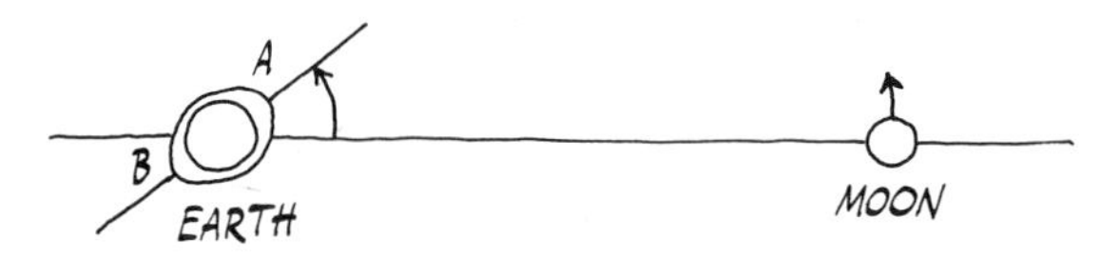

Figure 3-8
Tidal bulges (*A* and *B*) in the earth are dragged off the earth–moon line by the earth's rotation.

much easier to observe, as beach people know well. Second, the earth rotates faster than the earth–moon system revolves, so the side toward the moon is always changing. An observer at a given place on the turning earth sees the ocean tide rise and then fall as the tidal bulge sweeps past him. In actual fact high tide does not come when the moon is overhead. As suggested by Figure 3-8, the sun, the shapes of shorelines, and other complications modulate the tides so that they are not controlled solely by the moon's direction.

Tidal Evolution of Orbits

Because the earth is turning, the tidal bulges *A* and *B* are dragged around off the earth–moon line. This happens to both the body and ocean tides, although the amount differs because of the different amounts of friction involved. Bulge *A* is thus "out in front" of the moon as it moves in its orbit and bulge *B* is "behind" the moon. The bulges are, in fact, concentrations of mass, and so by Newton's law of gravity they exert forces on the moon. Bulge *A* is closer to the moon and thus stronger; the net effect is to pull the moon ahead. It is as if the moon had a small rocket motor pushing it steadily ahead in its orbit. The result is that the moon is slowly spiraling outward, away from the earth. This effect was actually detected late in the nineteenth century when observers noticed that the moon tended to depart from the position predicted by ordinary, nontidal, two-body Keplerian laws. Another way of stating the case is to say that tidal effects cause a transfer of angular momentum between two orbiting bodies.

In the same way, the moon has a net effect of pulling backward on bulge *A*, slowing the rotation of the earth. In the future, millions of years from now, the day will be noticeably longer than 24 hours and the moon will be farther away. The angular momentum gained by the moon is given up by the earth, thus satisfying the law of conservation of angular momentum in the two-body system.

In the past the earth turned faster and the moon was closer. Mathematically, we can trace the motions back in time to find that the moon may have once been very close to the earth, in which case the day would have lasted only about 4 to 5 hours. George Darwin (1845–1912), son of the famous naturalist, worked out this result (1898) and surmised that the moon might indeed have originated near the earth, or have broken off the earth, and has been moving out ever since. It is not possible to compute from tidal theory how long ago this might have occurred, because the extent of the ocean and the size and hence effect of the ocean and body tides are not known accurately for past ages. The idea that the moon broke off from the earth gave rise to the popular speculation that the Pacific Ocean basin might be the moon's natal scar. The idea remains only a hypothesis. The majority of post-Apollo opinion does not favor it, but it may be proved or disproved by further studies of composition of samples collected from different parts of the moon, and particularly by comparisons with terrestrial rocks.

★ MATHEMATICAL THEORY ★

Newton's law of gravity states that

$$F = \frac{GmM}{r^2}. \tag{1}$$

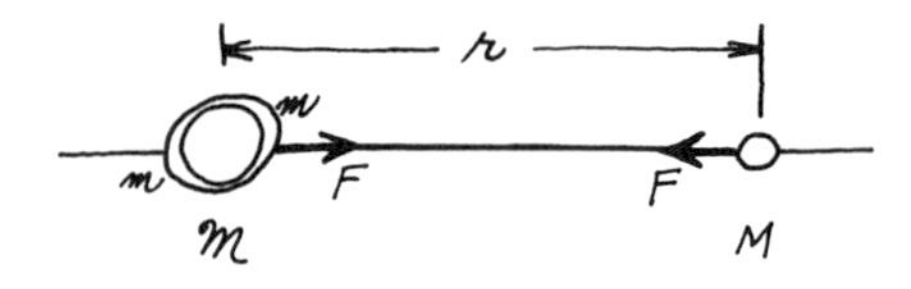

Figure 3-9
Properties related by Newton's law of gravitation in the case of tidal forces.

Defining masses as shown in Figure 3-9, we have

$M =$ mass of moon

$m =$ mass of tidal bulge

$\mathcal{M} =$ mass of earth

Thus the differential force on the earth caused by the moon is

$$dF_{\oplus} \propto r^{-3}\,dr \qquad \text{differentiating (1).}$$

The mass m of the tidal bulge raised by this force must be proportional to the force, so

$$(2) \qquad m \propto dF_{\oplus} \propto r^{-3}\,dr.$$

We want to know the net force on the moon caused by the two bulges. This is also a differential force, dF, proportional to the mass of the tidal bulge:

$$(3) \qquad dF_{☾} \propto mr^{-3}\,dr \qquad \text{differentiating (1),}$$

or

$$dF \propto r^{-6} \qquad \text{substituting (2) into (3).}$$

What we have shown is the important fact that *tide-raising forces are proportional to the inverse cube power of distance,* and that the net effect of tidal forces depends on the inverse sixth power of distance. These are much stronger dependences on r than the inverse-square law of gravity. Therefore, tidal effects can be very strong if the two bodies are close together but weak when the two bodies are somewhat farther apart.

ROCHE'S LIMIT

In discussing the mathematical theory of tides we showed that the force raising a tidal bulge obeys a $1/r^3$ law (whereas the net tidal interaction obeys a $1/r^6$ law). This means that as the distance r decreases, the stretching force gets very large. There exists a critical distance between two bodies within which the tide-raising force on the smaller body is strong enough to tear it apart. This critical distance is called *Roche's limit,* after its discoverer Edouard Roche (1850), a French mathematician.

To understand Roche's limit more clearly, imagine the earth–moon system. Suppose that we represent the moon by two particles just touching each other (Figure 3-10). The only force holding the particles together is their own mutual gravity; the only force tending to separate them is the tidal force. If, for any reason, this "moon" approaches the earth, the tidal force will increase much faster than the gravity, and at some point it will exceed gravity, causing the two particles to drift apart.

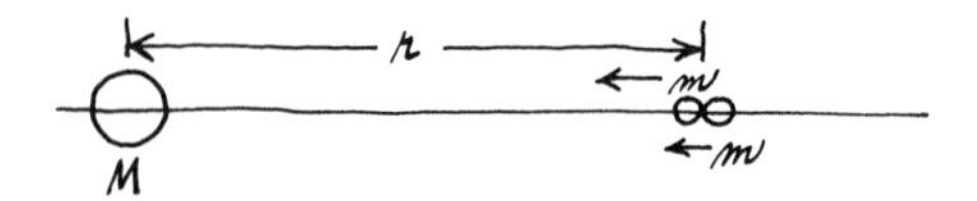

Figure 3-10
If two masses, m and m, orbit around a planet, M, Roche's limit occurs where the excess attraction of the closer mass is equal to the mutual gravitational attraction of m and m for each other.

In reality, of course, the moon is made of more than two particles; its many particles are bonded together in the form of rock, with high *tensile strength* (strength against rupture by stretching). But the principle is the same. For mathematical convenience, Roche's limit is defined as the critical distance at which a body with *no* tensile strength would be torn apart by tidal forces. A solid-rock satellite would have to pass inside Roche's limit before it would actually fragment.

Roche's limit is often invoked in discussions of Saturn's rings. The rings are inside the Roche limit for Saturn, and so they are regarded either as debris of an ancient satellite that was perturbed and passed too close to Saturn, or as a swarm of particles that never coalesced into a single satellite because they were too close. Similarly, if our own moon came within roughly 18,000 km (11,000 miles) of the earth, it would break apart and produce a ring of debris.

★ MATHEMATICAL THEORY ★

The mutual gravitational attraction for two touching particles is (see Figure 3-10)

$$F = \frac{Gmm}{(dr)^2}. \tag{1}$$

The disruptive tidal force is the differential gravity force:

$$dF = \frac{GMm}{r^3}\,2dr \qquad \text{see discussion of tides.} \tag{2}$$

At Roche's limit, r_R, these two forces are equal:

$$\frac{Gm^2}{(dr)^2} = \frac{2GMm}{{r_R}^3}\,dr \qquad \text{equating (1) and (2).}$$

Therefore,

$$\text{Roche's limit (for two touching particles)} \equiv r_R = \left(\frac{2M}{m}\right)^{\frac{1}{3}} dr.$$

More generally, *dr* can be interpreted as the dimension of the satellite body.

RADIATION PRESSURE

Sunlight, like all light, is a form of electromagnetic radiation. Electromagnetic radiation consists of pulses transmitted through electric and magnetic fields; it has some of the properties of wave motion yet some of the properties of particles. We commonly speak of radio *waves* or the *wave*length of light, but the particle-like properties are not so familiar. One of these properties is that the pulses of light, called *photons,* carry momentum. Like BB's, they transmit this momentum to whatever they strike. When a photon strikes an object, an impulse is transmitted away from the light source.

Suppose a body is orbiting in space, exposed to the sunlight. The body is being struck by photons that transmit momentum (i.e., exert pressure on the body). This is called *radiation pressure.* It has the effect of a force pushing the body away from the sun. If the cross-sectional area exposed to the sun is very large and the mass of the body is low, the radiation force can exceed the gravitational pull of the sun, causing the body to be "blown" out through the solar system.

One application of this has been the idea of a radiation-pressure-driven spacecraft, which would consist of an enormous thin sail. No such spacecraft has yet been built, but it may be a possibility for long solar-system voyages, which require low power and low fuel consumption.

★ QUESTION ★

How would you navigate such a spacecraft *toward* the sun?

Answer

Turn it sideways.

★ QUESTION ★

In a mixture of spherical particles of different sizes, which would be most affected by radiation pressure: large or small particles?

Answer

If you said "large" in order to maximize cross section, guess again. Remember, it is not the cross section but the ratio of cross section to mass that is important. The cross section of a sphere is the circular area πr^2. The mass will depend on the volume, $4\pi r^3/3$. So the ratio of cross section to mass varies as $1/r$, which gets large as r gets small. The effect of radiation pressure is therefore greatest for small particles.

Radiation pressure thus has another application—in the study of small particles in the solar system (e.g., meteoroids). Such particles are continually being sprayed into space by comets, collisions among asteroids, and other processes. The smallest particles are blown away from the sun and, in fact, are being continually blown out of the solar system.

★ MATHEMATICAL THEORY ★

Pressure can be expressed as the rate of transfer of momentum to a unit surface. The momentum carried by an individual photon is $h\nu/c$, where h is Planck's constant (see Table 2-2), ν is the frequency of the light, and c is the velocity of light. The total rate of momentum transfer is this momentum per photon times the number of photons per square centimeter per second for all frequencies:

$$P = \sum_{\nu} \frac{h\nu}{c} \frac{dn}{dt}\bigg|_{\nu} = \frac{\mathscr{F}}{c},$$

where $\mathscr{F}$ is the total flux of radiation, in ergs per square centimeter per second. The substitution of $\mathscr{F}$ can be made because $h\nu$ is the well-known energy per photon, so the total flux of energy is $h\nu(dn/dt)$, summed over all frequencies. If we want to know the force caused by radiation pressure on a small particle, we recall that pressure is force/area. Thus

$$\text{radiation force on a spherical particle} = F = \frac{\mathscr{F}}{c}\,\pi a^2 Q\,,$$

where a is the particle radius and Q is a correction factor on the cross section since small particles may have an absorption cross section different from the geometric cross section:

$$Q = \text{correction factor} = \frac{\text{effective absorption cross section}}{\text{geometric cross section}}.$$

Q depends on the particle size, particle composition, and wavelength of the light, but typically has values of 0.1 to 1.0.

★ PROBLEM ★

For particles of what size would the outward force of radiation from the sun be just balanced by the inward force of gravity? Assume that the particle is in the earth's orbit and that it is a stone particle of density $\rho = 3$ g/cm^3 and that $Q = \frac{1}{2}$.

Solution

Equating radiation and gravity forces, we have

$$\frac{\mathscr{F}\pi a^2 Q}{c} = \frac{GMm}{r^2} = \frac{GM}{r^2}\frac{4\pi a^3 \rho}{3}.$$

Solving for a and substituting the values for the constants (from Chapter 2), we have

$$a = 1 \times 10^{-5} \text{ cm}.$$

This is about 1000 angstrom units or about 0.1 micron. Particles of this size and smaller would thus be blown out of the solar system when exposed to sunlight.

★ PROBLEM ★

Prove that the ratio between radiation and gravity forces is independent of distance from the sun. (*Hint:* Note that in the equation above, the flux of sunlight at distance r is

$$\mathscr{F} = \frac{\text{total luminosity of sun}}{\text{area of sphere of radius } r \text{ centered on sun}} = \frac{L}{4\pi r^2}.$$

In the preceding problem, therefore, was it necessary to specify the distance from the sun?)

SOLAR WIND AND INTERPLANETARY GAS MOTIONS

Analyses of comets, especially by Biermann (1951), showed that material expelled from them was being accelerated away from the sun faster than could be accounted for by radiation pressure, suggesting that the interplanetary gas itself is moving away from the sun and carrying the cometary material with it. This led to the concept of the *solar wind*—an expanding, low-density *plasma,* or ionized gas, emanating from the sun. Theoretical and observational studies indicate that in the vicinity of the earth the interplanetary medium is composed of a plasma with average electron densities of the order two electrons/cm^3, temperatures about 2×10^5 °K (200,000°K), and expansion velocities about 600 km/sec, ranging on occasions up to 1000 km/sec. Theoretical models treat the solar wind in the first approximation as an expansion of the hot gases of the sun's outer atmosphere, or *solar corona* (temperature about 2 million °K at three solar radii from the sun's center).

Spacecraft have obtained good observations of the interplanetary gas in the vicinity of the earth. Experimenters are anxious to extend these data to the inner and outer solar system. One current problem is the nature of the interface of the solar wind with the interstellar gas. This interface may occur within the solar system somewhere beyond Mars. An alternative hypothesis places the interface beyond Pluto's orbit, at roughly 50 A.U. from the sun. Space probes into this region and beyond are planned for the decade of the seventies and may give us interesting new data on the interplanetary gas.

POYNTING–ROBERTSON EFFECT

The *Poynting–Robertson effect* perturbs the orbits of small particles (of centimeter dimensions and smaller, roughly), producing a tendency to spiral inward toward the sun.

Consider a small particle moving in a circular orbit around the sun. Sunlight, which can be thought of as a steady stream of photons, flows outward from the sun, while the particle moves at right angles to the photon stream. Thus the photons strike the particle preferentially on its leading side, just as a car driven through a rainstorm is struck on its front side even though the rain may be falling vertically. This effect is shown in Figure 3-11. The displace-

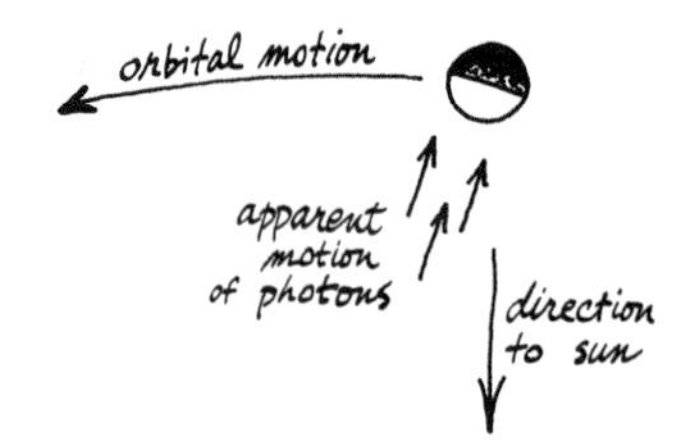

Figure 3-11
The Poynting–Robertson effect is caused by "drag" due to apparent displacement of photons from the solar direction (see the text).

ment of the apparent source of the light is called *aberration of light;* the apparent positions of stars seen from earth are actually displaced up to 20.5 seconds of arc due to the aberration of light, as discovered by the English astronomer James Bradley in 1727.

As a result of the aberration of light, then, the particle absorbs photon momentum preferentially on its leading edge, causing a resistive force. This resistance amounts to a net loss of energy and the particle slowly settles into a smaller and smaller orbit, spiraling in toward the sun.

This effect was first predicted by the British physicist Poynting (1903) and was later amended by the American physicist Robertson (1937) to take relativity into account.

Mathematical theory enables us to calculate the history of such a particle of any given size, starting in any given orbit. For example, if we consider the history of particles of various sizes starting in the asteroid belt, we find that particles of radius less than 10 cm (4 in.) will spiral into the sun in less time than the age of the solar system. The Poynting–Robertson effect, in summary, is very efficient in clearing small particles from the solar system. (Recall that radiation pressure drives much *smaller* particles *outward*. Particles affected by radiation pressure are about 0.00001 cm in radius.) Thus if we encounter any particle in space with dimensions less than about 1 cm, we can expect that it was created "recently," not at the beginning of the solar system. Such a particle might be a bit of ejecta from an impact on the moon, or material ejected from a comet, or a fragment of an asteroid, to mention a few possibilities.

★ MATHEMATICAL THEORY ★

If one writes the equations of motion, taking into account gravity and radiation forces, one can derive the time scale for spiraling of the particles into the sun. The result for near-circular initial orbits is*

$$t = 7.0 \times 10^6 \, r^2 \rho a$$

where t = lifetime of the particle (years)
r = initial distance from the sun (A.U.)
ρ = particle density (g/cm^3)
a = particle radius (cm)

TURBULENCE

Turbulence is familiar to anyone who has watched the eddies in a flowing river, or an explosion, or the dust swirling behind a passing car. It is the opposite of streamline, or *laminar,* flow. In planetary science turbulence is important in a number of cases. The clouds of gas and dust from which stars form are turbulent; formation of planets has been attributed to occurrence of turbulent eddies; turbulence characterizes the motions of clouds in some parts of planetary atmospheres.

The *Reynolds number* is a useful concept in discussing turbulence; it tells whether turbulence will occur on a certain scale in a given gaseous medium.

* The spiraling is so slow that any single orbit can be taken as circular if the initial orbit is near-circular.

The Reynolds number is a ratio of two numbers. The numerator represents the forces that promote turbulence and the denominator represent viscous forces that tend to damp out turbulence. If the Reynolds number is greater than 1, turbulence will grow in the medium, and if it is less than 1, turbulence will decline.

★ MATHEMATICAL THEORY ★

In the mathematical theory of fluid mechanics there is a quantity known as *vorticity,* which is a measure of the amount of rotary circulation of the medium per unit area. An equation called the *vorticity equation* gives the rate at which vorticity increases or decreases. This equation has several positive terms that increase vorticity and a negative term that decreases it. The ratio of these terms is the Reynolds number, which can be expressed as

$$R \equiv \frac{vL}{\nu}$$

where v = flow velocity in the medium
L = dimension of possible turbulent eddies
ν = kinematic viscosity = absolute viscosity/density

In looking up viscosities, the reader should be aware of the difference between absolute viscosity and kinematic viscosity as given above. Some typical absolute viscosities are water 10^{-2}, air 10^{-4}, hydrogen (100°K) 4×10^{-5}, in cgs units.

★ PROBLEMS ★

With normal wind conditions would you expect turbulence in the earth's atmosphere over dimensions like those of a thunderstorm?

Solution

We must use cgs units. An air speed of 36 km/hr equals 10^3 cm/sec for v. If we take 10 km for the storm dimension, $L = 10^6$ cm. The kinematic viscosity will be the viscosity 10^{-4} over the density of air, about 10^{-3} g/cm^3, or 10^{-1} cgs. Thus the Reynolds number is of the order 10^{10}, which is considerably greater than 1, and we correctly predict turbulence in the air.

Other possible problems would be to predict whether there is turbulence in the interstellar medium (see Chapter 5), or to calculate the minimum size of turbulent eddies if you swish your hand through your bath water at 10 cm/sec.

References

Biermann, L. (1951) "Kometenschweife und solare Korpuskularstrahlung," *Z. Astrophys., 29,* 274.

Bruman, J. R. (1969) "A Lunar Libration Point Experiment," *Icarus, 10,* 197.

Darwin, G. H. (1898) *The Tides and Kindred Phenomena in the Solar System* (San Francisco: W. H. Freeman and Company, Publishers, 1962).

Kordylewski, K. (1961) Summary of Letter to Editor, *Sky and Telescope, 22,* 63.

Newton, Isaac (1687) *Principia,* A. Motte, trans. (1729) (Berkeley, Calif.: University of California Press, 1962).

Poynting, James (1903) "Radiation in the Solar System: Its Effect on Temperature and Its Pressure on Small Bodies," *Monthly Notices Roy. Astron. Soc., 64,* App. 1.

Robertson, H. P. (1937) "Dynamical Effects of Radiation in the Solar System," *Monthly Notices Roy. Astron. Soc., 97,* 423.

Roche, Edouard (1850) "La Figure d'une masse fluide soumise à l'attraction d'un point éloigné," *Mém. Acad. Montpellier, 1* (Sciences), 1847.

Roosen, R. G. (1968) "A Photographic Investigation of the Gegenschien and the Earth–Moon Libration Point L_5," *Icarus, 9,* 429.

THEORIES OF THE ORIGIN OF THE SOLAR SYSTEM: HISTORICAL REVIEW 4

THE ANCIENTS

Man's philosophic, emotional, intuitive concern with the nature of the world around him is as old as man himself. Forty thousand years ago Neanderthal men were constructing what can best be described as "shrines," consisting of stone cists and geometric arrangements of animal bones, and they were systematically burying their dead in sleeping positions in prepared graves with stone tools and even food. This suggests, at least by analogy with practices of contemporary primitive and advanced cultures, that the Neanderthals had developed some sort of concern for the causes and destinies of things. These early pseudo-religious stirrings may have been engendered as much by social impulses, such as emotional attachments to friends, the euphoria of gatherings around the campfires, and the fear of the hunt, as by inquisitiveness into the cause of natural, physical processes. Probably there was not yet much of a *cosmogony*, or theory of the origin of the physical world.

Ten thousand years ago man discovered how to cultivate plants, which enabled him to begin living in substantial permanent villages and to begin systematizing his thought. By this time he had speculated on the origin of his world. It is important to note that primitives are concerned with explaining the world *around them,* not yet having invented the conception of a stellar universe. An example of this kind of speculation in primitive societies is a Tahitian hymn of the creation, given by Radin (1927) in his book *Primitive Man as Philosopher.* A common feature of primitive explanations of the world is the assumption that there had to be, in the beginning, some kind of creative force or spirit, which is usually given animistic character. We see this in the Tahitian chant, here translated and paraphrased:

> He abides—Taaroa by name—in the immensity of space. There was no earth . . . no heaven . . . no sea . . . no mankind . . . Taaroa is the root, the rocks, the sands, the light, within, below, enduring, wise. He created the land of Hawaii, great and sacred Hawaii, as a shell for himself. The earth is moving. O foundations, O rocks, O sands! Here, press together the earth: press and press again! Stretch out the seven heavens, let ignorance

cease. Let anxiety cease within . . . Complete the foundations. Complete the rocks and the sands.

In 2000 B.C. in the Egyptian empire the identification of the creative forces as animistic gods had reached an extreme. This assigning of "personalities" to natural forces was not entirely misguided. Wilson (1951) points out that the Egyptians had achieved a level of abstract thought, but needed to give concrete expression to the concepts they talked about. Although a number of competing cosmogonies and theologies were current, a recurring idea describes a creator-god of all-pervading nature. An example would be Amon, "associated with the air as an invisible, dynamic force . . . invisible, born in secret." The cosmogony promoted in the city of Memphis was particularly important in that it sought after first causes and ascribed creation to an intelligence operating under rational, purposeful principles.

Such ideas were apparently transmitted to tribes of nomads forced to work in Egyptian society about 1300 B.C. These nomads left Western culture its most famous and influential creation story, in which the world was created in seven stages:

In the beginning God created the heavens and the earth. At first everything was formless and dark. In the first stage of his creation God created light to make the day separate from night. In the second stage he created the sky, and in the third, the dry land, called Earth, and the oceans and vegetation. In the fourth stage, the sun, moon, and stars were put in the sky. In the fifth God put fish and sea monsters in the ocean and birds in the air, and in the sixth stage he created the land animals, and man and woman. In the seventh God put the finishing touches on his creation and rested.

In the interval from 600 to 200 B.C. the Greeks began to look at things from an entirely different viewpoint. Greek philosophers thought of things in terms of their natural physical "elements"—earth, air, fire, and water—and they looked for physical rather than metaphysical relations among phenomena. By 600 B.C. men could predict eclipses, and Thales of Miletus (ca. 636–ca. 546 B.C.) preached to his skeptical countrymen that the sun and the stars were not animate gods but balls of fire. Although we are intrigued to note that this is virtually correct, we should note that it was not a deduced fact derived by applying the modern scientific method but a philosophic construct. Although we would prove the statement with a spectroscope, Thales had no more proof than did his detractors, who said that the sun was a living god. The achievement is not that Thales came up with what we could call a right answer but that he and the other Greeks set men to thinking in terms of physical processes.

Heraclitus of Ephesus (ca. 535–ca. 475 B.C.), who was a member of the school founded by Thales, said

> This ordered cosmos, which is the same for all, was not created by any one of the gods or of mankind, but it was ever and is and shall be ever-living Fire, kindled in measure and quenched in measure . . . The fairest universe is but a dust-heap piled up at random.

Democritus of Abdera (ca. 460–ca. 370 B.C.) conceived the idea that "In reality there are only atoms and empty space." Again, we must caution that Democritus did not really know that matter is made of atoms in the modern sense, but he conceived that there should be ultimate, indivisible particles. His idea was widely celebrated 50 years ago when physicists proved the existence of what *we* call the atom, but it is perhaps once again subject to doubt as we now delve continually deeper into a hierarchy of subatomic particles.

Aristotle (384–322 B.C.) founded modern science or at least the idea of modern science. Among other conceptions he argued that the world was cyclical, composed of endless instances of rebirths and destructions. To illustrate that the old masters were not always correct, we might note that Aristotle rejected the notion of Pythagoras (ca. 582–ca. 507 B.C.) that the sun was the center of our system; Aristotle put the earth at the center—an error that was destined to stunt the thinking of medieval Aristotelian philosophers for several hundred years.

Finally, the Greeks made a beginning at buttressing their philosophic speculation with direct observations—the keystone of the modern scientific method. Aristarchus of Samos (ca. 310–230 B.C.) measured the length of the solar year and theorized that the earth rotated and moved in an orbit around the sun. Eratosthenes (ca. 275–194 B.C.), who served in the famous library at Alexandria, not only knew that the earth was round, but proceeded to measure its diameter to an accuracy within 20 percent by observing the difference in latitude between Alexandria and another city a known distance away. Hipparchus of Nicaea (ca. 160–ca. 125 B.C.) made charts of the sky and observed the precession of the equinoxes.

The Greeks provided a flowering of human thought that lasted 400 years and was perhaps never matched before or since until our own times, 1600–2000 A.D. Unfortunately, their thought was lost to the Western World with the destruction of the Alexandrian library. The reintroduction of Aristotle to the medieval world in the twelfth century via Arab translations was, more than any other single event, the spark that touched off a resurgence of rational thought in the present millenium.

THE PHILOSOPHERS—1300–1800

The Renaissance—the flowering of thought that ushered in our times—is said to have begun about 1300, not long after the writings of the Greeks began to be recirculated in the medieval world. Before we trace the further development of cosmological theories, we should note two characteristics of the period.

First, the development of science as we know it was a parallel stream to the development of philosophy. In Chapter 3 we traced the development of celestial mechanics from Copernicus (ca. 1500) to Newton (ca. 1700). This may be termed the scientific stream of thought, which fed ideas into the philosophic stream. It was usually the philosophers, not the scientific observers, who gave thought to the origins of things. Of course, these two streams merge at points since science was at that time known as "natural philosophy."

The second characteristic of the period is that the place of the sun in the stellar universe had not been discovered. Although Kepler, Galileo, and Newton had shown that a system of planets in addition to earth attended the sun, it was not realized that this "solar system" was a tiny, isolated unit moving independently in a much larger system, the Milky Way galaxy, which in turn was an independent part of the whole universe. Not until 1800 did the English astronomer William Herschel succeed in getting observational evidence of this fact by counting the numbers of stars visible in each direction (see North, 1965). Therefore, theories of origin before about 1800 tend to be theories of origin of the entire universe. Not until the late 1700s do we find recognition of the possibility that the solar system could have originated in events separate from the origin of the galaxy or of the universe.

Since the theories came from philosophers trying to conceive of the origin of the universe, we can correctly anticipate that they were derived mainly from mental constructs rather than physical observations. That is, the early "natural philosophers" tended to base their theories on intuitive mental pictures, reinforced by observations of nature; whereas we find science more successful if it is based directly on observations reinforced by intuition.

This is the difference between the medieval mysticism and modern science. The philosophers of the fourteenth and fifteenth centuries were trying to sort out the meaning and significance of material objects, spirits, motion, change, and constancy. Some people thought that by mastering certain mystical words or magical zodiacal signs, men could gain power over each of these properties. So alchemy—the forerunner of chemistry—came about. Even today, magical principles such as those found in astrology are more appealing to some people than the scientific method, in spite of the fact that science has had vastly more demonstrable success in telling us about nature.

Matter, spirits, space—these were hot issues of the times. Even as late as 1600, when Galileo was preparing to make his telescopic observations, the Italian astronomer Giordano Bruno was burned at the stake for philosophizing that the universe is a single, infinite, imperishable entity composed of ceaselessly changing parts. Aristotle had written, "It is clear that there is neither place nor void nor time beyond the heavens," and the sixteenth-century establishment was not about to listen to someone speculate about worlds beyond the visible sky—even Galileo was forced under threat of death to promise that he would not teach his belief that the sun was at the center of the "universe." Perhaps we should pause and be thankful that antiintellectualism has not in the last few decades reached such heights.

The man credited with the first modern view of the origin of things was the French philosopher René Descartes (1596–1650). Descartes's major contribution was to bridge the gap between medieval and modern views by separating the objective physical world from the world of the mind and sense perception. "I think, therefore I am," he said. In this way he pictured humans as observers in an independent world of physical reality, leaving his successors free to discuss the evolution of the physical world completely independent of human concepts or activities. He also proposed that if God had created the universe in a chaotic state, the incorporated laws of nature would have led to predictable, regular phenomena. Thus he separated scientific from theological inquiry.

Descartes pictured the universe as full of moving matter. Motion was indestructible and there must have been a certain amount of motion since the very beginning (note the foreshadowing of Newton's laws, which came about 40 years later). Therefore, Descartes theorized that the initial condition of the universe was that it was filled with gas, in which "vortices" swirled, as depicted in Figure 4-1. The vortex of Descartes can be likened to an eddy that

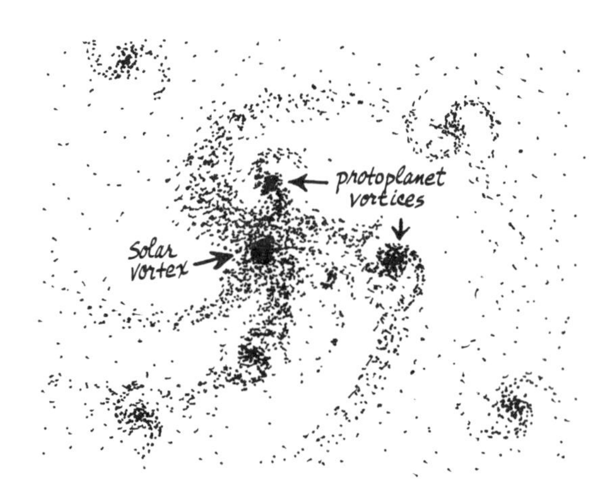

Figure 4-1
Vortex motions and condensations producing the sun and planets, as envisioned by Descartes (1644).

forms on the surface of a stream. In 1644 Descartes attempted to explain the laws of planetary motion found by Kepler about 1620 by suggesting that the sun condensed out of such a vortex of gas, thus gaining its own rotational motion and imparting circular orbital motions to its family of planets.

Theories of the origin of the solar system can be divided into classes, of which Descartes's exemplifies the first, the *evolutionary theories*. Evolutionary theories picture the formation of planets as the result of normal evolutionary process that may happen to most or all stars. The other theories are the *catastrophic theories,* in which the solar system is viewed as a unique or nearly unique accident. As a result of modern observations, evolutionary theories have come to be regarded as closer to the truth.

The first of the catastrophic theories was that of the French scientist Buffon (1745), who suggested that a "comet" had hit the sun and knocked off sufficient material to form planets, as shown in Figure 4-2. A glancing blow

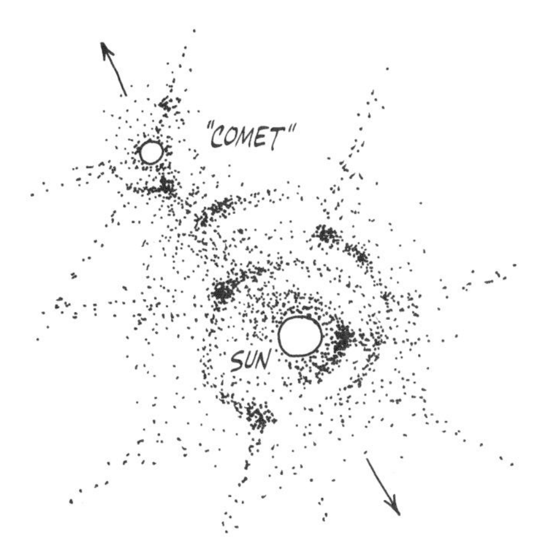

Figure 4-2
Condensations in a cloud of material knocked out of the sun by a passing "comet" (star), as envisioned by Buffon (1745).

could have caused the sun's rotation and planetary rotation. The reader should be warned of a semantic problem here, because a "comet" to Buffon was far from the modern conception of a comet. Eighteenth-century scientists did not know what comets are, and Buffon's "comet" was supposed to have had a mass comparable to that of the sun—over 1 billion times more massive than the mass of true comets. Buffon's "comet" thus more nearly resembled a star.

The great Newton, who preceded Buffon, did not develop a complete theory of solar-system origin, but he did point out one problem, which was the thorn in Buffon's side. According to Kepler's laws and Newtonian theory, planets could move in elliptical orbits of any eccentricity. Why then should all the planetary orbits be special cases of ellipses—that is, almost perfect circles? Buffon's theory did not explain this.

In the post-Newtonian years of the 1700s theorists and their theories grew more sophisticated. The great German thinker Immanuel Kant (1755), in addition to overturning philosophy and theology, offered some thoughts on the formation of the solar system. He assumed that at some time the universe was filled with gas—a starting point the same as that of Descartes. Newtonian gravitation would then cause contraction of the densest regions. If there had been any initial rotating vortices of the sort imagined by Descartes, the rotation would be preserved and the contracting system would flatten.

Secondary condensations would arise within the cloud to form the planets, as indicated in Figure 4-3. Kant even considered the motions of individual particles inside one of these rotating, flattening clouds and tried to show by Kepler's laws that the collisions between these particles would create the direct sense of rotation observed in planets. Kant's is clearly one of the evolutionary theories.

Figure 4-3
Condensations in a rotating solar nebula, as envisioned by Kant (1755).

The French mathematician Laplace (1796) independently hit upon a model that very much resembles Kant's, but because it was more clearly presented it became more famous. This model (as well as Kant's) is called a *nebular hypothesis* because it deals with the evolution of a *nebula*—or cloud of gas and dust—surrounding the sun. By the time of this hypothesis, the distinction between mere mental speculation and the scientific method of observation and analysis was clearly recognized, as is shown by an apology with which Laplace prefaced his theory: Since he had no physical proof, he offered the theory "with that diffidence that always ought to attach to whatever is not the result of observation or calculation."*

Laplace pictured his nebula as a hot extension of the early sun. He as-

* It is disquieting to note that today's scholars would be apt to dismiss such verbal models as pure "arm waving." Perhaps today's model makers are not as articulate as Laplace.

sumed that as it cooled it would contract. It would spin faster as it contracted, just as an ice skater spins faster when she pulls in her arms. Therefore, centrifugal force around the edge of the disk-shaped nebula would increase. Laplace reasoned that a ring of material thus might break off the extreme outer edge of the disk and be left behind in orbit as the rest of the disk continued to shrink. Soon the shrinking disk, illustrated in Figure 4-4, would be spinning still faster

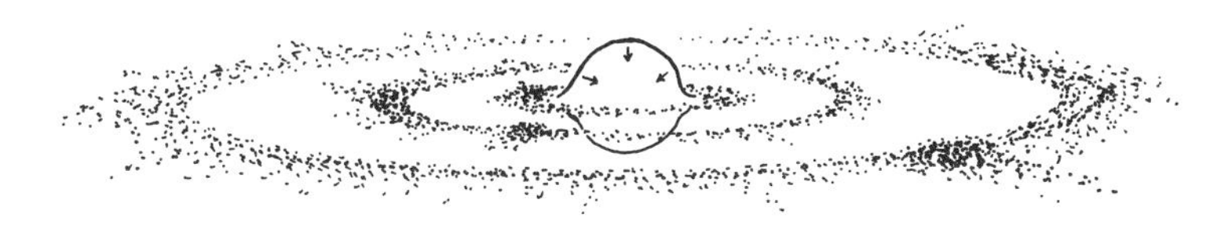

Figure 4-4
Condensations in rings of material shed by a contracting, rotating sun, as envisioned by Laplace (1796).

and another smaller ring would break off in the same way. The nebula would thus break up into a series of concentric rings surrounding the sun; these hypothetical rings came to be known as the *Laplacian rings.* Laplace hypothesized that each would condense into a separate planet. Thus he had apparently accounted for the flattened shape of the solar system, the sun's rotation, the planets' circular orbits, and the spacing of the planets in successive zones outward from the sun.

Among the objections to the Laplacian model are that it never explained why the rings should appear in discrete steps rather than in a continual series of flattenings that would leave an extended continuous disk, and that it predicted that the sun should be the fastest rotating body of all, which is not true. Nonetheless, we shall see that the Kant–Laplace nebular hypotheses are the foundations of modern theories.

NINETEENTH-CENTURY THEORIES

The 1800s should be viewed as a time of development for celestial mechanics and theoretical physics; there was little direct attack on the origin of the solar system. Maxwell and others worked out the theory of electric and magnetic fields, Darwin proposed evolution of biological systems, and Helmholtz and Lord Kelvin (W. Thomson) calculated the rate of energy radiation from a contracting body. Some thought that this so-called *Helmholtz contraction* might be the source of the sun's radiation.

One principle related to the origin of the solar system was proposed by Babinet in 1861. Newtonian physics had produced the principles of conservation of mass and conservation of angular momentum, which said that in a closed system, both mass and angular momentum are constant. Each must retain its initial value. *Babinet's criterion* applied this idea to the solar system and said that the present-day total mass of the system and its present-day angular momentum must equal the initial values. We shall see, however, that this principle is rejected in twentieth-century theories because the solar system is not necessarily closed; it can eject material into interstellar space.

CHARACTERISTICS TO BE EXPLAINED BY A SUCCESSFUL THEORY

We have now gone far enough in our review of theories that we should stop for a moment and ask what it is that we are really trying to explain.

The contemporary theorist H. Alfvén (1954) has remarked, "To trace the origin of the solar system is archeology, not physics." By this he meant that we do not start out, as in most physics problems, with a ready-made set of initial conditions, because we do not know what the initial conditions were. (In Chapter 5 we shall use astrophysics to get a better grip on that problem.) Instead, we have to work backward through time like an archeologist, using available orbits, meteorite samples, comet spectra, and similar data in much the way an archeologist uses a fortuitous charred bone or pottery shard.

We shall now list some observed solar-system properties that a theorist is obliged to explain. The rationale of this list is quite important. These are the observational facts for which there is no *obvious* explanation in terms of present-day processes. Kepler's laws would not be included since they can be explained by Newtonian mechanics. The characteristics in this list must have come about as a result of the process of origin. Theory must predict these characteristics as its end products.

1. *Planetary orbits are coplanar.* The solar system has the shape of a very flat disk. The planets whose orbits are most inclined to the ecliptic are

Pluto	17°.2 (escaped satellite of Neptune?)
Mercury	7.0
Venus	3.4

2. *All the planets orbits lie near the plane of the sun's rotation.* The sun's inclination to the ecliptic plane is only 7°.

3. *Planetary orbits are nearly circular.* The three greatest eccentricities are

Pluto	0.249 (escaped satellite of Neptune?)
Mercury	0.206
Mars	0.093

4. *The directions of planetary revolution are all direct and the same as the solar rotation.*

5. *With a few remarkable exceptions the rotations of planets are direct with small obliquities* (angle between rotation axis and pole of ecliptic). The three greatest obliquities are*

Venus	174°	(retrograde, Carpenter, 1970)
Uranus	98°	(retrograde)
Neptune	29°	(prograde)

6. *Planetary distances can be given by a relatively simple relation* (Bode's law). Each planet tends to be roughly twice as far from the sun as its next inward neighbor. An important result which would have removed this characteristic from our list was claimed by the Soviet theoretician Molchanov (1968). He argued that Bode's law was not a result of a primeval origin process but a resonance effect caused by the normal workings of the laws of celestial mechanics. However, Molchanov's argument has been strongly criticized by various workers (see *Icarus, 11,* No. 1, 1969; and Dermott, 1969).

7. *Satellite systems mimic the planetary systems.* Among the multisatellite systems of the outer planets, the major satellites form systems lying in the plane of the planets' rotation, with direct revolution and other properties similar to those of the planets. Even Bode's law finds its analog among the spacings of satellites (Alfvén, 1954; Dermott, 1968).

8. *Planets together contain much more angular momentum than the sun.* (The failure of Laplace's nebular ring theory to predict this was its major flaw.) Satellite systems, on the other hand, contain angular momentum less than or comparable to that of their parent planets.

9. *Comets form a class of objects with highly discordant, almost randomly distributed orbits, as opposed to the concordant planet, satellite, and asteroid orbits.*

10. *The various planetary bodies have different compositions as evidenced by different mean densities and by terrestrial, lunar, and meteorite samples.* This includes not only different abundances of chemical elements but also certain differences among isotopes of single elements. There may be a systematic decrease in percentage of heavy elements from Mercury out to Saturn.

11. *Periods of rotations of planets are related in such a way as to suggest an initial tendency to produce the same period for all planetary bodies* (Brosche, 1963; MacDonald, 1963; Alfvén, 1964; Hartmann and Larson, 1967; Fish, 1967).

12. *In addition to the systematic differences in composition remarked in property 10, there is a noticeable division between the "giant" or "Jovian planets"* (Jupiter, Saturn, Uranus, and Neptune) *and the "terrestrial planets."* The former are much larger and occupy a distinct zone of the solar system.

* Pluto uncertain, Mercury probably smaller than any of these (Goldstein, 1970).

EARLY-TWENTIETH-CENTURY THEORIES

In the first decades of our century it seemed that the nebular hypothesis of Kant and Laplace had died a natural death. For one thing, there was the seemingly insurmountable problem of how to have the nebula contract to form the sun without ending up with the sun rotating very fast and containing a large fraction of the total angular momentum. There was still the problem of how to generate Laplace's rings, and on top of that, astrophysicists had concluded that hot rings with the mass of the present planets would not condense into planets but would expand and dissipate into space.

Therefore, catastrophic theories of the Buffon type were revived. The Chamberlin–Moulton hypothesis was one of the most popular of these. Chamberlin (1901) and Moulton (1905) proposed that another star passed so near the sun that it raised high tides and drew filaments out of the sun, as shown in Figure 4-5. These formed clouds which subsequently condensed

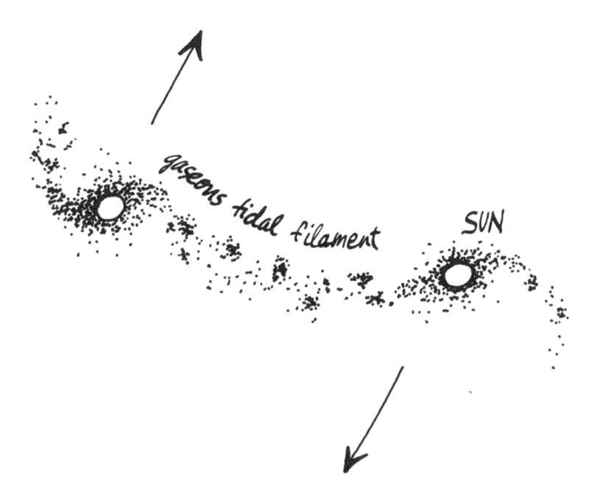

Figure 4-5
A subfragmenting tidal filament drawn from the sun by a passing star, as envisioned by Chamberlin and Moulton (1905).

into a multitude of small particles which accreted to form the planets. Since the attracting star passed by the sun, it was supposed to impart a lateral motion to the ejected filaments, putting them into orbit around the sun. Jeffreys (1916) criticized this model on the grounds that the small particles, or planetesimals, would collide at such high velocities that they would vaporize instead of sticking together. Jeffreys, together with Jeans (1917), produced alternative theories in which the approaching star made a grazing collision with the sun and drew out a fairly massive, cigar-shaped filament, which condensed into the separate planets. The densest region was supposed to be in the central part of the cigar, so the giant planets formed at an intermediate distance from the sun, as shown in Figure 4-6.

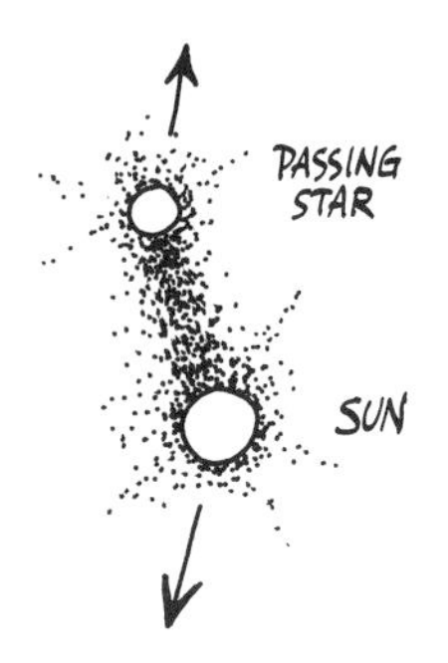

Figure 4-6
A cigar-shaped filament torn from the sun by a grazing collision with a passing star, as envisioned by Jeffreys (1916) and Jeans (1917).

Clearly these catastrophic theories involve a lot of ad hoc assumptions, improbabilities, and physical difficulties. The frequency of stellar collisions in our galaxy can be calculated, and it turns out to be extremely rare. If the solar system were formed by such an event, it might be one of only a few such systems among the hundred billion of stars of our galaxy. Of course one can always answer, a posteriori, that ours is that one system! But astrophysicists soon showed that the theory was unattractive on more substantial grounds: It did not predict planetary rotations, it did not predict the right distribution of angular momentum in the solar system, and it did not adequately explain how hot material torn directly from the sun could contract rather than expand into space and dissipate.

Theorists looked for another way to generate a dense circumsolar cloud of gas from which planets could form. Russell (1935) suggested that the sun might once have been half of a binary system, and that the other star exploded to form a dense cloud of gas. (Such exploding stars are called novae and are a late stage in the evolution of at least some stars.) Both theory and observation, however, suggest that novae produce giant, expanding clouds rather than dense clouds that could contract or self-gravitate to form planets. Lyttleton (1941) tried to salvage this idea by making the sun originally a member of a triple-star system, but his theories met the same difficulties. The same thing happened to still another attempt to salvage the idea on the part of Hoyle (1946), who made some assumptions about the binary undergoing an asymmetric *supernova* explosion.

The result of the growth of electromagnetic theory and the increasing dissatisfaction with the catastrophic theories led to a new class of theories which invoked electromagnetic fields. An early example was a quantitative theory by Birkeland (1912), who assumed that the atoms in the nebula had become ionized. The motions of the different atoms would be affected not only by their masses but by their total charge. In this way, atoms of different charge/mass ratios would be separated by the action of an assumed solar electric field. Unfortunately, Birkeland argued himself into a situation where each planet was to be made from a different isotope; the earth was supposed to be mainly phosphorus! The theory, with what a later critic called its "many weird points," was a failure, but it did introduce the important idea of electromagnetic fields influencing motions in the solar nebula.

Alfvén (1954), who was ultimately to win the Nobel Prize for his work on magnetohydrodynamics, published a series of papers stretching from 1942 to the present and made more fruitful use of this idea. He used not an assumed electric field of the sun but the magnetic field. Alfvén went so far as to assume that the magnetic field was the dominant phenomenon in the formation of the system. Ionized atoms (atoms that have lost or gained one or more electrons and thus have an electric charge) in the earth's vicinity would experience a greater force from the sun's magnetic field than from its gravitational field. Alfvén assumed that ions would have been plentiful in the solar nebula and that therefore its evolution was governed not by simple dynamical effects but by *magnetohydrodynamic* effects—effects of magnetic fields acting on ionized gasses. The principal problem with this theory is that most modern investigators do not believe that significant numbers of atoms in the outer parts of the solar nebula would have been ionized. Neutral atoms would move unaffected by the magnetic field, and as long as the nebula is neutral, magnetohydrodynamics does not apply.

However, one exceedingly important magnetic effect has been adopted by most contemporary authors. The sun itself is composed of ionized atoms, and disturbances on the solar surface continually spew these out into space. At the present time, such a process feeds solar wind, the field of ions moving outward from the sun. During the origin of the solar system, the early sun might very well have had a high density of ions nearby, even if the rest of the nebula was not ionized. If the sun originally had a strong magnetic field and was surrounded by ionized gas in its immediate vicinity, the gas would be controlled by the motions of the magnetic field. In the parlance of modern physics, we express this by saying that the nebula's ionized particles would be *frozen* to the magnetic field, which means that the magnetic field carries the ions along with it. If the sun formed from a contracting nebula, the sun and its magnetic field would be rapidly turning, and the magnetic field would "stir" the gas. Figure 4-7 shows what would happen next. As a result of the contraction process, the sun would be spinning faster than the nebular particles. Therefore, the sun's magnetic field would tend to grip the nebula and speed it up, while the nebula in turn would cause a drag on the magnetic field and slow down the solar rotation. *The magnetic field is thus the probable solution to the old problem of why the sun is rotating much slower than was expected by Laplace.* It seems impossible to solve this problem by pure mechanics, but magneto-

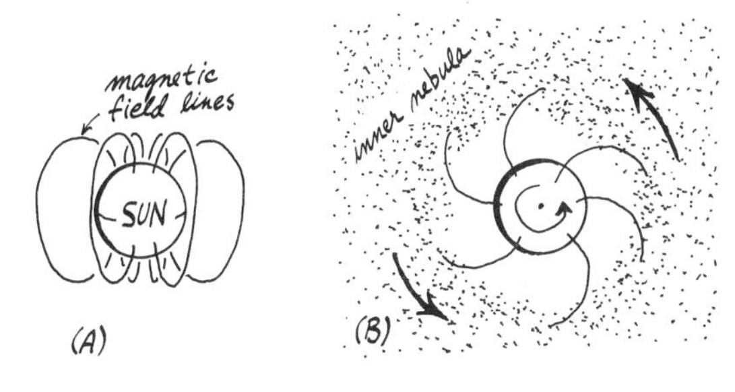

Figure 4-7
Magnetic braking of the early sun's rotation. A: "Side" view of the sun's magnetic lines of force. B: "Top" view of the rotating sun, showing effect of drag by ionized gas on the magnetic lines of force.

hydrodynamics at once gives a solution. As we shall see in Chapter 5, observations of young stars support this theory.

Having shed at least a glimmer of light on the problem of the sun's rotation, the nebular hypothesis and the evolutionary theories were quickly adopted by most researchers. The problem of the origin of the solar system became a problem of working out the details of the evolution of the nebula. How could planets form in a disk-shaped cloud of gas rotating around the sun?

Before proceeding to theories that attack this question we should comment on the origin of the assumed solar nebula. *Solar nebula* is a term used to designate the cloud of dust and gas around the sun. The nebula was for the most part probably a residual cloud of the same material from which the sun was formed. We shall see in Chapter 5 that modern theories of the origin of stars suggest such a leftover cloud. An alternative source for part of the nebula, however, would be material ejected from the sun (Hoyle, 1946).

THREE POSSIBLE MODES OF PLANET ORIGIN

Let us return to the question of how the planets grow in the nebula. There are three possible processes.

1. *Gravitational collapse* is the process by which sufficiently dense regions in the nebula have enough self-gravitational force to begin to contract. Once the contraction starts, barring external energy sources or disruptions, the region can collapse all the way to planetary dimensions. This was essentially Kant's view of the formation of the solar system.

2. *Accretion* is the process by which small particles collide and stick together to make larger particles that ultimately grow into planet-sized objects. A familiar example is the sticking of snowflakes as they fall through the air, producing occasional snowflake clusters.

3. *Condensation* is growth by the sticking of individual atoms and molecules to a growing particle. Condensation is the process by which raindrops and *individual* snowflakes grow.

INTRODUCTION TO CONTEMPORARY THEORY

The first group of current theories concentrated on mass motions. In 1944 and 1946 the German physicist C. F. von Weizsäcker studied *turbulence* in the nebula (see the excellent summary in English by Chandrasekhar, 1946). Von Weizsäcker's work indicated that turbulence was indeed a major factor in the motions of the nebula. It created local whirlpool-like eddies—or vortices reminiscent of those postulated 300 years before by Descartes. Von Weiz-

Figure 4-8
The solar nebula divided into turbulent eddy-zones, as envisioned by von Weiszäcker (1946).

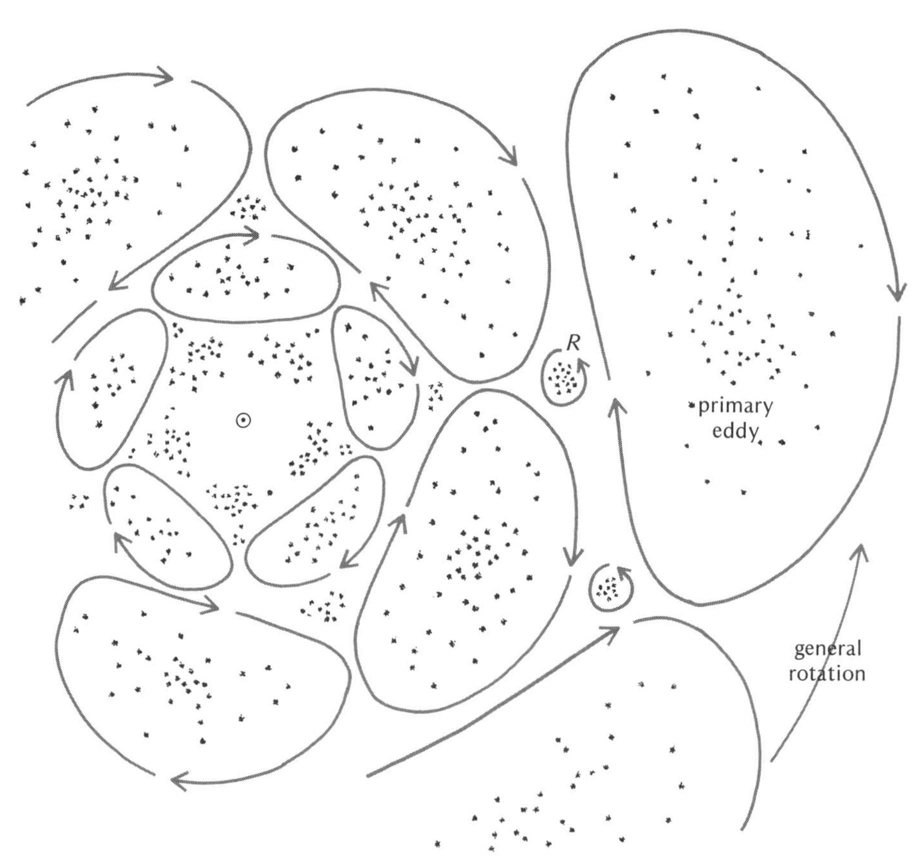

sacker argued that these "primary" eddies would tend to arrange themselves in a certain "packing order." The result would be that the nebula would automatically partition itself into zones. Von Weizsacker showed that under certain conditions these zones could have just the spacing predicted by Bode's law. Since the primary eddies would have a retrograde sense of rotation, von Weizsäcker assumed that planets would grow in the smaller secondary—or "roller-bearing"—eddies located at the contact points of the primary eddies, as shown by Figure 4-8. This would ensure direct rotation. A weak point in the von Weizsäcker theory was that his model of turbulent eddies was derived by mentally following the relative motions of adjacent atoms in Keplerian orbits around the sun. In reality, if the nebula were dense enough to spawn the planets, the atoms would not have long enough free paths to move in complete Keplerian orbits. Rather they would be constantly perturbed by many collisions. Thus von Weizsäcker's turbulent theory was internally inconsistent.

Ter Haar (1948, 1950) pointed out that von Weizsäcker's turbulence model was suspect and concluded by means of a revised turbulence model that the nebula would be too low in density for turbulent eddies to have any important gravitational effect. Rather, ter Haar concluded that small grain-like particles would condense out of the cooling nebula and that they would grow by accretion. Ter Haar's most important contribution was probably the recognition of the importance of condensation in providing particles with different compositions, varying with temperature of the gas and hence with distance from the sun. Ter Haar concluded that heavier metallic particles would be more abundant in the inner regions of the nebula while hydrogen-rich icy particles would dominate in the outer part. This would account for the composition variations from the terrestrial planets outward to the Jovian planets.

The American astronomer Kuiper (1951) enlarged upon von Weizsäcker's and ter Haar's work in an important synthesis of views. First, Kuiper used a revised turbulence theory developed by the Russian physicist Kolmogorov. This indicated that the turbulent eddies would be more chaotically distributed but that the major "primary" eddies would dominate zones at difference distances from the sun, hence leading to Bode's law. Next, Kuiper assumed that grains would condense and begin to grow by accretion, but he found that collisions between grains in the "roller-bearing eddies"—shown in Figure 4-9—would be too violent to be satisfactory. Therefore, he was forced to accept the *retrograde*-rotating primary vortices as the source of planets. There were two difficulties with this: The planets do not rotate in a retrograde sense and the von Weizsäcker theory showed that individual eddies in the vortex pattern had lifetimes much shorter than the 10^7 to 10^8 (10,000,000 to 100,000,000) years that Kuiper found necessary for accretion to work.

To escape these difficulties, Kuiper next postulated that the nebula was dense enough that the individual eddies would be stabilized by their own self-gravitation. Therefore, instead of forming and dissipating they would form and

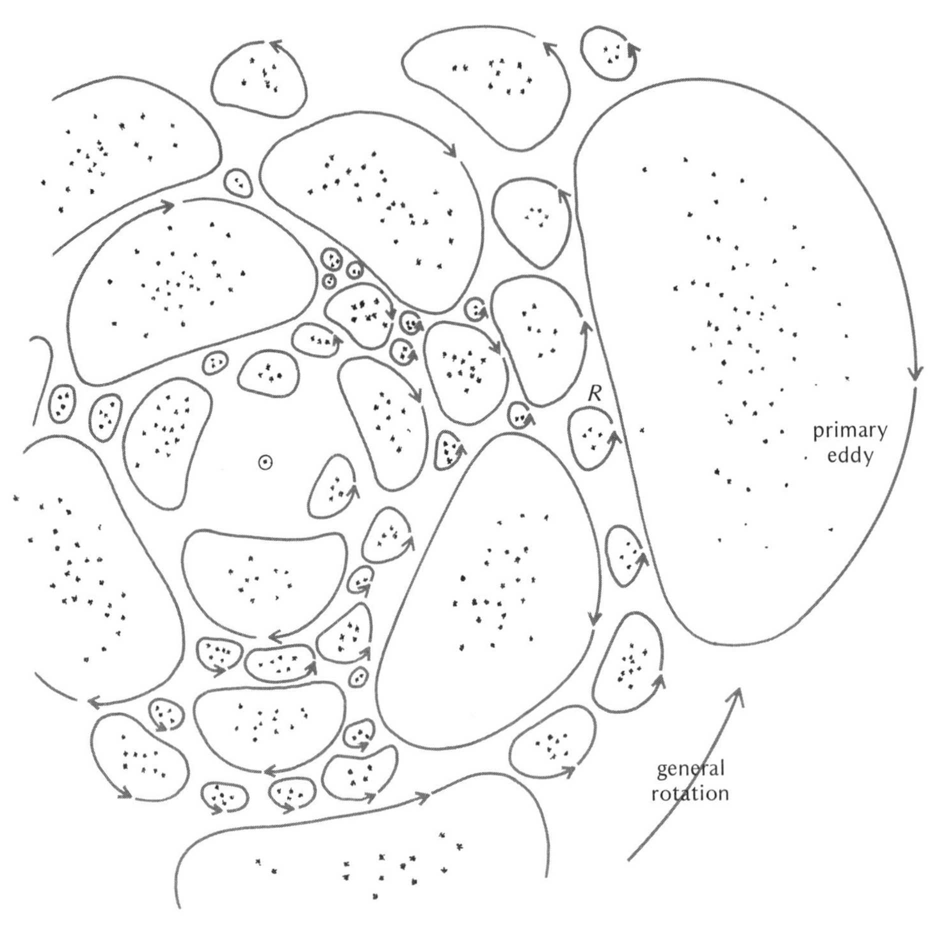

Figure 4-9
The solar nebula divided by turbulence into protoplanets, as envisioned by Kuiper (1951). Smaller turbulent eddies act as "roller bearings" (*R*).

divide the nebula into large gravitationally stabilized, long-lasting units which Kuiper called *proto-planets*. A proto-planet could be several hundred times more massive than the final planet it would create, but it would contain most of the material that would ultimately form the given planet and its satellites. Because the proto-planets were long-lived portions of the nebula, the sun would raise tides on them and lock their rotation into direct sense rotation—thus solving the embarrassing problem of the initial retrograde rotations mentioned above.

Kuiper's proto-planet theory thus combined self-gravitation, accretion, and condensation. It had the advantage of using modern turbulence theory and predicting the proper spacings of planets, the proper sense of rotation, and other properties. The proto-planet theory was perhaps the most widely quoted theory during the last two decades.

Nonetheless, the proto-planet theory, like others, had drawbacks. It was criticized on several grounds:

1. A complete explanation of the role of gravitation was not provided. If gravitational collapse of the proto-planets had been effective, the accretion and condensation processes become irrelevant, since all the mass of the proto-planet is gravitationally trapped and a planet will form whether these processes operate or not.

2. For the nebula to divide into self-gravitating units an unattractively high density is required.

3. If the planets form by gravitational collapse, we would expect all of the heavier elements, such as xenon, to be incorporated in them in their cosmic abundances, but this is not observed.

Not widely known to Western scientists until the late 1950s, the Russians had been developing their own school of cosmogony, principally under the guidance of O. Schmidt, who published an English summary of his work in 1958. In the 1940s, Schmidt developed a number of cosmogonic ideas independent of Western scientists. Subsequently, Schmidt studied the Western work and questioned several aspects of the proto-planet idea. How could the earth and other planets have lost most of the mass that was originally gravitationally tied to the immense proto-planet? Is it not artificial to assume that there will be just one proto-planet identified with each finished planet? Schmidt argued that the "central problem of planetary cosmogony" was to define the state of material in the nebula before the planets formed, and he argued that one could do this only by studying the properties of the planets as we see them.

Schmidt pointed out that the circular orbits and other orbital characteristics of the planets can be explained by the dynamical averaging of a large number of small particles in a resisting medium, not by motions of a few large entities. "When large numbers of bodies were joined into a single planet their orbits were, naturally, averaged, and as a result they could only be . . . circular, and close to (the ecliptic) plane." Schmidt thus parallels ter Haar and Kuiper in picturing a multitude of small dust grains in the nebula, but he departs from von Weizsäcker and Kuiper in discounting the importance of self-gravitating proto-planets. Schmidt pictures the planets as growing by collisions between dust grains in the nebular medium through which they are moving. The dispersed material is thus not just a resisting medium, he says; it is also a "feeding medium."

Another branch of scientific observation with important consequences for solar-system origin was from the study of the chemistry of meteorites. Especially due to the work of the geochemist V. Goldschmidt, the chemist and Nobel laureate H. C. Urey (1952), and other workers, scientists realized that the meteorites were primeval material, dating back to 4.6 or 4.7 $\times 10^9$ (4.6 or 4.7 billion) years ago, when the planets formed. Urey's work in particular emphasized that the meteorites contained clues to the duration of the forming process, the temperature, and chemical composition of the original material. Urey also showed that chemical models of the condensation of different elements and compounds in the solar nebula could help explain differences in compositions of the planets.

Many contemporary workers are contributing to our view of the solar-system's origin. Before proceeding with a synthesis of modern views of how the solar system formed, let us turn to a different branch of science—astronomy—to see what *observational* data are available on the problem of formation of stars and planetary systems. We shall discuss this in Chapter 5 and then return to the solar system in Chapter 6.

References

Alfvén, H. (1954) *On the Origin of the Solar System* (New York: Oxford University Press, Inc.).

______ (1964) "On the Origin of the Asteroids," *Icarus, 3,* 52.

Birkeland, K. (1912) *Compt. rend. acad. sci., 155,* 892.

Brosche, P. (1963) "Über das Masse-Drehimpuls Diagramm von Spiralnebeln und anderen Objekten," *Z. Astrophys., 57,* 143.

Buffon, G. L. L. (1745) *De la formation des planètes* (Paris).

Carpenter, R. L. (1970) "A Radar Determination of the Rotation of Venus," *Astron. J., 75,* 61.

Chamberlin, T. C. (1901) "The Genesis of Planets," *Astrophys. J., 14,* 17.

Chandrasekhar, S. (1946) "On a New Theory of Weizsäcker on the Origin of the Solar System," *Rev. Mod. Phys., 18,* 94.

Dermott, S. F. (1968) "On the Origin of Commensurabilities in the Solar System—II," *Monthly Notices Roy. Astron. Soc., 141,* 363.

______ (1969) "On the Origin of Commensurabilities in the Solar System—III," *Monthly Notices Roy. Astron. Soc., 142,* 143.

Descartes, R. (1644) *Principia Philosophiae* (Amsterdam).

Fish, F. F. (1967) "Angular Momenta of the Planets," *Icarus, 7,* 251.

Goldstein, M. R. (1970) "Mercury: Surface Features Observed during Radar Studies," *Science, 168,* 467.

Hartmann, W. K., and S. Larson (1967) "Angular Momenta of Planetary Bodies," *Icarus, 7,* 257.

Hoyle, F. (1946) "On the Condensation of the Planets," *Monthly Notices Roy. Astron. Soc., 106,* 406.

Jeans, J. H. (1917) "The Motion of Tidally-Distorted Masses, with Special Reference to Theories of Cosmogony," *Mem. Roy. Astron. Soc., 62,* 1.

Jeffreys, H. (1916) "On Certain Possible Distributions of Meteoric Bodies in the Solar System," *Monthly Notices Roy. Astron. Soc., 77,* 84.

Kant, I. (1755) *Allgemeine Naturgeschichte und Theorie des Himmels.*

Kuiper, G. P. (1951) "On the Origin of the Solar System," in *Astrophysics*, J. A. Hynek, ed., p. 404 (New York: McGraw-Hill, Inc.).

Laplace, P. S. (1796) *Exposition du Système du Monde, Paris.*

Lyttleton, R. A. (1941) "On the Origin of the Solar System," *Monthly Notices Roy. Astron. Soc., 101,* 216, 349.

MacDonald, G. J. F. (1963) "The Internal Constitution of the Inner Planets and the Moon," *Space Sci. Rev., 2,* 473.

Molchanov, A. M. (1968) "The Resonant Structure of the Solar System," *Icarus,* 8, 203.

Moulton, F. R. (1905) "On the Evolution of the Solar System," *Astrophys. J., 22,* 165.

North, J. D. (1965) *The Measure of the Universe* (New York: Oxford University Press, Inc.).

Radin, Paul (1927) *Primitive Man as Philosopher* (Appleton-Century-Crofts; New York: reprinted by Dover Publication, Inc.).

Russell, H. N. (1935) *The Solar System and Its Origin* (New York: The Macmillan Company).

Schmidt, O. (1958) *A Theory of the Origin of the Earth* (Moscow: Foreign Language Publishing House).

Ter Haar, D. (1948) *Kgl. Danske Videnskap. Selskab. Mat.-Fys. Medd., 25,* 3.

______ (1950) "Further Studies on the Origin of the Solar System," *Astrophys. J., 111,* 179.

Urey, H. C. (1952) *The Planets: Their Origin and Development* (New Haven, Conn.: Yale University Press).

von Weizsäcker, C. F. (1944) "Über die Entstehung des Planetensystems," *Z. Astrophys., 22,* 319.

______ (1946) "Über die Entstehung des Planetensystems," *Naturwissenschaften, 33,* 8.

Wilson, J. A. (1951) *The Culture of Ancient Egypt* (Chicago: University of Chicago Press).

THE FORMATION OF STARS

5

IN THE PRECEDING CHAPTER we reviewed the history of one method of attack on the origin of the solar system. This was the traditional approach: looking at the earth, sun, and other planets, listing their characteristics, and then trying to explain them. This approach is solar-system-oriented and involves interdisciplinary studies by celestial mechanicians, physicists, and geochemists—all pursuing planetology. In a sense, this approach starts with the earth and works outward.

In this chapter we study a different approach—parallel but separate from planetology—which starts with the stars and works back toward the planetary system. This approach is through stellar astronomy and astrophysics. For 50 years scientists in this area have been developing increasingly reliable theories of how stars form. They have worked independently of the planetary evidence, for only recently has it been widely accepted that the planets could give evidence of star-forming processes.

This gives us an interesting scientific situation. The two independent disciplines (planetology and astrophysics), both investigating the same problem (how a star formed), should have come up with answers that dovetail perfectly. In actual practice the two lines of thought can be made to dovetail only with effort. There is scarcely an astrophysical theory of star formation that predicts the existence of planets, not to mention the detailed characteristics that we listed in Chapter 4. The reason is that traditionally, astrophysicists have not set out to predict planetary systems. Their models have been designed to predict stars of various masses. It appears that planets form from residual gas and dust surrounding the star during and after the formation process, and astrophysical theories for the most part have not been sufficiently detailed to trace the evolution of this "leftover" material.

Therefore, there exists a fertile interdisciplinary area for new work, lying halfway between planetology and astrophysics. We shall now explore the astrophysical side of this gap.

THE SOLAR-SYSTEM'S PLACE AMONG THE STARS

The solar system is only a minute detail in the vast system of stars. To show this, let us review the structure of the universe in a series of progressively finer and finer stages.

The *universe* is everything that exists. The universe is filled, as far as we

Figure 5-1

A spiral galaxy of form similar to that of the Milky Way. The bright nucleus and spiral arms are composed of billions of stars, and the dark patches and lanes are clouds of dust particles.

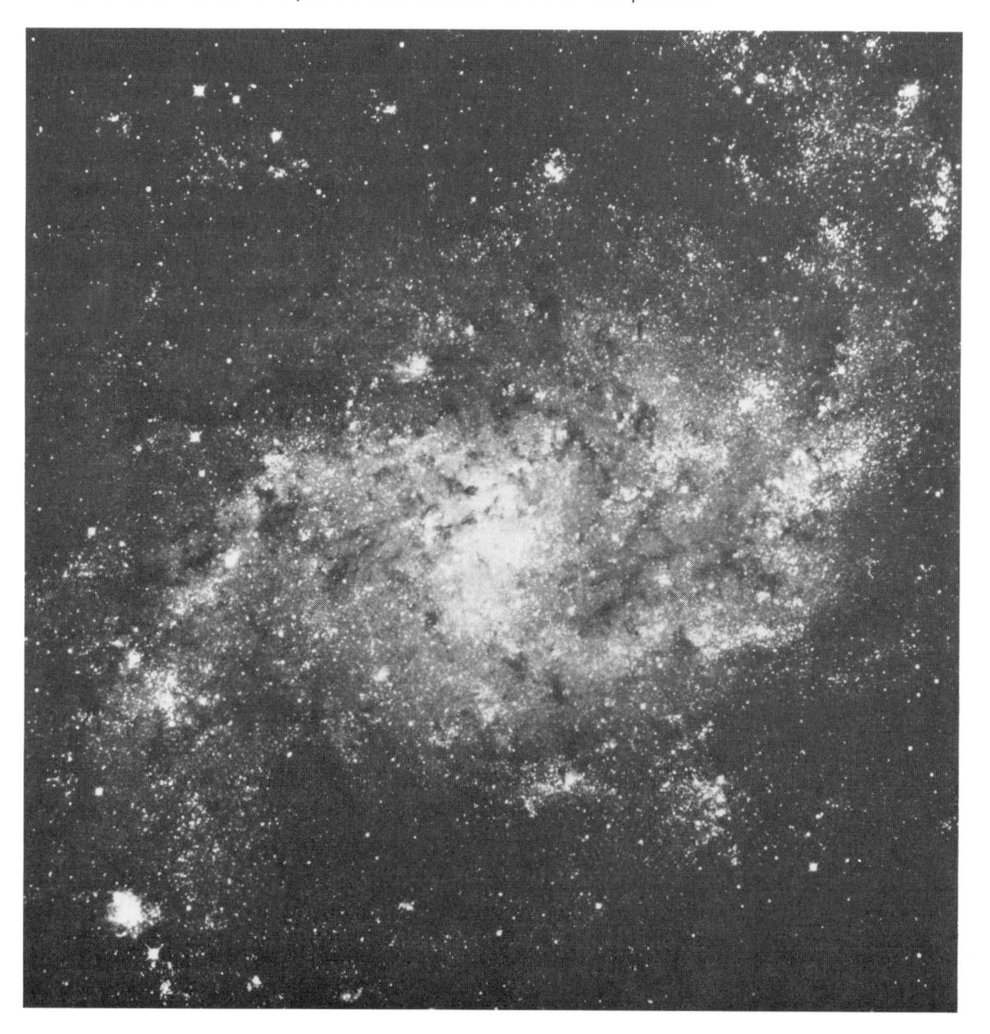

can see, with *galaxies*. Galaxies are not randomly distributed through the universe, but often occur in local groupings. For example, our own galaxy, known as the Milky Way because of its visual appearance from earth, is a member of a grouping that also contains the famous Andromeda galaxy and its small companion galaxy, two nearby irregular galaxies known as the Magellanic Cloud (visible from the earth's Southern Hemisphere), and two recently discovered galaxies known as Maffei 1 and Maffei 2.

Each galaxy consists of billions of stars. The stars may be arranged in different ways, forming galaxies with a sequence of different shapes, including tightly packed elliptical galaxies, irregular galaxies, and the familiar spiral galaxies (Figure 5-1). The Milky Way is a spiral galaxy.

Spiral galaxies are disk-shaped, with a central bulge called the *nucleus*, and a more spherical, thin halo of stars and globular star clusters. The stars of the galactic disk are arranged in spiral arms, loosely or tightly winding around the nucleus, and the whole system rotates. In our own spiral galaxy there are about 100 billion stars, and one rotation of the sun around the center takes 200 million years. Mixed with the stars in the spiral arms are immense clouds of dust and an ambient field of gas. This interstellar medium will be discussed later in greater detail.

Some idea of the relative sizes of the galaxy and solar system can be gained if we introduce one of the astronomical units of distance, the *parsec*, which is equal to 3×10^{18} cm or 3×10^{13} km. The sun is about 10,000 parsecs from the center of the galaxy. The distance between stars in the sun's neighborhood is of the order of 1 parsec, and the diameter of the solar system is of the order of 1/10,000 parsec. When William Herschel discovered in the 1790s that the solar system was but a minute part of the Milky Way galaxy, there was no way of knowing that many other galaxies existed outside the Milky Way. Indeed, it was thought for more than a hundred years thereafter that all the stars of the universe belonged to only one disk-shaped galaxy, the Milky Way. No one knew how or when the stars formed. Modern knowledge of stars and star formation traces back to the first decades of this century, when the principles of spectroscopy and nuclear sources of energy were recognized and applied.

EVIDENCE FOR CONTINUING FORMATION OF STARS

If we are to use stars as examples in reconstructing the history of the solar system, the first question is whether all the stars of the galaxy were formed at once, or whether star formation is an ongoing process, active even at the present day. There is a good bit of evidence for the latter view.

Youth of the Solar System

Recently evidence has accumulated that all the galaxies formed after some kind of primeval, explosive creative event that occurred probably 8 to 24×10^9 (8 to 24 billion) years ago. This was the creation of the universe, as best we can tell. Yet our solar system formed only 4.7×10^9 years ago, as determined from radioisotopic studies of meteorites (see Chapter 6). This shows that formation of stars within our own galaxy continued long after the creation.

Short-Lived Massive Stars

Astronomical measurements show that the most massive, hottest, and brightest stars are using up their nuclear energy at a very fast rate. We can measure the rate of energy loss and calculate the amount of energy available, and so calculate how long the star can last. It turns out that such stars have lifetimes of only a few million years. This is very short by astronomical standards, only a hundredth of a percent of the age of the universe. Therefore, these stars have formed very "recently." The brightest star in the sky, Sirius, is an example of one of these stars. Its bluish color is a consequence of its high temperature. (The hotter an object gets, the bluer its emitted light. The reddish stars are the coolest.)

Short-Lived Star Clusters

Many young, massive stars are associated with groupings called *open star clusters*. Dynamical studies show that open star clusters cannot last very long, because during the galaxy's ponderous rotation, each star individually orbits around the nucleus. For example, while the sun takes roughly 2×10^8 years to complete one revolution, stars closer to the nucleus require less time. This is a result of Kepler's laws (just as on a much smaller scale, the inner planets Mercury and Venus take less time than the earth to go around the sun). Because each star moves in a separate orbit around the galactic nucleus, there is a shearing effect which tears apart star clusters. Star clusters are also subject to tidal distortions caused by the gravitational attraction of the galaxy's nucleus, and the individual stars are subject to perturbations caused by interaction with other nearby stars in the cluster. All these effects tend to disrupt star clusters, and calculations show that star clusters rarely last more than a few hundred million to a billion years. Thus, when we look at the night sky and see clusters of stars, we know that these groupings must have formed comparatively recently.

The Pleiades, or "seven sisters," a familiar sight in winter skies, is an example of a star cluster. The bright stars in the Pleiades are hot blue massive stars of the kind discussed above. The age of the Pleiades is estimated to be about 50 to 100 million years. In a direction near the Pleiades is another familiar star cluster, the Hyades, forming the bull's face in the constellation Taurus. The age of the Hyades is estimated to be several hundred million years.

Expanding Gas and Runaway Stars

In the same regions where star formation seems to be occurring we often find violent motions. Clouds of gas are being pushed away from the center of activity—apparently by the radiation pressure from the most massive, superbright blue stars. A few individual stars, called *runaway stars,* are moving out fast enough that we can track their motions by comparing photographs made in different years. These are probably stars that have been ejected from young star clusters by perturbative interactions or disruption of fast-orbiting binary stars (pairs that orbit around their mutual center of mass). Motions can be projected backward to calculate how long ago the gas and stars were ejected from the star-forming regions.

The example *par excellence* of expanding material and runaway stars is the region of the constellation Orion. The middle "star" in Orion's sword is revealed by the telescope to be a luminous gas cloud with a central tight star cluster known to astronomers as the "trapezium." The gas and dust density is

Figure 5-2

The region of Orion—a view down one of the Milky Way's spiral arms. This picture is a mosaic of six exposures made with an ordinary 35-mm camera, hand-guided for 5 minutes each at f/2 on Tri-X film.

much higher in the Orion region than in most of interstellar space, and star formation seems to be proceeding there at a great pace. The region is a hot-bed of activity. Expansion of great clouds of hydrogen has been observed. Some clouds were blown out only 10^4 to 10^5 (10,000 to 100,000) years ago (Blaauw and Morgan, 1954). Star-forming activity has been going on in the Orion vicinity for several million years.

The stars and star clusters that we have mentioned all lie in the same direction from the sun. Sirius, Orion, the Hyades, and the Pleiades lie at different distances along one of the spiral arms of the galaxy, and when you look out toward Orion on a February evening—a view shown in Figure 5-2—you are looking right down that spiral arm. If you swing your head around toward different parts of the sky, you quickly become aware of the great concentration of bright stars in the direction of Orion—a consequence of the large number of young, massive stars formed in the dust clouds of that region of our local galactic spiral arm.

Association of Young Stars with Gas and Dust

Theoretical analysis of star formation inevitably shows that if stars are to form, they must form in regions where the gas density is higher than usual, with enough self-gravitation and enough raw material to supply the formative process. In support of this we find that the stars thought to be young are associated with clouds of gas and dust. The Pleiades, for example, are embedded in such a cloud, which reflects the light of the local stars and is hence called a *reflection nebula*. This nebulosity is shown in Figure 5-3, which is a telescopic photograph of the Pleiades.

These five pieces of evidence of star formation assure us that we can get observational evidence of the star-forming process in action if we know what to look for. Before starting such a search, let us consider what is known about the interstellar medium from which stars are born.

THE INTERSTELLAR MEDIUM

The *interstellar medium* is the gaseous environment between the stars. Since stars are forming from this material, an understanding of its properties must precede any understanding of star formation. The medium includes gas, tiny dust grains, and magnetic and radiation fields.

Interstellar Gas

The composition of the interstellar medium by mass is roughly $\frac{2}{3}$ hydrogen, $\frac{1}{3}$ helium. In addition, there is about 2 percent carbon, nitrogen, and oxygen;

Figure 5-3
The Pleiades—an open star cluster. (Steward Observatory.)

and a fraction of 1 percent of heavy elements, sometimes loosely lumped together under the heading "metals": magnesium, silicon, iron, nickel, chromium, and so on. Material with such elemental abundances is said to have *cosmic composition*.

The interstellar medium is not uniform in its properties. Astronomers often divide the interstellar gas into two regimes: the cool *HI regions,* and the scattered, hot *HII regions*. *HI* is the symbol for neutral hydrogen gas; *HII* is the symbol for ionized hydrogen in which each atom has lost its electron. The HI regions comprise the larger part of the interstellar medium, where the temperature lies in the range 25 to 125°K, too cool to ionize the hydrogen. HII regions, on the other hand, are regions surrounding hot stars, where the interstellar gas is heated to several thousand °K and the gas is ionized.

In regions where star formation has been occurring, very hot stars may have heated the original HI medium, creating new, expanding HII regions which introduce turbulence as gas rushes out into the normal interstellar medium. Thus, even if we could magically calm the interstellar gas, hot stars would cause new HII regions which would introduce new motions.

The density of the interstellar medium is variable. In most regions there are about 1 to 5 atoms/cm^3, but in the denser clouds densities may be much higher. Recent measures of clouds in the Orion complex (Werner and Harwit, 1968) indicate hydrogen *molecule* densities as high as 10^3 to 10^4 (1000 to 10,000) molecules/cm^3. This indicates the high gas density in star-forming regions as well as the tendency to form molecules when the gas density is high. Complex molecules have been found. Recent discovery of organic interstellar molecules, such as formaldehyde (H_2CO), has heightened interest in the possibility of life elsewhere in the galaxy (see Chapter 13).

Interstellar material thins out away from the plane of the Milky Way disk. Measurements of halos of other galaxies, bordering on intergalactic space, suggests gas densities less than 10^{-3} ($\frac{1}{1000}$) atoms/cm^3 (Roberts, 1966). When the void of intergalactic space is taken into account, the mean density of matter in the entire universe is found to average about 10^{-8} atom/cm^3.

Interstellar Dust Grains

Floating among the atoms and molecules of interstellar gas are tiny grains of dust. They have been detected primarily because they redden the light of distant stars, much as dust in our own atmosphere helps redden the light of the setting sun. Calculations based on the amount of reddening show that the grains are about 10^{-5} (1/100,000) cm in diameter.

No one knows the exact composition of the grains. Ices, such as frozen water, methane, or ammonia, have been suggested as constituents, because such compounds are composed of common, chemically active elements in the interstellar gas (H, O, C, and N). Icy grains are also favored by observations of grains in reflecting nebulae, whose properties indicated that the grains had relatively high albedo. When it was found that the grains are apparently affected by magnetic fields in space,* iron was suggested as a constituent. Other suggested constituents include diamond (carbon), graphite (carbon), and silicates similar to those that form the minerals in common rocks.

Many contemporary models of the grains involve composite compositions, such as a silicate–iron core surrounded by an icy mantle.

No one is certain of the origin of the grains, although theories of origin have been proposed and have influenced the estimates of the grains' compositions. When it was shown by early Dutch investigators (Oort and van de

* The grains are not all spherical, and polarization studies show that they tend to lie with their long axes aligned in the same direction—possibly determined by the galactic magnetic field.

Hulst, 1946) that there were not enough atoms in interstellar space to allow grains to nucleate and grow, other investigators looked for places in the universe where the gas would be dense enough and cool enough to allow grains to form. Hoyle and Wickramasinghe (1962) suggested that graphite grains would condense in the atmospheres of certain cool stars and then be blown out into interstellar space. Kamijo (1963) predicted that quartz (SiO_2) grains might originate from such a source.

The origin of interstellar grains might seem a subject far removed from planets, but a relationship is emerging from current research. Larimer (1967) and Larimer and Anders (1967) investigated the condensation of elements in the "solar nebula" believed to surround the sun during the formation of the planets. They found that iron, silicates, and ices would condense in that order. Herbig (1970) and Hartmann (1970) suggested that the interstellar grains might be "planetesimals" of iron, silicate, and possibly icy composition that formed in nebulae around young stars and were then blown clear of the star by radiation pressure. Independently, Wickramasinghe (1967) found that an admixture of silicates to the graphite grains he had earlier postulated would satisfy polarization observations that graphite alone would not explain, and a number of observers (Woolf and Ney, 1969; Fertel, 1970; Knacke and Gaustad, 1969) reported observational evidence of silicates in interstellar and circumstellar grains near various stars. Thus interstellar grains may be analogs or direct examples of the planetesimals produced early in the planet-forming processes.

Dust grains are distributed irregularly in space. The dust is even more highly concentrated toward the galactic plane than is the gas. It is also concentrated in nebulae: for example, the Orion nebula, where star formation is occurring, and in dark nebulae such as the famous "Coal Sack," which blots out part of the Milky Way in the Southern Hemisphere.

Reviews of the physics of interstellar grains and their behavior near stars are given by Wickramasinghe (1967) and Greenberg (1968).

Interstellar Magnetic Fields

Although intermagnetic fields are very weak (roughly 10^{-5} to 10^{-6} gauss) they may have an important bearing on star formation. Stars must form by the contraction of large clouds of dispersed interstellar material. Each such cloud contains a certain amount of energy locked in the magnetic field that passes through it. During the contraction that leads to the formation of the star, the magnetic field may become concentrated and thus become a source of energy important in governing the duration of the star's formation. The significance of this is that a *complete* theory of star formation must deal with physics more complicated than simple nonmagnetic celestial-mechanical effects, even though mechanics may give a first-order description of the formative process. A recent review of the interstellar magnetic field is given by Burbidge (1969).

Turbulence

The irregular forms of many nebulae, plus the observation of great clouds of gas expanding into the interstellar medium, suggest that conditions in the medium are not static or regular. Application of simple turbulence theory, such as calculation of the Reynolds number for the medium, also suggests that turbulence will be induced, even by expansion of HII regions and the normal shear introduced by galactic rotation. Thus fluid mechanics as well as magnetohydrodynamics become important for a thorough description of the prestellar formative process.

THE PROCESS OF GRAVITATIONAL COLLAPSE

Stars are formed by the contraction, or "collapse," of large clouds of interstellar gas and dust. The only way for an interstellar cloud of gas and dust to collapse from nebular to stellar dimensions is to be dense enough so that the self-gravitational attraction of its particles, one for another, is strong enough to start it contracting.

★ QUESTION ★

Why shouldn't a cloud collapse regardless of its density?

Answer

There are competing mechanisms, or energy sources, that tend to keep the cloud in an expanded state. One is the energy density in the magnetic field and another is the kinetic energy involved in turbulent motion, but probably the most important source is thermal energy. Heat acts to make the cloud expand. In terms of atomic theory, each atom has a certain mean velocity that increases with the temperature of the cloud. In a given cloud at a given temperature, the atoms are darting about, colliding with each other, and exerting an outward pressure. The only way that cloud can collapse is for the density of atoms to become so great that the interatom attractions overcome the outward pressures.

This concept is expressed by a crucial astrophysical theorem called the *virial theorem*. This theorem, which can be derived by the methods of statistical mechanics, shows that gravitational collapse of a proto-star will begin only if the total gravitational potential energy of the cloud exceeds twice the total thermal energy (neglecting other energy sources such as magnetic fields).

The fact that the entire interstellar medium is in motion, with turbulence and expanding HII regions adding to the chaos, helps assure that some portions of the interstellar gas will be compressed and become dense enough—have enough mass and gravitational potential energy—for collapse to start.

The virial theorem has the further consequence that as long as the contraction is very slow (strictly speaking, as long as the collapsing proto-star is nearly in hydrostatic equilibrium), energy will be released in a specific way: Half the "gravitational" energy released by the infalling material will be used in heating the gas of the proto-star and the other half will be radiated away into space.

★ THEORETICAL NOTE ★

The opposite extreme would be a gas cloud collapsing in free fall with all the atoms moving radially inward. In this case, there is no hydrostatic equilibrium. The "gravitational" energy released (i.e., the change in gravitational potential energy) would be completely transformed into kinetic energy of the atoms as they fall toward the center. In real life the atoms do not fall radially inward because of their erratic thermal motions. Throughout the collapse, the atoms are colliding and the collapse is slowed. Some of the released gravitational potential energy is converted into heat, and some is radiated away.

The virial theorem has many important applications. It can be used to estimate the temperature and the amount of radiation coming from a collapsing proto-star, and these estimates can be used to identify proto-stars in space. We can also use the virial theorem to compute the size and density of interstellar clouds that could collapse to form a star such as the sun.

Are nebulae of such size and density common in interstellar space? Paradoxically, the answer is no. There are no known nebulae in the interstellar medium small enough and dense enough to collapse into single stars.

How, then, do stars form? To resolve the paradox, we must recall the evidence in the first part of this chapter that stars are forming in clusters. A typical star cluster has a mass of several hundred solar masses. Calculations using the virial theorem reveal that even the smallest interstellar clouds dense enough to collapse have a mass of several hundred to a thousand solar masses. Further study has indicated that during the collapse of these *proto-clusters,* local condensations form within the cloud and begin to collapse as independent entities. The cloud thus *subfragments* rapidly into a cluster of individual stars. The process is reviewed in detail by Fowler and Hoyle (1963), Layzer (1963), Gaustad (1963), Mestel (1966), and others.

In summary, *the virial theorem predicts that stars form not in isolation but in clusters of several hundred members, which indeed are observed.* An example was shown in Figure 5-3. These clusters, as we noted above, have a relatively short lifetime before they disperse, and the stars, like the sun, become lone objects circling the galactic nucleus.

★ MATHEMATICAL THEORY ★

According to the verbal statement above, the cloud will contract if the gravitational potential energy exceeds twice the thermal energy. The gravitational energy of a homogeneous sphere (which we can assume to represent the proto-star or proto-cluster) is

$$\text{potential energy} = \frac{3}{5}\frac{GM^2}{R},$$

where G = gravitational constant
M = mass of cloud
R = radius of cloud

The thermal energy is the thermal kinetic energy of *one* particle, $\frac{3}{2}kT$, times the total number of particles, $M/\mu M_H$,

where k = Boltzmann's constant
T = temperature of gas
μ = mean molecular weight $\simeq 1$
M_H = mass of hydrogen atom

The inequality stated by the virial theorem is thus:

$$\frac{3}{5}\frac{GM^2}{R} > 2 \times \frac{3}{2}\frac{kTM}{\mu M_H} \qquad \text{for collapse.}$$

Canceling common factors and substituting for the mass of a sphere $M = \frac{4}{3}\pi R^3 \rho$, we have

$$\rho > \frac{15kT}{4\pi G R^2 \mu M_H} \qquad \text{for collapse to start,}$$

where ρ = density of gas.

★ PROBLEM ★

Find the density required in an interstellar cloud of hydrogen at 10^2 °K to allow collapse to begin if the cloud has dimensions somewhat larger than a typical star cluster, say 10^{20} cm (an astronomer would express this distance as roughly 30 parsecs). The answer should be equal to the density cited for the denser regions of interstellar space.

★ PROBLEM ★

In a collapsing body in hydrostatic equilibrium, the above relation may be treated as an equality. Show that the interior temperature in a collapsed body of the mass and dimensions of the sun ought to be of the order of a few million degrees. Assume that the sun is in hydrostatic equilibrium. Recall that

$$R = 7.0 \times 10^{10} \text{ cm} \qquad M = 2.0 \times 10^{33} \text{ g}.$$

STAR CLUSTERS AND ASSOCIATIONS

In addition to *open star clusters,* which are relatively obvious groupings such as shown in Figure 5-3, there exist looser, larger, and more groups of stars known as *associations*. As is the case with clusters, associations contain young stars. Some associations are called *O associations,* because they contain a number of very luminous, hot, young, bluish stars known to astronomers as "O stars"; some associations are called *T associations* because they are characterized by a certain type of unstable young star known at T Tauri stars (see the next section). The Orion region, for example, is both an O and a T association. The Orion nebula, near the center of the star-forming region, is shown in Figure

Figure 5-4
The Orion nebula, center of recent star-forming activity in Orion. The nebula forms the central "star" in the "sword" of the constellation Orion. (Steward Observatory.)

5-4. The vicinity of this nebula contains virtually all the above-discussed features that indicate recent formation of stars.

Associations may represent regions of recent star formation, from which young stars are now dispersing. Estimates of the outward motions of some groups of such stars indicate that they were closely associated only about 10^7 years ago.

PROTO-STARS: THEORY

A type of graph called the *Hertzsprung–Russell* diagram is widely used among astronomers who discuss stellar evolution. Named after its originators, the Hertzsprung–Russell diagram (often called the *H–R diagram*) plots two observable properties of stars: the brightness versus the temperature.* The brightness is usually called the *luminosity* (the total rate of energy output of the star, in ergs per second). Because of astronomical traditions, the luminosity increases upward on the diagram and the temperature increases to the *left.*

When a random selection of typical field stars is plotted on an H–R diagram, it is found that they define a diagonal band known as the *main sequence,* as shown in Figure 5-5. Hot, luminous stars (upper left corner) are the most massive ones, as we have already noted. Cool, faint stars (lower right) are the least massive. The sun lies approximately in the middle of the main sequence. The main sequence is the position of stars that are in equilibrium; these are the stars which, like the sun, are in a stable configuration and are "burning" their nuclear fuel. Until a star reaches this stable configuration, it evolves rapidly. After its fuel is exhausted, it also evolves rapidly, away from the main sequence.

★ QUESTION ★

Where on the Hertzsprung–Russell diagram would you expect to find an evolving proto-star?

Answer

Since the proto-star is heating as it forms, it must have a lower temperature than its main sequence descendent. Therefore, proto-stars must lie to the right of the main sequence.

Let us consider the formation of a star like the sun. The star begins as a very rapidly contracting gas cloud. Collapse from a radius of about 1000 A.U.

* The temperature is usually represented by a more readily observable quantity, either color or spectral class—a classification based on spectrum of the star.

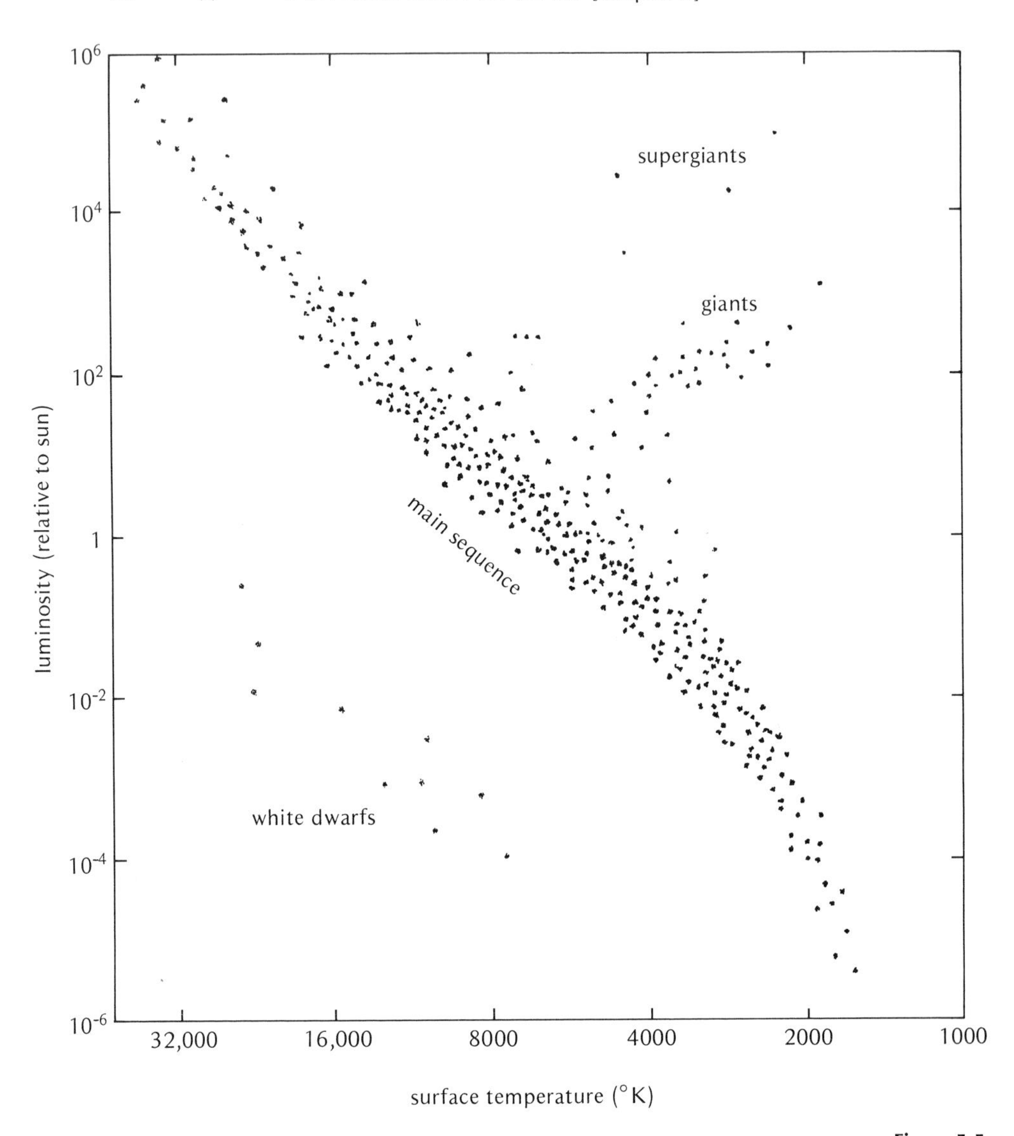

Figure 5-5

The H–R diagram. Named after the astrophysicists Hertsprung and Russell, this plot of stellar luminosity versus stellar temperature provides many clues to stellar evolution. Each point represents a star.

down to about 10 A.U. may take only 5000 years (Ananaba and Gaustad, 1968). At a radius of about 6 A.U. (~1300 solar radii) the star suddenly adjusts its internal structure in order to transport the internally liberated heat to the surface. Convective motion sets in (much as heated water begins to circulate before it boils), and the star suddenly (within a hundred years?) flares up and may become about 100 to 1000 times as bright as the sun. Now it contracts slowly, shining by the energy liberated during its contraction. So far there have been no nuclear reactions. The star grows considerably smaller and fainter as it begins to approach the main sequence.

Figure 5-6 shows this stage of the evolution in terms of the Hertzsprung–Russell diagram. The behavior of the star during this stage was first determined by the Japanese astrophysicist Hayashi (1961). The path followed by the star

Figure 5-6

Evolutionary tracks for formation of stars of three different masses (see the text), plotted on the H–R diagram. See page 111 for a time scale of formation for a star of 1 solar mass.

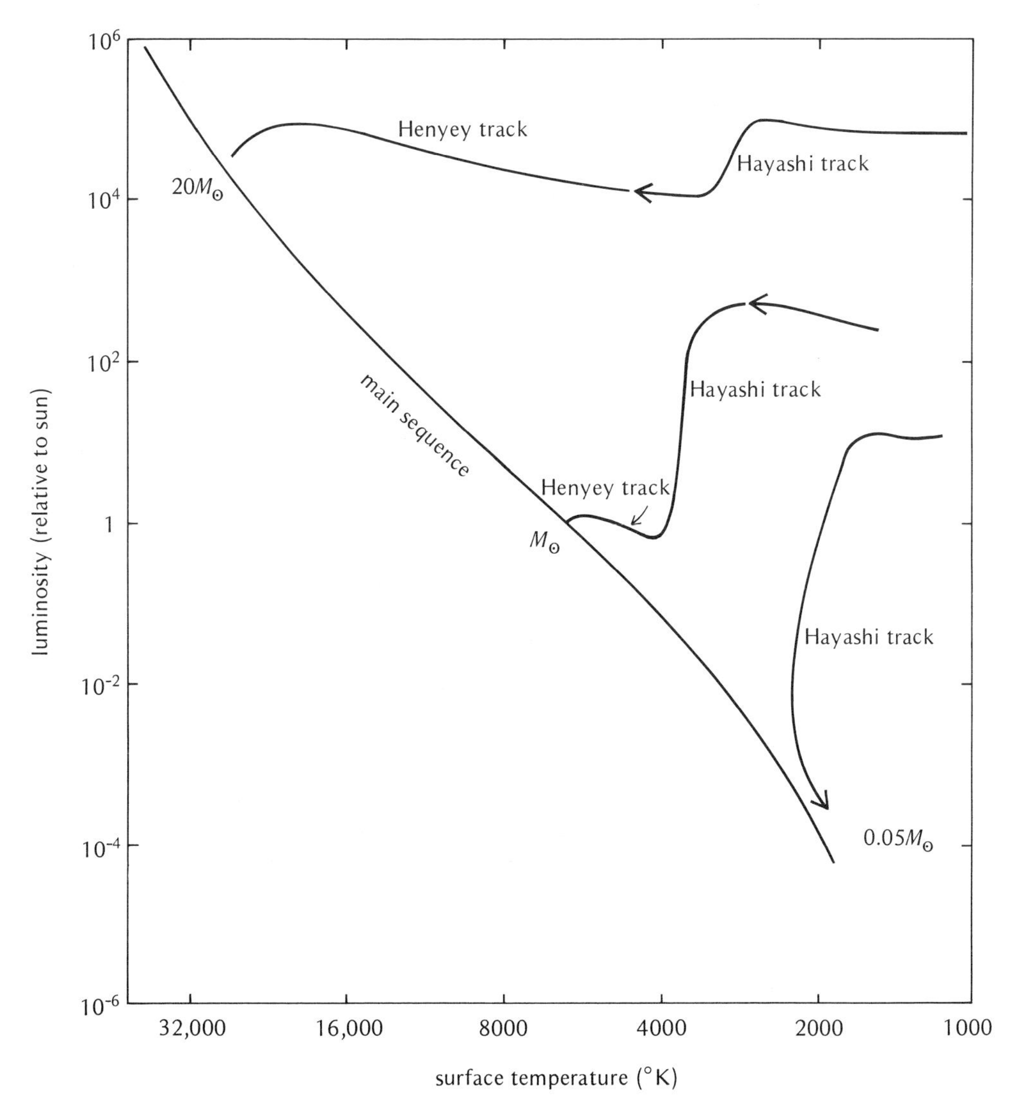

to this point is known as the *Hayashi track*. Stars of different masses will follow somewhat different Hayashi tracks. Examples are given by Bodenheimer (1968).

Our solar-type star has now reached a radius about twice the radius of the sun. There are still no nuclear reactions inside, and the internal structure now undergoes another change. The energy output is sufficiently low that convective motions are no longer needed to transport it; rather the energy is now carried outward from the interior by radiation. The photons of radiative energy travel from atom to atom until they reach the surface and are radiated into space. An astrophysicist would say that the star has changed from "convective transport" to "radiative transport" of energy. In the case of a sun-like star, this stage occurs about 1 million years after the star begins to form. The star has reached the bottom of the Hayashi track.

Now the star evolves along a different path, to the left, toward the main sequence, as shown in Figure 5-5. This new path is called a *Henyey track* after Henyey, Le Levier, and Levee (1955), who first derived its position. After some 10^7 (10 million) years the temperature in the central regions becomes high enough to trigger self-sustaining nuclear reactions. The star makes a final adjustment of its internal structure and "drops" slowly from the Henyey track onto the main sequence, where it has achieved a structural stability that will last for some 10^{10} (10 billion) years until the nuclear "fuels" are exhausted.

From this description alone, we might expect that we could detect protostars in the sky by watching for the appearance of superluminous stars located at the top of the Hayashi track. However, there is a complication. We have not considered the residual nebular material that might surround the star and shield its light. A series of studies of such nebular material by Woolf (1961), Poveda (1965), and others was climaxed by the work of Davidson and Harwit (1967), who coined the term *cocoon nebula* to describe the cloud of dust-laden gas which was predicted to surround a newly-born star. The star was pictured as enveloped in a "cocoon" which dims or entirely obscures its visible radiation. The dust in the cocoon may absorb even the large amount of visible radiation from the star during its Hayashi phase. The dust is thereby heated and reradiates the star's radiation. Whereas the star is radiating with a temperature of 1000°K or more, the dust is heated to only a few hundred degrees (being a large distance from the star), and hence radiates only in the infrared. Thus the nebula may not be visible, except by special infrared detectors.

It is interesting to note that the cocoon nebula predicted theoretically in the 1960s was just the same cloud of dust and gas which planetologists had long theorized as surrounding the early sun, merely on the grounds that the planets had to have a medium in which to form. Cross-talk between the disciplines was sufficiently slight, however, that the astrophysical result was independent of the planetary theory and gave new physical insight into the nature of the nebula.

The theory of cocoon nebulae, then, predicts that newly born proto-stars are surrounded by opaque clouds radiating brightly in the infrared. As the cocoon nebula evolves and breaks up, the visible radiation begins to leak through directly from the star, and the new star bursts forth upon the scene like a young butterfly emerging.

PROTO–STARS: OBSERVATIONS

Cocoon Nebulae and Infrared Stars

In 1966 Low and Smith announced the discovery of what they interpreted to be a preplanetary cocoon nebula surrounding a young star. Their infrared observations apparently showed a young star surrounded by a nebula rich in dust grains and heated to several hundred °K. The object in question was the "star" R in the constellation Monocerotis, known to astronomers as R Mon. What was actually observed was a spectrum showing (1) a small amount of visible radiation leaking out from the obscured star, and (2) a larger amount of infrared radiation matching that which would be given off by a large number of solid bodies heated to different temperatures in the range of a few hundred °K (Figure 5-7). Herbig (1968a) studied the structure and spectrum of this object and found that no star was visible but only a very small nebula, just discernible in the largest telescopes. The brightness of the nebula varies from year to year, as does the spectrum. Inside the nebula is apparently a variable and unstable young star surrounded by a short-lived dusty shell.

Since these early observations, a host of other stars and nebulae with excess infrared radiation have been found. A number are discussed by Mendoza (1968). Among them are several small infrared nebulae in Orion (Ney and Allen, 1969), a highly irregular variable star, FU Orionis, located in or near a dust cloud in Orion, and an infrared object known as NML Cygnus (Gillett, Stein, and Low, 1968), all discussed as possible young stars obscured by dust clouds. Some, such as FU Orionis, are located in the Hertzsprung–Russell diagram near the point where stars are predicted to appear at the top of the Hayashi track. Marginally observed absorptions in the spectra of such objects have given clues about the composition of the dust grains surrounding them. Frozen water or ammonia has been suggested for NML Cygnus, and silicates resembling meteoritic material have been reported in the dense trapezium of the Orion nebula (Stein and Gillett, 1969).

The difficulty with interpreting these objects as proved cases of star birth and analogs of the solar nebula is that *dying* stars are also known to evolve into the upper right corner of the Hertzsprung–Russell diagram. Although all the stars mentioned were first interpreted as instances of star formation, some have recently been reinterpreted as possible very old stars (Hyland et al., 1969). Until this is settled, we will not know which, if any, objects represent analogs of the solar nebula and possible forerunners of planetary systems.

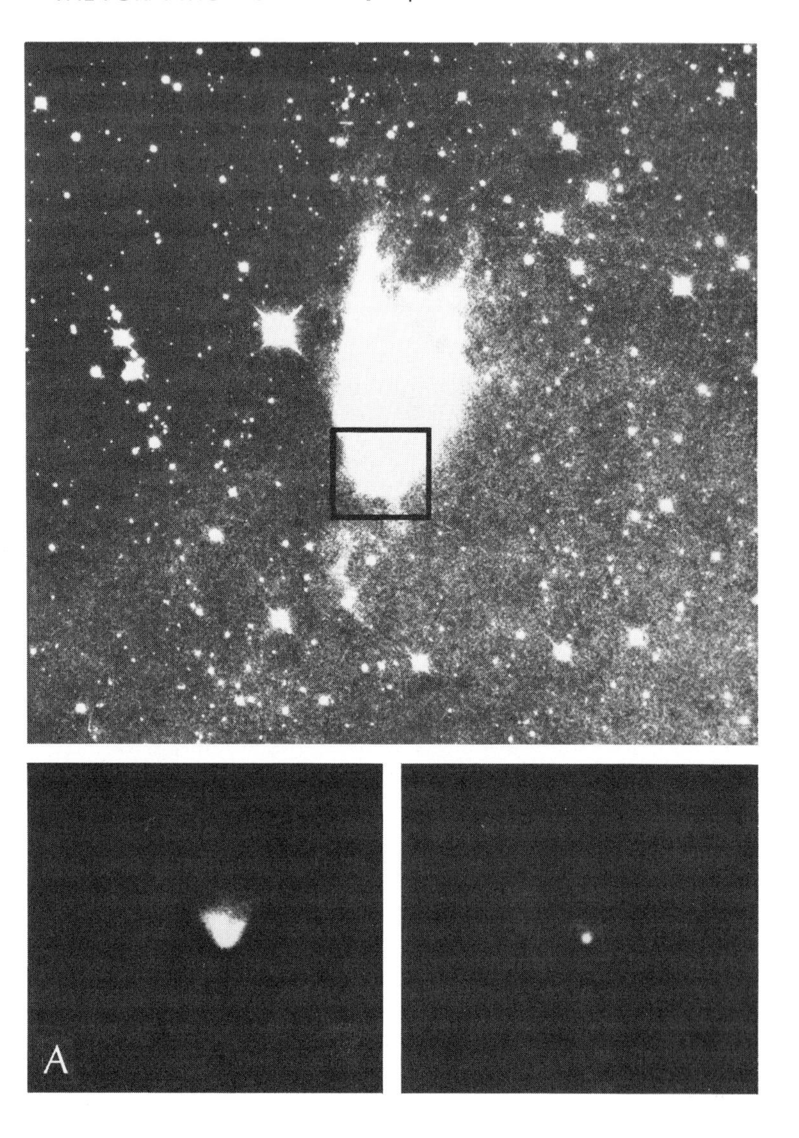

Herbig–Haro Objects

These form another class of objects which may be related to the earliest stage of stars. They are small, highly variable nebulae, often in clusters. A Herbig–Haro object may brighten without notice, remain visible for a few years, and die out again. An example of this behavior is shown in Figure 5-8. In this sense Herbig–Haro objects are related to the highly variable objects with infrared excesses, such as R Monocerotis. Indeed, several possible Herbig–Haro objects lie near R Mon.

No star is visible in any of the Herbig–Haro objects, yet their spectra show light of certain colors (emission lines) characteristic of the next class of star

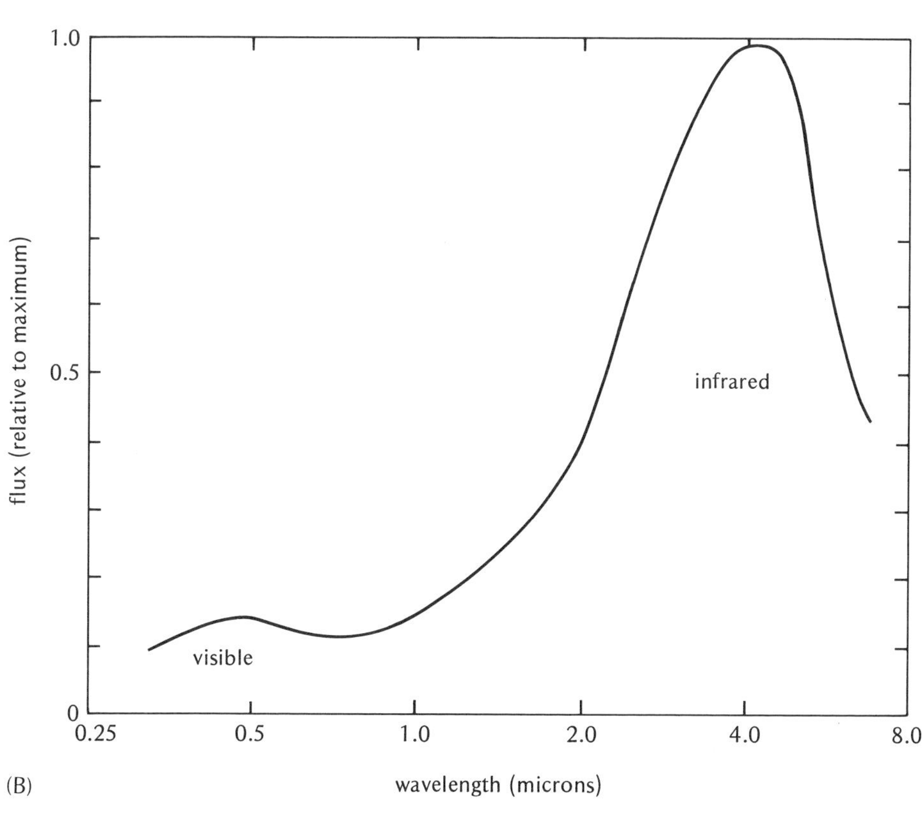

(B)

Figure 5-7

A (*left*): Three photographs of the "star" R Monocerotis and its associated nebula. The top photograph is a long exposure showing the nebula; the bottom photographs are short exposures at high magnification of the rectangular area indicated above. Successively shorter exposures indicate a star-like object at the nebula's tip. (Lick Observatory.) B: Spectrum of R Monocerotis. Most of the light is infrared radiation emitted by dust particles surrounding R Mon, but some visible light leaks through the nebula.

Figure 5-8

Four photographs of Herbig–Haro object number 2, showing changes over 22 years. The 1960 exposure is an enlargement showing detailed structure. The other three are taken about 11 years apart and show emergence of at least two new "stars" on the left side of the cluster. (Lick Observatory.)

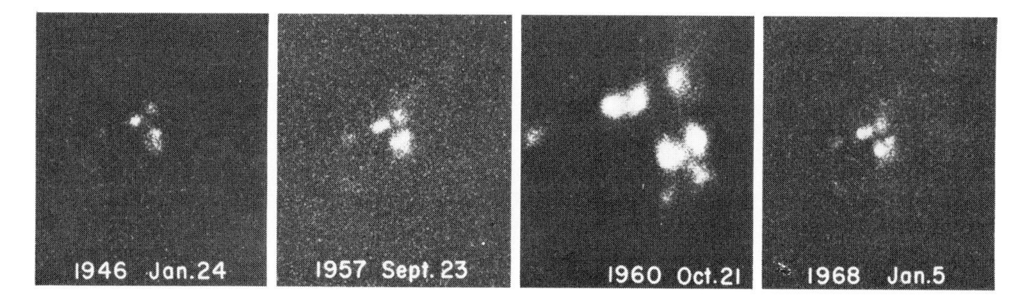

forerunners, the T Tauri stars (see below). Also, they have a distribution in space similar to that of T Tauri stars. They are discussed by Herbig (1968b).

T Tauri Stars

These stars are named after the type example, the star T in the constellation Taurus and were first recognized as a class by Joy (1945). If infrared stars and Herbig–Haro objects really do represent an early stage of evolution, the T Tauri stars probably represent the next stage of evolution, when the cocoon nebula clears and the star draws closer to the main sequence.

T Tauri stars are highly and irregularly variable in brightness, accompanied by variable nebulae and strong magnetic fields, and found in regions with abundant dust clouds and young, massive stars. They have fast rotation (predicted by Laplace's nebular hypothesis; see Chapter 4) and have expanding gas clouds around them (Figure 5-9). They occupy a band in the Hertzsprung–

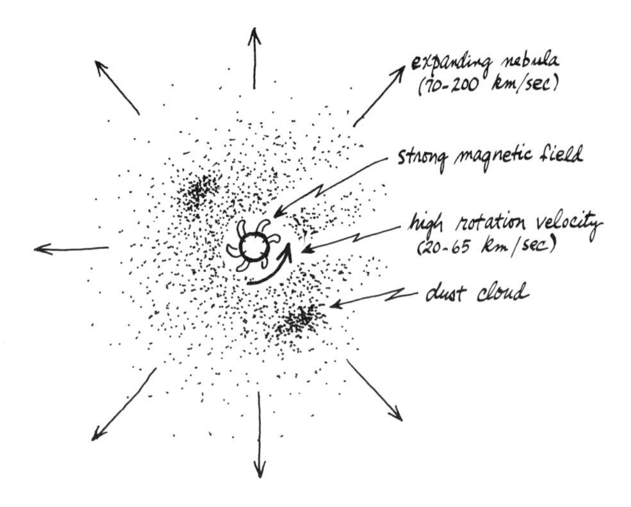

Figure 5-9
Schematic view of a T Tauri star, showing the expanding nebula and features inferred from observations.

Russell diagram just above the main sequence. This is the region that would be reached stars that have just evolved to the bottom of their Hayashi tracks or are moving onto their Henyey tracks. Many T Tauri stars have infrared excesses suggesting accompanying dust grains embedded in their nebulosity (Mendoza, 1968).

A review of their status as young stars is given by Kuhi (1966), who studied the expansion and dispersal of their nebulae. Kuhi's study of the rate of mass loss of six T Tauri stars suggested that they will lose as much as 0.4 solar mass by the time they reach the main sequence. This loss of nebular mass could represent the shedding of the cocoon nebula, with the residual nebular material being blown away from the sun by radiation pressure and/or strong "solar wind" activity. Thus, Kuhi's upper limit of 0.4 solar mass serves as an approximate upper limit for the possible mass of the solar nebula.

The reader will note other striking similarities between the T Tauri stars and the postulated condition of the sun during the formation of the planets described in Chapter 4. First, there is a surrounding nebulosity with evidence of the formation of solid particles. Second, there is fast rotation, as predicted by the old nebular hypothesis of Laplace and Kant. Third, there is evidence of magnetic activity that could act to slow the rotation.

The evidence for strong magnetic fields is particularly instructive and involves the abundance of the element lithium. Bonsack and Greenstein (1960) and Bonsack (1961) showed that T Tauri stars had as much of the element lithium as the earth and meteorites but much more than the sun and normal main-sequence stars. In view of the destruction of lithium by nuclear reactions inside stars, it was surprising to find that T Tauri stars contained so much lithium. It was also observed that there was more lithium in the T Tauri stars than in the interstellar medium or in the nebulae associated with the T Tauri stars. If the stars form from a low-lithium nebula, yet produce planets with high lithium abundances, how and when is the lithium created? The interpretation is that during the T Tauri stage stars create light elements such as lithium by energetic reactions in their atmospheres. Bonsack and Greenstein and others inferred that these reactions involve acceleration of atomic particles in strong magnetic fields near the surfaces of the T Tauri stars. This indicates that newly forming stars have strong magnetic fields which might serve to brake their rotation in the manner described in Chapter 4 and at the same time explains some aspects of planetary composition.

"Bok" Globules

Certain regions of dust concentration, star formation, and nebulosity show small, round dark "clots" of opaque nebular material silhouetted against brighter background nebulae (Figure 5-10). Spitzer (1941) and Bok and Reilly (1947) suggested that these *globules,* or *Bok globules,* as they came to be called, might mark a stage of star formation. Radii of the globules lie in the range 10^3 to 10^5 (1000 to 100,000) A.U., many times bigger than the solar system. Early observers suggested that the globules were proto-stars, but it seems likely that they are compact nebular debris left over after star-forming activities. Two supports for this point of view are

1. There is no theoretical or observational evidence that globules are contracting; some estimates of mass suggest that they are too low in density to contract gravitationally.

2. Globules tend to lie in regions of much dust but less gas than would be expected for star-forming regions. The dust may be blown out of cocoon

Figure 5-10

Dark dust globules and bright nebulosity associated with the open star cluster NGC 6611. The dark globules are foreground dust clouds silhouetted against background nebulosity. (Steward Observatory.)

nebulae after star formation has exhausted the original interstellar gas in a region.

BINARY AND MULTIPLE STARS

Our description of star formation has one major flaw: We have not yet accounted for the fact that about half of all stars are binary or multiple. A *binary star** is a pair of stars orbiting around a common center of gravity; a *multiple star* is a system of three or more stars coorbiting. Since binaries and multiple stars are so common, their formation must be a common process, yet there is no satisfactory theory to account for them.

One reason for the lack of understanding of binaries is the lack of good statistics. Selection effects are devastating. If we study only nearby stars, which

* The term *double star* applies to two stars that appear close together in the sky by fortuitous geometric alignment but without any proved gravitational connection.

we can see well, we have too small a number to be a good sample. If we study distant stars the data will be biased in several ways (e.g., toward binaries with equal brightness because in very unequal pairs, the faint member will lose in the glare of the brighter). *Visual binaries* are binaries with members far enough apart to be seen separately in a large telescope; their statistics are biased toward large separation distances. *Spectroscopic binaries* are so close together that their orbital velocities are very high (recall Kepler's third law), causing measurable, periodic *Doppler shifts* that identify the star as a binary. Statistics of spectroscopic binaries thus favor very close separations. Because of these selection effects in detection and cataloging, no list can be guaranteed to give a statistically valid sample for all binary and multiple stars.

A third type of binary is called the *astrometric* binary and is detected by observations of stellar motions with incredible precision. These are called *astrometric observations.* If a star is accompanied by another star or planet, the two bodies will be orbiting around each other. Therefore, the star's drifting motion across interstellar space will not be a straight line but will have a "squiggly" shape, because it is always orbiting back and forth around its companion. The companion may be an unseen, tiny body, but if the star is close enough to the sun, the sinuosities will be detectable. It may take decades to get enough good data to detect any periodic sinuous motion.

Fortunately, this type of observational work was started at several observatories by far-sighted astronomers 60 years ago. Analysis of the accumulated observations has shown that a number of seemingly single stars actually have very, very tiny sinuous motions. They are so tiny that statistical methods must be used to sort out the unavoidable inaccuracies in measurement. They are also so tiny that the invisible companions must be much, much less massive than the sun, and at least one, a companion to Barnard's Star, is of nearly planetary mass.

Table 5-1 gives a census of the nearest stars and shows the importance of binary and multiple systems detected by all available methods. Including the sun, 5 of the nearest 6 stars, and 13 of the 28 in the whole list are binary or multiple. Binary and multiple stars are hardly rare!

What is the importance of binary and multiple systems to the study of the planets? There are two answers. First, Table 5-1 shows that some of the binary companions have masses comparable to the mass of Jupiter—in other words, some of them may actually *be* planets! Second, the origin of binaries may be related to the origin of planets, since secondary stars and planets both are examples of small bodies orbiting around larger stars.

Let us consider the question of origin first. Binary and multiple stars are so common that their mode of origin must be a commonplace process, not a rare or unique event. There are three theories for their formation.

The first theory is that binaries form in cases where the proto-star has a higher amount of angular momentum than usual. The completed star therefore spins too fast to be stable and *fissions,* or splits, into two stars. This hypothesis accounts for some, but not all, properties of multiple stars. In par-

Table 5-1
Survey of the nearest star systems[a]

Star or system	Distance (light-years)	Mass of known components (solar masses)		
		A	B	C
Sun	0	1.0	0.001	0.0003
Alpha Centauri	4.3	1.0	0.9	0.1
Barnard's Star	5.9	0.2	0.0016	0.001?
Wolf 359	7.6	0.1		
Lalande 21185	8.1	0.3	0.1	
Sirius	8.6	2.2	0.9	
Luyten 726-8	8.9	0.14		
Ross 154	9.4	0.2		
Ross 248	10.3	0.1		
Epsilon Eridani	10.7	0.8		
Luyten 789-6	10.8	0.1		
Ross 128	10.8	0.6		
61 Cygni	11.2	0.7	0.7	0.008
Epsilon Indi	11.2	0.7		
Procyon	11.4	1.8	0.6	
Σ 2398	11.5	0.4	0.4	
Groombridge 34	11.6	0.4	0.2	
Lacaille 9352	11.7	0.3		
Tau Ceti	11.9	0.9		
BD +5° 1668	12.2	0.3	0.1?	
Lacaille 8760	12.5	0.4		
Kapteyn's Star	12.7	0.5		
Krüger 60	12.8	0.27	0.16	
Ross 614	13.1	0.14	0.08	
BD −12° 4523	13.1	0.2		
van Maanen's Star	13.9	0.3?		
Wolf 424	14.2	0.1	0.1	
CD −37° 1549	14.5	0.3		

[a] This table includes all known stars out to 15 light-years (1 light-year is about 10,000,000,000,000 km) from the sun, based on a list by van de Kamp (1969), with addition of new data on Luyten 726-8. Total single stars and systems, 28; no. of multiple systems, 13.

ticular, it explains the properties of certain very close binary pairs, such as the "W Ursa Majoris stars." The fission theory does not satisfactorily explain widely separated binary pairs (Kuiper, 1955).

A second theory is that binaries and multiple stars originate with *capture* of one star by another. This view is supported by the fact that binaries bear some resemblance to random pairings of stars of various types (van de Kamp, 1961). Some years ago this theory was discounted because statistical astronomers showed that encounters between ordinary stars like the sun, drifting in the vast emptiness of deep space, are far too rare to account for the observed number of binaries. However, the hypothesis has been revived by the more recent realization that stars form in crowded clusters, where they stay for 100 million years or more. In these crowded conditions, encounters and captures may occur, and it might be that by the time the cluster disperses, many of its stars have linked themselves into multiple systems.

The third theory, involves *breakup of the cocoon nebula.* According to this idea, as advanced by Kuiper (1951, 1955), the solar system is just an extreme case of a multiple star system. The cocoon nebula breaks up to form a group of tiny planets if it has high angular momentum and its density is too low to form another object comparable in mass to the central star. Depending on the angular momentum and density distributions, we might get a system with two stars of 1.0 and 0.9 solar mass, another with 1.0 and 0.1, and still another with 1.0 and 0.01 solar mass. The solar system would be a still more extreme example. Jupiter is by far the largest planet and has only $\frac{1}{1000}$ the mass of the sun. If it had roughly 100 times more mass, it would have been a star. This theory tempts one to view the solar system as an extreme sort of binary "star"—the sun and Jupiter, with 1.0 and 0.001 solar mass.

Which is the right theory? The fact is that the origin of binaries is just not known. There is an increasing tendency to suspect that different binaries have different origins. Thus, while the close W Ursa Majoris binaries may result from fission, wider pairs may be chance captures. This is not as intellectually satisfying as a unified theory for all binaries, but it appears to be gaining observational support.

The possible relation of binaries to planetary systems (theory 3) is tempered by the orbits reported for binaries. Binary orbits are usually highly elliptical, and in multiple systems the three or more orbits are often highly inclined to each other. This contrasts sharply with the solar system with its circular, coplanar orbits. In any case it is unlikely that planets develop within binary or multiple star systems, because the complex perturbations would cause instabilities in orbits.

In summary, we are not obliged to consider binary and multiple systems as an immediate part of the problem of planetary origin, yet we cannot ignore the possibility of such a relation.

STARS, BLACK DWARFS, AND PLANETS

So far we have sidestepped a delicate question. What is the difference between planets and stars? The obvious difference is that stars are large and give off visible light, while planets are small and dark. What happens in the transitional size range? Stars are radiating energy produced in their gaseous interiors by nuclear reactions and are defined by the existence of these nuclear reactions. In order to start the nuclear reactions, the star must be very hot and, in turn, very large. If a proto-stellar body is not massive enough, the interior temperature during its formation never gets hot enough to trigger the nuclear reactions. Instead, it would heat to some maximum temperature and then slowly cool. The transition from stars to planets thus is defined by the existence of nuclear reactions. Calculations (Kumar, 1964; Hoxie, 1969) show that the smallest object that can become a true star has a mass of only 0.085 times that of the sun. Objects smaller than 0.085 solar mass never achieve nuclear chain reactions, although they radiate some internal heat.* Astronomers call them "black dwarfs."

Notice the small masses of some of the "stars" in Table 5-1. Is the companion to Barnard's Star a black dwarf or a planet? Or is there any difference? Should Jupiter and the earth be considered black dwarfs?

We are skirting the edges of a jungle of semantic confusion. The large black dwarfs, say of mass 0.080 solar mass, will certainly not look like our ordinary conception of a planet because they have not had enough time within the history of our galaxy to cool off! The heat produced as they contract takes a long time to dissipate; large black dwarfs would still be radiating dull red or infrared light.

Another possible distinction involves orbits. Reported orbits for black dwarfs tend to be more elliptical than for solar-system planets, and most known black dwarfs cannot be proved to have the circular orbits characteristic of the solar system. Orbits for most of the small-mass objects in Table 5-1 are notoriously difficult to determine. By means of Fourier analysis any elliptical orbit can be broken down into a series of circular orbital components, reminiscent of the Ptolemaic system of epicycles by which the ancients tried to describe the solar system. If a given astrometric binary is assumed to have *one* black dwarf secondary, the orbit calculated from the observed periodic motions may turn out to be highly eccentric. But if we assume two or three companions, calculations show them to have more nearly circular orbits,* more like solar-system planets. The true nature of the orbits is rather uncertain.

The circular orbits of solar-system planets are a key factor in our theories of how the planets formed. They indicate that the planets formed from many small particles moving through a resistive medium of gas and dust that would damp out noncircular motions. If black dwarfs move in eccentric elliptical orbits (as an uncritical look at catalogued orbits would suggest), they would seem to be truly genetically different from planets. While planets may form from particles sticking together, eccentric black dwarfs might have formed by gravitational collapse, either within a cocoon nebula or in interstellar space. Only further observations and possibly new techniques will accurately determine the orbits of the astrometric binary companions.

* An example of this is the case of the possible planets circling around Barnard's star, which is discussed in Chapter 13.

* In Chapter 10 we shall see evidence that Jupiter is radiating significant heat, possibly due to a slow contraction.

In summary, a true planet is distinguished from a star or black dwarf not solely by mass but probably also by the regularity (low eccentricity, low inclination) of its orbit, which reflects its mode of origin.

References

Ananaba, S. E., and J. E. Gaustad (1968) "The Early Evolution of a Protostar," *Astrophys. J., 153,* 95.

Blaauw, A., and W. W. Morgan (1954) "The Space Motions of AE Aurigae and μ Columbae," *Astrophys. J., 119,* 625.

Bodenheimer, P. (1968) "The Evolution of Protostars of 1 and 12 Solar Masses," *Astrophys. J., 153,* 483.

Bok, B. J., and E. Reilly (1947) "Small Dark Nebulae," *Astrophys. J., 105,* 255.

Bonsack, W. K. (1961) "The Abundance of Lithium in T Tauri Stars: Further Observations," *Astrophys. J., 133,* 340.

——— and J. L. Greenstein (1960) "The Abundance of Lithium in T Tauri Stars and Related Objects," *Astrophys. J., 131,* 83.

Burbidge, G. R. (1969) "The Galactic Magnetic Field," *Comments Astrophys. Space Sci., 1,* 25.

Davidson, K., and M. Harwit (1967) "Infrared and Radio Appearance of Cocoon Stars," *Astrophys. J., 148,* 443.

Fertel, J. F. (1970) "Silicon Monoxide Bands in Some Low-Temperature Stars," *Astrophys. J. Lett., 159,* L7.

Fowler, W., and F. Hoyle (1963) "Star Formation," *Roy. Observ. Bull.* No. 67.

Gaustad, J. E. (1963) "The Opacity of Diffuse Cosmic Matter and the Early Stages of Star Formation," *Astrophys. J., 138,* 1050.

Gillett, F., W. Stein, and F. Low (1968) "The Spectrum of NML Cygnus from 2.8 to 5.6 Microns," *Astrophys. J. Lett., 153,* L185.

Greenberg, J. M. (1968) "Interstellar Grains," in *Nebulae and Interstellar Matter,* B. M. Middlehurst and L. Aller, eds. (Chicago: University of Chicago Press).

Hartmann, W. K. (1970) "Growth of Planetesimals in Nebulae Surrounding Young Stars," in *Evolution stellaire avant la séquence principale,* Proc. Liège Symp., June 30–July 2, 1969, Liège Collec. in −8°, 5th Ser., *19,* 215.

Hayashi, C. (1961) "Stellar Evolution in Early Phases of Gravitational Contraction," *Publ. Astron. Soc. Japan, 13,* 450.

Henyey, L., R. Le Levier, and R. Levee (1955) "The Early Phases of Stellar Evolution," *Publ. Astron. Soc. Pacific, 67,* 154.

Herbig, G. H. (1968a) "The Structure and Spectrum of R Monocerotis," *Astrophys. J., 152,* 439.

——— (1968b) "The Light Variations in the Nuclei in Herbig–Haro Object No. 2, 1946–1968," Contr. Lick Observ. 282, in *IAU Colloq. on Non-periodic Phenomena in Variable Stars,* Budapest.

——— (1970) "Introductory Remarks," in *Evolution stellaire avant la séquence principale,* Proc. Liège Symp., June 30–July 2, 1969, Liège Collec. in −8°, 5th Ser., *19,* 13.

Hoxie, D. (1969) "The Structure and Evolution of Stars of Very Low Mass," Ph.D. Dissertation, University of Arizona.

Hoyle, F., and N. Wickramasinghe (1962) "Graphite Particles as Interstellar Grains," *Monthly Notices Roy. Astron. Soc., 124,* 417.

Hyland, A., E. Becklin, G. Neugebauer, and G. Wallerstein (1969) "Observations of the Infrared Objects, VY Canis Majoris," *Astrophys. J., 158,* 618.

Joy, A. H. (1945) "T Tauri Variable Stars," *Astrophys. J., 102,* 168.

Kamijo, F. (1963) "A Theoretical Study on the Long Period Variable Stars. III. Formation of Solid and Liquid Particles in the Circumstellar Envelope," *Publ. Astron. Soc. Japan, 15,* 440.

Knacke, R., and J. E. Gaustad (1969) "Possible Identification of Interstellar Silicate Absorption in the Infrared Spectrum of 119 Tauri," *Astrophys. J. Lett., 155,* L189.

Kuhi, L. V. (1966) "T Tauri Stars: A Short Review," *J. Roy. Astron. Soc. Can., 60,* 1.

Kuiper, G. P. (1951) "On the Origin of the Solar System," in *Astrophysics,* J. A. Hynek, ed., p. 404 (New York: McGraw-Hill, Inc.).

______ (1955) "On the Origin of Binary Stars," *Publ. Astron. Soc. Pacific, 67,* 387.

Kumar, S. S. (1964) "On the Nature of Planetary Companions of Stars," *Z. Astrophys., 58,* 248.

Larimer, J. W. (1967) "Chemical Fractionations in Meteorites—I. Condensation of the Elements," *Geochim. Cosmochim. Acta, 31,* 1215.

______ and E. Anders (1967) "Chemical Fractionations in Meteorites—II. Abundance Patterns and Their Interpretation," *Geochim. Cosmochim. Acta, 31,* 1239.

Layzer, D. (1963) "On the Fragmentation of Self-Gravitating Gas Clouds," *Astrophys. J., 137,* 351.

Low, F., and B. Smith (1966) "Infrared Observations of a Preplanetary System," *Nature, 212,* 675.

Mendoza V., E. E. (1968) "Infrared Excesses in T Tauri Stars and Related Objects," *Astrophys. J., 143,* 1010.

Mestel, L. (1966) "Problems of Star Formation—I," *Roy. Astron. Soc. Quart. J., 6,* 161.

Ney, E. P., and D. A. Allen (1969) "The Infrared Sources in the Trapezium Region of M42," *Astrophys. J. Lett., 155,* L193.

Oort, J. H., and H. C. van de Hulst (1946) "Gas and Smoke in Interstellar Space," *Bull. Astron. Inst. Neth., 10,* 187.

Poveda, A. (1965) "The H-R Diagram of Young Clusters and the Formation of Planetary Systems," *Bol. Observ. Tonantz. Tacubaya, 4,* 15.

Roberts, M. S. (1966) "Upper Limit to the Neutral Hydrogen Density in the Halo Regions of Spiral Galaxies," *Phys. Rev. Lett., 17,* 1203.

Spitzer, L., Jr. (1941) "The Dynamics of the Interstellar Medium. II. Radiation Pressure," *Astrophys. J., 94,* 232.

Stein, W., and F. Gillet (1969) "Spectral Distribution of Infrared Radiation from the Trapezium Region of the Orion Nebula," *Astrophys. J. Lett., 155,* L197.

van de Kamp, P. (1961) "Double Stars," *Publ. Astron. Soc. Pacific, 73,* 389.

______ (1969) "Stars Nearer than Five Parsecs," *Publ. Astron. Soc. Pacific, 81,* 5.

Werner, M., and M. Harwit (1968) "Observational Evidence for the Existence of Dense Clouds of Interstellar Molecular Hydrogen," *Astrophys. J., 154,* 881.

Wickramasinghe, N. C. (1967) *Interstellar Grains* (London: Chapman & Hall Ltd.).

Woolf, N. J. (1961) "On the Star Herschel 36 Near M8," *Publ. Astron. Soc. Pacific, 73,* 206.

______ and E. P. Ney (1969) "Circumstellar Infrared Emission from Cool Stars," *Astrophys. J. Lett., 155,* L181.

THE GROWTH OF THE PLANETS

THE HISTORICAL DEVELOPMENT of a number of theories, reviewed in Chapter 4, indicated that the planets grew in a dust-laden, disk-shaped nebula surrounding the early sun. Chapter 5 discussed astrophysical evidences that every newly formed star may have such a nebula. These two lines of evidence do not prove that production of planets accompanies formation of every star, but they do suggest that during star formation, grain-like or larger planetesimals will form and that planets can grow from these small bodies. This chapter attempts to use a mixture of current theoretical and observational data to draw a consistent picture of this process and to bridge the evident interdisciplinary gap between astrophysical and planetary research. There is no guarantee that this chapter gives the correct theory, but it presents a view at least compatible with modern research.

To review: A fixture of theories of solar-system origin is the solar nebula. The concept of such nebulae surrounding all solar-type young stars during their pre-main-sequence evolution is supported by theoretical predictions of "cocoon stars" wrapped in opaque dusty nebulae with visible radiation dimmed or obscured entirely. The discovery of just such infrared cocoon stars, apparently surrounded by dusty nebulae radiating at a few hundred degrees Kelvin, is taken to establish the reality of cocoon nebulae as a stage in the formation of solar-type stars. The cocoon nebula around the sun is identical to the solar nebula predicted by early planetary theorists.

Our first task is to describe the properties of the solar nebula. But first, let us formally define some terms we have been using and will use.

Solar nebula: disk-shaped cloud of gas and dust surrounding the early sun.

Cocoon nebula: extrasolar analog of the solar nebula.

Proto-planet: hypothetical large mass of gas and dust, of larger mass than a planet, but representing the total material of cosmic composition that contributed to the construction of a given planet.

Planetesimals: hypothetical small solid bodies, ranging from microscopic to multihundred kilometer dimensions and contributing to the construction of planets.

EARLY CONDITIONS IN THE SOLAR NEBULA

At the end of the sun's rapid, unstable collapse, as it started down the Hayashi track, its radius was some tens or a hundred times its present radius. During most of the Hayashi phase, therefore, the sun would have been well inside Mercury's orbit, and the solar system would have been dynamically similar to its present state, to the first order. The solar nebula during the Hayashi contraction can thus be viewed as surrounding a central, gravitationally distinct star of approximately 1 solar mass, although the nebula itself may have formed before the Hayashi track was reached.

Origin of Nebula

It is not clear whether the solar nebula should be considered simply a residue of the interstellar material from which the sun condensed or material spun off from the sun's equator by rotational instability setting in during the contraction. Models based on the latter view are inconclusive, with the estimated radius of the sun at the time the nebula was created varying from greater than 100 A.U. (Cameron, 1962, p. 59) to 0.2 A.U. ($40R_{\odot}$) (Hoyle, 1960).

Composition of Nebula

Meteorite compositions have long been used to argue that, basically, the nebula had the same composition as the solar material and the rest of the universe. Concise examples of comparisons between meteorites and solar abundance are given by the geochemists Ringwood (1966) and Wood (1968). Observations of young stars, such as T Tauri stars, give no evidence that their cocoon nebulae markedly differ from cosmic composition.

It seems clear, therefore, that the nebular material initially had solar composition. Nonetheless, abundances of certain isotopes may have been altered subsequently by nuclear reactions, and many of the lighter elements have escaped during or after formation of the planets. Thus current planetary matter may differ substantially in some ways from the primordial material.

Mass of Nebula

Some early theorists incorrectly assumed that the mass of the nebula was equal to the total mass of the planets. However, the mass of the nebula must have been many times the planetary mass, since the planets are composed of only residual elements from a nebula originally of cosmic composition. The terrestrial planets are composed mostly of silicates and iron; the giant planets are composed mostly of ices of water, methane, ammonia, and hydrogen. In the case of each planet, but especially the terrestrial planets, many of the elements originally present were not incorporated, or were incorporated and then lost.

By mass, most of these elements, especially the gases, such as hydrogen, helium, the inert gases, water, and so on, are described as *volatiles*. The volatiles are the elements that, under the given conditions, are not solid and either do not combine chemically to form a planet or escape easily from a finished planet; volatiles have high vapor pressures.

Consider water as an example. Urey (1952), and more recently Watson, Murray, and Brown (1963) have shown that when exposed to sunlight, ice crystals quickly evaporate or sublime in the region of the terrestrial planets, while beyond the asteroids, frozen water is stable. Thus water could easily have been trapped and incorporated in the giant planets but was more highly volatile in the case of the terrestrial planets. Even frozen water on the surface of a "terrestrial planetesimal" would have quickly dissipated, if exposed to sunlight.

A minimum mass for the solar nebula can be calculated if we can estimate the total amount of missing volatiles that must be restored to each planet to produce a proto-planet of cosmic composition. Since silicates and iron constitute only a small fraction of cosmic material, the mass of earth must be multiplied by a large factor to restore the lost mass.

Table 6-1 summarizes the results of four authors who have attempted this calculation. Jupiter apparently had the largest proto-planet mass, closely followed by the other outer planets. The total nebular mass contributed by

Table 6-1
Estimated masses of nebular matter required to form planets[a]

Planet	Assumed composition	Present mass (g)	Estimated ratio: required mass/present mass	Estimated nebular mass (g)
Mercury	Silicates, rich in iron	3×10^{26}	500	2×10^{29}
Venus	Silicates, iron	5×10^{27}	450	2×10^{30}
Earth	Silicates, iron	6×10^{27}	450	3×10^{30}
Mars	Silicates, iron	6×10^{26}	450	3×10^{29}
Asteroids	Silicates, iron (ices?)	$\sim 10^{24}$	250	3×10^{26}
Jupiter	Hydrogen, ices	2×10^{30}	5	1×10^{31}
Saturn	Hydrogen, ices	6×10^{29}	10	6×10^{30}
Uranus	CH_4, NH_3, H_2O, ices	9×10^{28}	70	6×10^{30}
Neptune	CH_4, NH_3, H_2O, ices	1×10^{29}	70	7×10^{30}
Pluto	?	7×10^{26}?	100?	7×10^{28}?
Comets	CH_4, NH_3, H_2O, ices	$>10^{30}$?	5	$>5 \times 10^{30}$?
			Total:	$>4(10^{31})$ g or >0.02 solar mass

[a] Adapted from Kuiper (1956a), Cameron (1962), Hoyle (1963), and Whipple (1964).

comets is uncertain, since we do not know the total number of comets, and the amount of mass lost by the giant planets is also uncertain. It is interesting to note that the masses that contributed to the formation of the various planets were much more nearly equal than the final masses of the planets themselves. The calculated total mass is a lower limit for the mass of the nebula because we do not know how much additional interplanetary material may have been blown away without contributing to the planets. The minimum mass for the solar nebula is thus found to be $0.02M_{\odot}$ ($M_{\odot}$ represents units of solar masses).

A maximum mass of the solar nebula comes from Kuhi's (1966) upper limit on the mass of material being expelled from T Tauri stars, $0.4M_{\odot}$. Consistent with this, it seems improbable that a nebula left over after formation of the sun would be as massive as the sun itself, although higher masses, ranging up to $3M_{\odot}$ have been proposed (Cameron, 1962, p. 63).

An independent rough estimate of the nebular mass comes from Herbig's (1970) model of a young star (VY Canis Majoris), which requires a cocoon nebula of about $0.15M_{\odot}$. Another estimate, $0.05M_{\odot}$, comes from a review of solar nebula physics by the Soviet theoretician, Safronov (1967). Still another estimate comes from Schwartz and Schubert (1969), who calculate that the sun could have been slowed to its present rate of rotation by the action of magnetic fields during loss of a nebula of $0.15M_{\odot}$.

We thus infer that the mass of the solar nebula was within an order of magnitude of $0.2M_{\odot}$.

Shape of Nebula

Von Weizsäcker (see Chandrasekhar, 1946), Kuiper (1951), and others have used the hydrostatic equation to model the shape of a rotating nebula encircling the early sun. The nebula reaches an equilibrium disk-shape under the conditions shown in Figure 6-1, when centrifugal forces and the

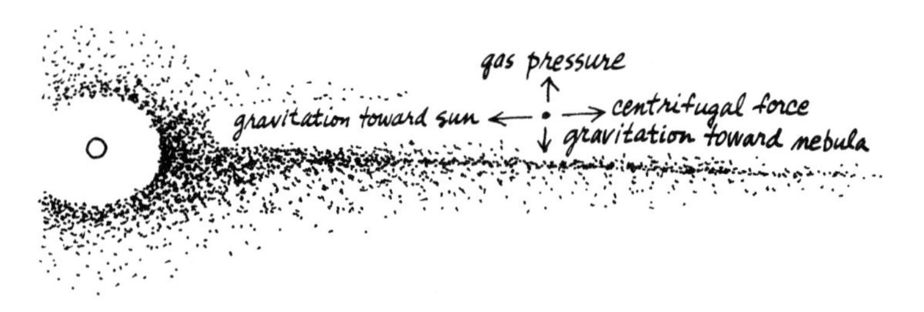

Figure 6-1
Balanced forces on a particle in the solar nebula.

outward gas pressure match the inner gravitational force toward the sun, and when the gravitational attraction toward the plane of the nebula matches the gas pressure away from the plane.

These conditions allow an estimate of the distribution of mass in the nebula. The results show that the dust and gas would be highly confined to the ecliptic plane. The result would be strikingly similar in appearance to one of the spiral galaxies, which accords with the historical fact that the distant galaxies were originally interpreted as stars in formation. That the planets all lie in a flat system with coplanar orbits also tells us that the original nebula must have been highly flattened.

Density of Nebula

A density can be estimated from the volume and mass distribution of the nebula. $0.2M_{\odot}$ distributed over a lenticular cloud of radius 20 A.U. and thickness 0.1 A.U. would give a mean density of about 10^{-9} g/cm^3. This is a minimum for our purpose, since the density *in* the ecliptic plane, where the planets formed, would have been higher. Kuiper (1956a) adopted a value 10^{-8} g/cm^3 in the plane 4 A.U. from the sun. Cameron (1962), who adopted a high mass and modeled the compression toward the ecliptic, estimated densities as high as 10^{-5} g/cm^3 in the plane at the earth's position, increasing toward the sun.

Temperature of Nebula

Urey (1952) showed that the final process of accumulation of planetary material into planetesimals occurred at low temperatures (hundreds of degrees). Urey's work helped to overturn the older idea that planetary masses formed directly from gases at solar temperatures torn catastrophically out of the sun.

However, recent evidence, such as the theoretical models of Cameron (1962) and the chemical studies by Larimer (1967), suggests that the nebular gas started out at high temperatures (2000°K) after its collapse to solar-system dimensions. The nebula then cooled before accumulation of planets began. Final temperatures probably reached the lower values found by Urey, who based his estimates on studies of meteorites (see Chapter 9).

Summary of Nebular Conditions

In short, the solar nebula is viewed at the outset of the planetary formation process as a flattened, cooling cloud of gas of density at least 10^{-9} g/cm^3, with dimensions comparable to those of the present solar system.

TIME LIMITATIONS ON THE FORMATION OF PLANETS

Date of Solar-System Formation

Careful dating of meteorites shows that they formed into solid bodies about 4.5 to 4.7×10^9 years ago (see Chapter 9). Meteorites are the oldest rocks known in the solar system, and this well-determined age evidently dates the formation of the planets.

Duration of Planet Formation

Another type of dating of meteorites gives not the date of a specific event, but a time interval called the *formation interval.* The formation interval is the amount of time elapsed between two events: (1) the moment when the concentration of certain short-lived radioactive isotopes reached certain values, and (2) the time when the planetesimals formed.* Measured formation intervals for various meteorites range from 6×10^7 to 8×10^8 (60 to 800 million) years (Anders, 1963; Wood, 1968).

In the 1960s the interpretation of this seemed fairly straightforward. Researchers concluded that the process of planet formation lasted a few hundred million years. The event that formed the short-lived isotopes was widely supposed to have been a series of high-energy nuclear reactions in the twisted magnetic fields of the outer atmosphere of the early sun. Such reactions would have thus marked the start of the formation interval. However, attempts to construct detailed theoretical models of these reactions have not been entirely successful.

It is presently thought that the "beginning" of the formation interval involves some sort of averaging over various events that preceded the formation of the planetesimals. Some of these events may have been nuclear reactions in other stars outside the solar system. These stars may have dispersed their material into interstellar space prior to the formation of the sun, thus contributing matter that ultimately was incorporated into the solar system.

Because the interpretation of the formation interval depends on the nature of these presolar events, some researchers have interpreted only the *differences* among formation intervals of various meteorites—that is, the relative values rather than the absolute formation intervals.

For example, Podesek (1970) finds a spread in formation intervals of about 15 million years for material in meteorites of different classes, including an iron meteorite. Similarly, Papanastassiou and Wasserburg (1969) have used measurements of rubidium and strontium isotopes in a certain class of meteorites (basaltic achondrites) to show that all members of this particular class finished separating from the solar nebula within an interval of only a few million years (as little as 2×10^6 years) or less. Such studies have led recent researchers to conclude that the era of planet formation was not much longer than 15 million years.

* More specifically, the end of the formation interval is the time when the planetesimals began to retain the gaseous decay products, Xe atoms, of the radioactive isotopes. This is probably the time at which the planetesimals cooled to a certain critical temperature (see Chapter 9).

In summary, the best present evidence from meteorite ages and formation intervals is that the planets formed over a period probably between 15 and 100 million years, beginning about 4.7×10^9 years ago.

Evolutionary State of the Sun during Planet Formation

In the context of stellar evolution outlined in Chapter 5, where was the sun on the Hertzsprung–Russell diagram as the planets formed? There is quite a bit of evidence.

First, the formation interval of some 10^7 years is no longer than the time required to go from the top of the Hayashi track to a point on the Henyey track, as shown by Figure 6-2. Hence the planetesimals should have been forming

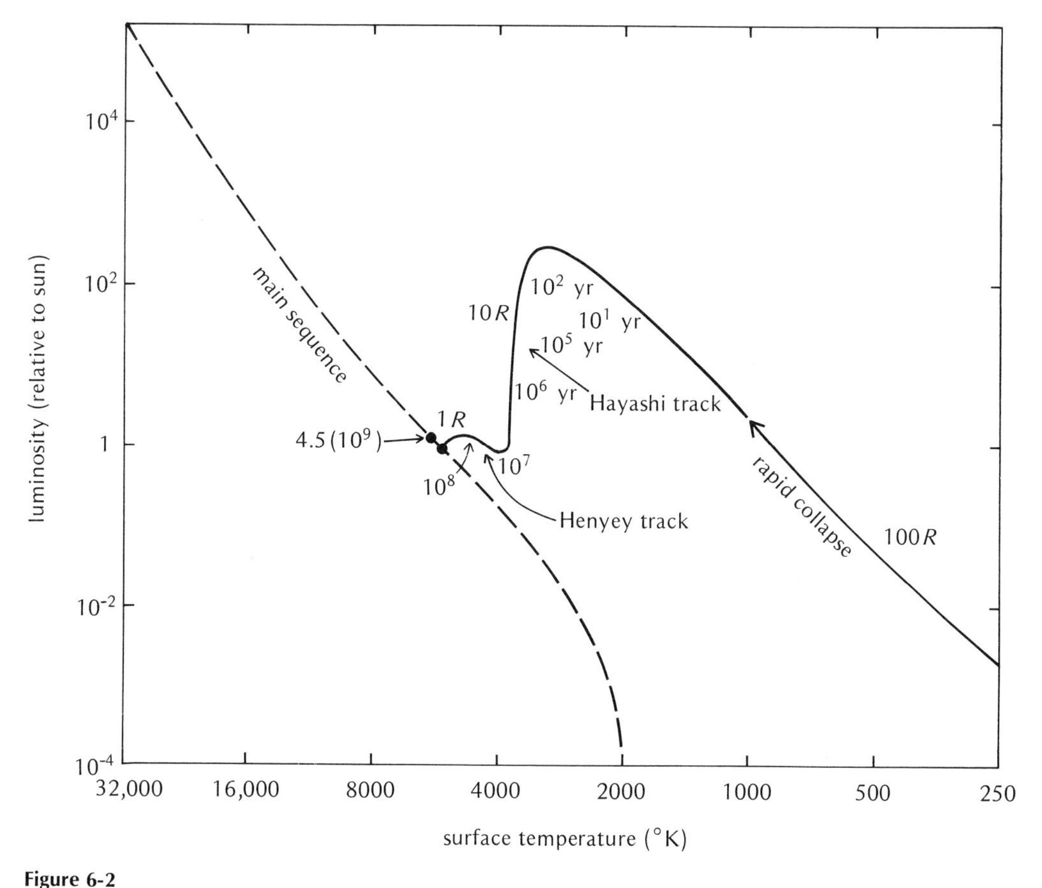

Figure 6-2

The computed evolutionary track of the early sun, plotted on the H–R diagram. Numbers above the track give the solar radius (relative to present value); numbers below the track give elapsed time (years).

by the time the sun approached the main sequence. Second, we would not expect planets to begin forming while the sun was in its rapid *pre-Hayashi* collapse, because the sun's radius was then so big that some planets would have had to form inside the body of the collapsing gas cloud. A third evidence is the flattened shape of the solar system, which indicates that the sun was already a dense, central body when the solar nebula was evolving, as noted at the beginning of this chapter. A fourth evidence is that the cocoon nebulae last no later than the T Tauri stage, which precedes arrival on the main sequence. Nebulae around T Tauri stars appear to be dispersing.

In summary, whatever processes began the formation of planets must have resulted from evolution of the solar nebula within a period of about 100 million years between the early evolution of the sun onto the Hayashi track and its subsequent evolution along the Henyey track as its solar nebula dispersed and it quit the T Tauri stage.

CONDENSATION OF THE HEAVY ELEMENTS

Suggestions in the 1940s and 1950s that solid microscopic grains would condense out of the solar nebula have been followed by detailed, physical-chemical studies of the condensation process (Urey, 1954; Wood, 1963; Lord, 1965; Larimer, 1967; Larimer and Anders, 1967; Blander and Katz, 1967; Blander and Abdel-Gawad, 1969). These studies have greatly aided our understanding of planet formation. Let us attempt to reconstruct the sequence of processes.

We have seen that the solar nebula probably started at high temperatures and proceeded to cool. Physical chemistry shows that in such a system, at the given temperatures and pressures, certain elements and compounds cannot remain in a gaseous state. Just as water vapor condenses out of a cooling high-altitude cloud to form snowflakes, certain solid particles condensed out of the cooling nebula.

Larimer (1967) and Larimer and Anders (1967) made important studies of this condensation process and found that iron was the first major constituent to condense out of the cooling gas, at about 1610°K, assuming a nebular pressure only a few thousandths of the air pressure at the earth's surface. (Had the nebula never got that hot, the iron grains would have formed eventually during the presolar collapse when the density reached a high enough point, or they might have already existed in the nebular material in the form of interstellar grains.)

The next major constituents to condense—beginning at about 1400°K—were the various silicates, similar to common rock-forming minerals. These probably used the preexisting iron grains as condensation nuclei, just as raindrops and snowflakes grow on preexisting dust particles.

An important method of research in this field is to test the theoretical predictions by examining the ancient, 4.5 billion-year-old meteorites, which are samples of the planetary material that formed in the primeval nebula. Thus guided, Blander and his coworkers showed that the problem of *nucleation*—the tendency to condense only on preexisting particles—might delay the condensation process until the nebula became supercooled. In this case the condensates, once they began to appear, would grow very rapidly in a way which can account for some properties of certain meteorites known as chondrites (see Chapter 9).

Larimer and Anders derived the sequence in which various elements would condense as the nebula cooled. They found that certain relatively volatile elements which condense at less than about 530°K are depleted from chondritic meteorites—indicating that these meteorites formed from material that had condensed by the time the temperatures got that low.

Larimer and Anders (1970) later studied meteoritic abundances of the relatively nonvolatile elements, such as silicon, iron, nickel, and so on, and found chemical evidence that they, as well as the volatiles, had undergone a substantial amount of segregation *in the nebula,* before the planetary bodies formed. This conclusion provides empirical support for the idea that chemical differences among the planets arose in part due to nebular processes. A possible mechanism for such a separation would be that the first condensates to grow into substantial particles would tend to settle through the gas toward the ecliptic plane, as a result of the gravitational attraction of the nebula. This would isolate and concentrate early condensates in the plane and leave them deficient in the later, lower-temperature condensates. Another possible separation mechanism would be selective condensation of high-temperature condensates in the warmer, inner nebula, and other material in the outer part of the nebula.

Thus, the theory of condensation not only is consistent with what we know about the solar nebula, but also illuminates the abundance patterns found in meteorites.

OPACITY OF THE SOLAR NEBULA

The grains that we have discussed are the first planetesimals—the forerunners of planets. Initially they are microscopic, like the particles that make up the smoke from a match. As soon as they become abundant, they obscure light from the sun. Because more and more grains form as the nebula cools, the nebula becomes more and more opaque. Indeed, it can be shown (Poveda, 1965; Hartmann, 1970) that the solar nebula became very rapidly opaque as the first grains formed.

Under such conditions the light from the central sun does not shine directly on most of the newly formed planetesimals. Rather, it is absorbed by grains in the inner nebula. Each grain must find some equilibrium temperature at which its rate of energy loss by radiation equals its rate of energy gain by absorption. Hence the grain temperatures rise until the grains radiate enough to reach equilibrium. This happens at temperatures of a few hundred degrees. Grains at that temperature radiate energy in the form of infrared, not visible, radiation.

The grains and gas in the innermost part of the nebula, where some sunlight filters through, are the hottest; the temperatures in the outer nebula are lower. Most of the planetesimals first form in near-darkness.

We have just repredicted the existence of a cocoon nebula around the sun. An observer on the outside would see not a bright star but a disk-shaped dark nebula, radiating in the infrared, with perhaps a small amount of visible radiation filtering through the thinnest part of the disk. Such nebulae have been observed, as we saw in Chapter 5.

CONDENSATION OF ICES

The region where the planets were forming was dark and cooling after the condensation of the heavy elements. Eventually the temperature reached values where the ices of cosmically common compounds—water (H_2O), methane (CH_4), and ammonia (NH_3)—could condense. The temperature at which frozen water condensed into solid form, under the nebula conditions postulated by Larimer and Anders (1967), was near 200°K, instead of the familiar 273°K at which H_2O ice freezes under terrestrial conditions.

Because most of the nebula was shielded from sunlight by dust, cooling was rapid. In the absence of heating from the sun, according to the Larimer–Anders model, it took less than 1000 years for the temperature of the nebula to fall to low-enough values for ices to condense.

A question remains as to how much ice was incorporated into the planets. The giant planets must have received a great deal, since their region of the solar nebula was cold and since they are observed to be rich in the volatiles CH_4 and NH_3. Some ices in the inner parts of the solar system may have condensed on planetesimals and later vaporized when exposed to sunlight as the nebula cleared.

The condensation of ices is an important milestone for two reasons: (1) it releases major constituents of the nebula (H, C, N, O) for conversion to solid material, and (2) it coats the preexisting grains with icy materials, which may facilitate the sticking together of the grains during later growth.

COLLISIONS OF THE PLANETESIMALS

The process of sticking together of the grains is called *accretion*. Without accretion being effective, we would end up with a cloud of microscopic particles instead of a system of planets. How does accretion proceed?

Accretion can occur only during collisions of grains. Calculations show that collisions are relatively frequent, but now the questions of whether and how the grains stick become crucial to the theory. If the grains simply hit and bounce apart, no growth will occur. Worse yet, if they hit too hard (at too high velocity) they may shatter each other, and instead of growth we would find fragmentation. This problem remains a topic of current research (Marcus, 1969).

It would seem that accretion would proceed only if the grains in the nebula move slowly with respect to one another. One might thus picture icy grains gently touching and sticking together like snowflakes. Nonetheless, if the velocities are *too* low, there will not be enough collisions to allow fast enough growth.

Fortunately, we have an independent way to estimate the velocities and motions of the grains: by considering the mass motions of the gas in the nebula, after the fashion of von Weizsäcker (see Chandrasekhar, 1946, and Kuiper, 1951, 1956a). The colliding grains are so small that they are caught up and swept along in the gas but cannot move independent of the gas. In the same way, snowflakes may be swept along together *in* the wind, but one snowflake cannot sail *through* the air at tremendous speeds with respect to the others. Thus, the grains move at slow velocities with respect to each other, perhaps a few centimeters per second, plus or minus an order of magnitude.

By calculating the number of grains we can estimate the collision rate. *If the grains stick when they collide,* the growth rate is found to be of the order 10^{-3} cm/year assuming a 1 cm/sec collision velocity. In 10,000 years we could get a large snowball.

The question remains—why should the grains stick together instead of bouncing apart? A possible answer is that the newly formed grains—especially the iron and silicate-coated grains—may be warm enough to coalesce in a semimolten fashion upon contact. At lower temperatures, ices may form loosely packed possibly slushy coatings conducive to sticking. A third suggestion is that certain organic molecules may form on the grain surfaces, making them sticky. The question requires further laboratory experimentation before it can be answered.

SUPPORT FROM OBSERVATIONS OF CIRCUMSTELLAR MATERIAL

In Chapter 5 we described two important sets of observations that support the theory derived so far. First are the observations of cocoon nebulae radiating in the infrared. The second is the accumulating evidence that certain stars have circumstellar grains composed in part by silicates. References to this research were given in Chapter 5.

GROWTH TO METRIC DIMENSIONS AND CLEARING OF NEBULA

A growth rate of 10^{-3} cm/year, quoted above, would produce planetesimals 10 meters (about 10 yards) across in about 1 million years. Several new processes occur during growth through this size range.

Stokes's law gives the maximum velocity at which a particle can move through a gas under the influence of a given force. Applications of Stokes's law to the planetesimals in the nebular gas show that as long as they are microscopic they will be swept along with the gas, but before they reach centimeter dimensions, they can begin to move independently. This suggests that the differences in velocity among the planetesimals (i.e., the collision velocities) begin to increase as the planetesimals grow larger.

At first sight, this helps to increase the growth rate of the planetesimals. However, we have as yet no theoretical guarantee against velocities getting so large that fragmentation ensues. Öpik (1958) calculated that fragmentation of rocky planetesimals would ensue at collision velocities of some 10^4 cm/sec—somewhat under the speed of a .22 bullet. The fact that planets exist, as well as other empirical evidence, suggests that collision velocities did not exceed this value. A reasonable collision velocity is 10 cm/sec, which would allow a growth rate of about 10^{-2} to 10^{-1} (0.01 to 0.1) cm/year, producing planetesimals as big as 10 km across in 10 million years.

As the planetesimals grow and begin to move independently from the gas, the tendency for them to settle toward the ecliptic plane, due to the gravitational attraction of the nebular disk, would increase the number of particles near the plane, encourage collisions, and speed the growth rate.

It is obvious that if all the accretable material coalesced into a few planets, the nebula could no longer obscure the sunlight. Calculations suggest that by the time most of the planetesimals became basketball- or meter-size objects, the solar nebula began to clear (Poveda, 1965; Hartmann, 1970). This clearing is important because it allows the sunlight to strike the planetesimals directly. Once this starts, the smallest grains will be blown away from the star into interstellar space by radiation pressure from the star (Wickramasinghe, 1967, p. 866). The larger planetesimals, which remain behind, will be heated by the sunlight and may loose some of their volatiles—especially their ices—leaving dense material in the planet. This may explain in part why the density among the inner planets increases toward the sun.

Although we cannot describe this stage in great quantitative detail, we can at least paint a qualitative picture. By the time the planetesimals reach multimeter dimensions, the sun has entered the T Tauri stage, the nebula is dispersing and growing less dense, and individual planetesimals no longer are entrapped in turbulent gas but move on Keplerian orbits. An imaginary view of conditions in the solar system at this time is shown in Figure 6-3. Many aspects of the growth process at this stage require further analysis and experimentation.

Figure 6-3

A scene during the clearing of the early solar nebula. As the sun becomes visible, abundant planetesimals define a swarm compressed toward the ecliptic plane.

RAPID GROWTH TO PLANETARY DIMENSIONS

At this time, 10^7 to 10^8 years after the formation of the sun, the planetesimals approached radii that were critical to the further development of the solar system. At a certain critical radius, a planetesimal will have a sufficient gravity field to affect the approach of other planetesimals. It will then begin to sweep up surrounding particles not with its geometric cross section (πr^2) but with a gravitational cross section that increases as r^4 instead of r^2. This means that the planetesimal will begin to grow very fast, because it gravitationally "pulls in" approaching particles. The bigger it grows, the more effective it becomes at accreting material—and the faster it grows. This feedback effect produces a dramatic change in the revolution of the system of particles.

Once one planetesimal exceeds the critical radius it will grow very rapidly, sweeping up mass at the expense of surrounding planetesimals. It then grows quickly toward planetary dimensions. Only the bodies that pass this critical size that can be called *true planets*. At a given distance from the sun, the first planetesimal that passes the critical radius will tend to sweep up all the rest. Several other planetesimals may pass the critical radius by the time the largest one has grown to planetary size. The critical radius depends on the distance from the sun and other factors but is on the order of a few hundred kilometers.

Thus during the final formation of the planets, each zone of the solar system may have contained one dominant, thousand-kilometer-sized planet, a few 100- to 1000-kilometer bodies, and many residual planetesimals some tens of kilometers in dimension.

This view of planet formation from a swarm of particles competing with each other in a collisional growth process draws from various contemporary references listed in Chapter 4, as well as references listed in this chapter. The concept of a critical radius, discussed by the German physicist von Weizsäcker (see Chandrasekhar, 1946, for a review in English), finds several supports in the asteroid belt. The numerous asteroids (see Chapter 8) appear to exemplify a planet-forming process that was halted before completion. Perturbations by the giant planet Jupiter may have been responsible for increasing collision velocities among the asteroids and hence preventing the growth of a planet in that zone of the solar system. Whatever the reason, asteroids appear to be remains of original planetesimals. By studying the asteroids' mass distribution, Anders (1965) found evidence that the most common radius among the original asteroids at the cessation of their growth was about 30 km—a size consistent with the theory outlined above. Using this theory, Hartmann (1968a) found that the critical radius for asteroids was about 175 km. There are three asteroids larger than this, and they are found to be abnormally large in comparison to the size distribution of the others. They are evidently the few asteroids which surpassed the critical radius and started to grow very fast. The largest of the three, Ceres, has a radius of about 400 km. There may have been more than three originally; meteorites—which are thought to be fragments of broken asteroids—show evidence of having come from parent bodies ranging up to this dimension, suggesting that many original planetesimals of the asteroid belt apparently collided and fragmented (see Chapters 8 and 9). In this view the asteroids provide a frozen tableau of the ancient solar system.

FATE OF PLANETESIMALS

What happened to the remaining nonasteroidal planetesimals? At any time during planet formation, four possible destinies awaited them.

Fragmentation

Some planetesimals collided with one another. As we have seen, collision velocities in the solar system were probably increasing during the final formative stages of planet growth. This came about because the planetesimals occasionally were perturbed by the newly forming planets. Their orbits thus became less and less circular, with the result that they no longer moved together but acquired intersecting orbits that favored high-speed collisions.

Such collisions would have produced fragments with size distributions that can be predicted. The predicted distributions agree with those observed among asteroids (Marcus, 1969; Hartmann, 1968b). The collision process apparently progressed in the asteroid belt to a point where almost all the smaller asteroids are merely fragments of original planetesimals (see Chapter 8 for further evidence).

The end result of collisions is to grind the planetesimals down into particles so small that they either spiral into the sun by the Poynting–Robertson effect or are blown out of the solar system by radiation pressure. A small fraction of the mass of the solar nebula may have been lost in this way.

Collision with Planets

A second fate that befell some of the planetesimals and their fragments is that they collided directly with one of the growing planets. This added to the planet's mass by accretion (as long as the planet was big enough not to be fragmented by the impact). Such collisions caused numerous impact craters on planetary surfaces.

The last such collisions occurred after the planets reached their present radius, and the craters are preserved on several planetary surfaces. The moon, Mars, and even the earth show such craters, although in the case of the earth, geological processes such as mountain building have erased almost all craters older than a few million years. The craters of the moon and Mars show a size distribution corresponding to the size distribution predicted for fragmented planetesimals. Even the two small satellites of Mars—about 12 and 22 km in diameter—display a population of small impact craters.

The largest craters on the moon, which range up to nearly 700 km (440 miles) across, were formed by bodies of the order of 50 to 200 km in diameter [depending on the impact velocity—probably on the order of 5 km/sec, as estimated from crater-energy relations published by Baldwin (1963)]. This dimension is consistent with the theoretical size estimates for the largest planetesimals left in orbit by the time the planets had grown, as discussed above.

Calculations by Öpik (1963, 1966) and Arnold (1965) show that planetesimals which followed orbits that crossed the paths of the terrestrial planets had mean lifetimes of only about 10^8 (100,000,000) years before colliding with one of the planets. In other words, the swarm of residual planetesimals and fragments of planetesimals would theoretically be used up several hundred million years after the formation of the planets. The resulting intense early bombardment of the planets was predicted by Urey (1952). The dating of the lunar surface during the Apollo program shows that the great majority of lunar impacts did indeed occur not more than 5×10^8 (500,000,000) years after the formation of the moon (Hartmann, 1970), although sporadic cratering has continued until the present.

The accretion process evidently produced near-zero obliquities among planets. Safronov (1966) has pointed out that the final impacts of planetesimals upon the nearly completed planets affected their obliquities. The greatest departure from zero (apart from Venus which may involve resonance effects) is that of Uranus, which Safronov attributes to the impact of a mass about 0.05 of that of Uranus itself. The largest bodies to hit the earth he finds to have had 10^{-3} earth mass, equivalent to the largest asteroids. This is consistent with our theoretical view of a few hundred-kilometer bodies near each planetary orbit.

These four lines of evidence—lunar and Martian crater size distribution, sizes of largest craters, crater ages, and planetary obliquities—are all consistent with the theory that the solar system was densely populated with planetesimals as the planets formed.

Ejection from the Solar System

The third possible disposition of the planetesimals applies to those which approached close to planets but experienced a near-miss. In such cases strong perturbations would occur, and some planetesimals would be ejected from the solar system. These have been discussed in detail by Öpik (1963, 1966) and Arnold (1965). Jupiter, being the most massive planet, was especially effective in ejecting planetesimals from the system. The planetesimals in the vicinity of Jupiter were likely to have been rich in ices, since ices are stable beyond the asteroid belt but are melted by sunlight in the region of the terrestrial planets (Watson, Murray, and Brown, 1963). It is thought that some of the icy planetesimals that originally populated the vicinity of Jupiter and the other giant planets were ejected to the outskirts of the solar system to become what we now know as comets. This will be discussed in Chapter 7.

Capture into Satellite Orbits

The fourth possible fate of planetesimals and their fragments was to be captured into orbits around planets. The complex history of planetary satellites will be discussed in the next two sections.

THE RINGS OF SATURN AS CLUES TO PLANETARY ORIGIN

Saturn is surrounded not only by 10 satellites (counting the recently reported but sometimes questioned Janus) but also by a unique system of rings described in Chapter 2. The rings are composed of a multitude of small particles of uncertain dimension. Such a phenomenon should be explicable in terms of a successful theory of solar-system origin.

Three important research questions are associated with the rings and relate to the formation of planets. First, of what are they composed? The answer to this question illustrates the vicissitudes of scientific research. Kuiper, as early as 1956, proposed that the ring particles are principally frozen water. However, later spectra by Kuiper and his coworkers showed absorption features that seemed to match that of frozen ammonia better than that of frozen water observed in the laboratory at −20°C; Kuiper thus announced that the rings were made of frozen ammonia. Simultaneously, however, Lebofsky, Johnson, and McCord (1970) pointed out that under special conditions (e.g., −190°C) the frozen water does match the spectrum observed. They concluded that ice particles or ice-covered silicate particles, with probable minor contaminants, are the main constituent of the Saturn rings. Kuiper, Cruikshank, and Fink (1970) concurred, thus confirming Kuiper's original prediction.

The second question is: How do Saturn's rings maintain their flatness and the inter-ring gaps? Alfvén takes the view that the present-day ring structure directly reflects primeval conditions among the Saturn satellite system during the formation of Saturn, and on the basis of this theory he predicts the existence of an undiscovered satellite between the inner satellites, Janus and Mimas. On the other hand, most theorists follow the model of analysts such as Reiffenstein (1968), who find that present-day perturbations and resonances among the known satellites are sufficient to account for the structure of the rings. The gaps between rings, according to this model, are produced solely by resonance effects with the inner satellites and do not reflect primeval conditions.

The third and most obvious question is: How did the rings form? The existence, if not the structure, of the rings must be a unique clue about the conditions attending the formation of planets, if we could but interpret it correctly. Alfvén (1968) and Kuiper (1956b) take the view that the rings are the remnant of the cloud of particles surrounding any newly forming planet, from which satellites may form. In Alfvén's model, "proto-Saturn" was surrounded by a magnetic field and a cloud of ionized gas. Certain parts of the ionized gas under these conditions moved inward to condense into separate grains and form the rings. Kuiper proposed that the rings formed from a cloud of uncharged particles that had too low a density to condense into a satellite. In each view, a satellite never formed from the particles in the rings because these particles are located inside Roche's limit for Saturn, and hence would not accumulate by gravitational forces. An older theory of the formation of the rings is also still widely held, and is adopted by Reiffenstein (1968). According to this model, the particles forming the rings were once part of a planetesimal or a satellite (or several such bodies) that approached so close to Saturn that it came inside Roche's limit. In this case, the planetesimal could disintegrate, producing a swarm of fragments that eventually assume a flat distribution due to tidal and satellite perturbations. Either theory, combined with the evidence of icy composition, is consistent with current theory of the growth of planets from multitudes of planetesimals. The ring particles could be viewed either as preserved planetesimals associated with Saturn or as fragments of a satellite that itself grew according to the accretion process. This raises the question of satellite origins, which is discussed next. Before leaving Saturn's rings, we should note that further observations and dynamical studies are clearly needed before these mysterious companions of Saturn can be fully understood.

ORIGIN OF SATELLITES

Complexities of Satellite Systems

In the 1800s many theorists thought that the satellite systems were simply scaled-down versions of the solar system, formed in the same manner. The satellites were even thought to hold the key to the origin of the solar system because they appeared to be more regular than the system of planets.

Satellites known at that time, such as the moon, the Galilean satellites of Jupiter, and certain large satellites of Saturn, had very low eccentricities, very low inclinations to their planet's equator, and regular spacings reminiscent of Bode's law. It was widely assumed that the satellite systems must have formed in disk-shaped nebulae somehow thrown off from the planets just as the solar nebula was supposed to have been thrown off from the contracting sun. Each system seemed to verify the theory as applied to the other systems, with the exception of Neptune's largest satellite Triton, which revolved in the retrograde direction. (The rotation direction of Neptune itself was not yet known and some theorists thought that it would also turn out to be retrograde, but this was later disproved.)

The twentieth century brought the discovery of a number of new satellites which upset this orderly scheme. The new satellites tended to be small and faint (hence their late discovery); many were in orbits outside those already known, and these were often in relatively high inclination orbits (about 30°) or in eccentric orbits or in retrograde orbits.

It thus became clear that the satellite systems are much more complex than mere copies of the solar system, although the early-known regularities are undoubtedly significant.

A number of complexities about satellite systems should be noted. First, we cannot trace the past history of satellite orbits by contemporary celestial mechanics. The contemporary theory and observational parameters are only accurate enough to trace the orbits through roughly the last 10^8 (hundred million) years—a few percent of solar-system history (Kuiper, 1956b). Thus we cannot rigorously prove that the eccentricities, inclinations, and other properties of satellites observed now are necessarily close to the values they had initially, when the environment may have been much different.

Second, the regularities in satellite orbits are associated with special perturbation conditions, known as *commensurabilities* (see Chapter 8), between pairs of satellites within each system. Such commensurabilities can produce spacings resembling Bode's law within each satellite system. There is evidence that the commensurabilities were effective early in the history of the systems (Dermott, 1968a, 1968b). Such commensurabilities are *not*, however, the dominant influence in establishing Bode's law among the planets; this has been claimed but has subsequently been refuted, as discussed in Chapter 4.

The third complexity is that satellites must have been lost by planets, recaptured, fragmented, and possibly even exchanged during the past history of the solar system. The best-known example is the case of Pluto. Lyttleton (1936) and Kuiper (1956b, 1957) suggested that Pluto might have originated as a satellite of Neptune, Pluto being the only "planet" in the solar system whose orbit overlaps another. Kuiper suggested that Pluto escaped because the Neptunian proto-planet (as described by Kuiper's theory of the origin of the solar system) was losing mass and the satellite Pluto therefore increased its semimajor axis until it escaped into its own orbit around the sun, overlapping Neptune's.

Tidal transfer of angular momentum from planets to satellites also tends to force prograde satellites outward toward eventual escape, as in the case of the moon (see Chapter 3). McCord (1966, 1968) has shown that retrograde satellites by the same mechanism will be efficiently drawn in toward planets and that in the past there may have been many more close retrograde satellites than we now see. Triton, for example, will either be disrupted by coming within Neptune's Roche limit, or will crash onto Neptune possibly within 10^7 to 10^8 years. Tidal effects also tend to reduce the inclination of a satellite's orbit toward zero, which makes it risky to assume that the regular satellites have always been so precisely in the planetary equatorial planes. These effects all *suggest* that ancient satellite systems could have been more erratic than those now observed.

Theories of Origin

Let us begin with the peculiar retrograde satellites. How can retrograde satellites have been born from prograde-spinning planets? It has become clear that the irregular satellites, especially the retrograde satellites, must not have come directly from the primary planet but must have been captured from interplanetary space (Kuiper, 1956b). These capture processes must have been common in the early days of the solar system. As illustrated in Figure 6-4, two groups of Jupiter's outer satellites—one prograde group and one retrograde—have such similar orbits that each group has been interpreted as the fragments from a collision between a satellite of Jupiter and an interplanetary planetesimal in the process of being captured or nearly captured (Aitekeeva, 1968; Bronshten, 1968). More recently, Bailey (1971) has suggested that the inner direct group

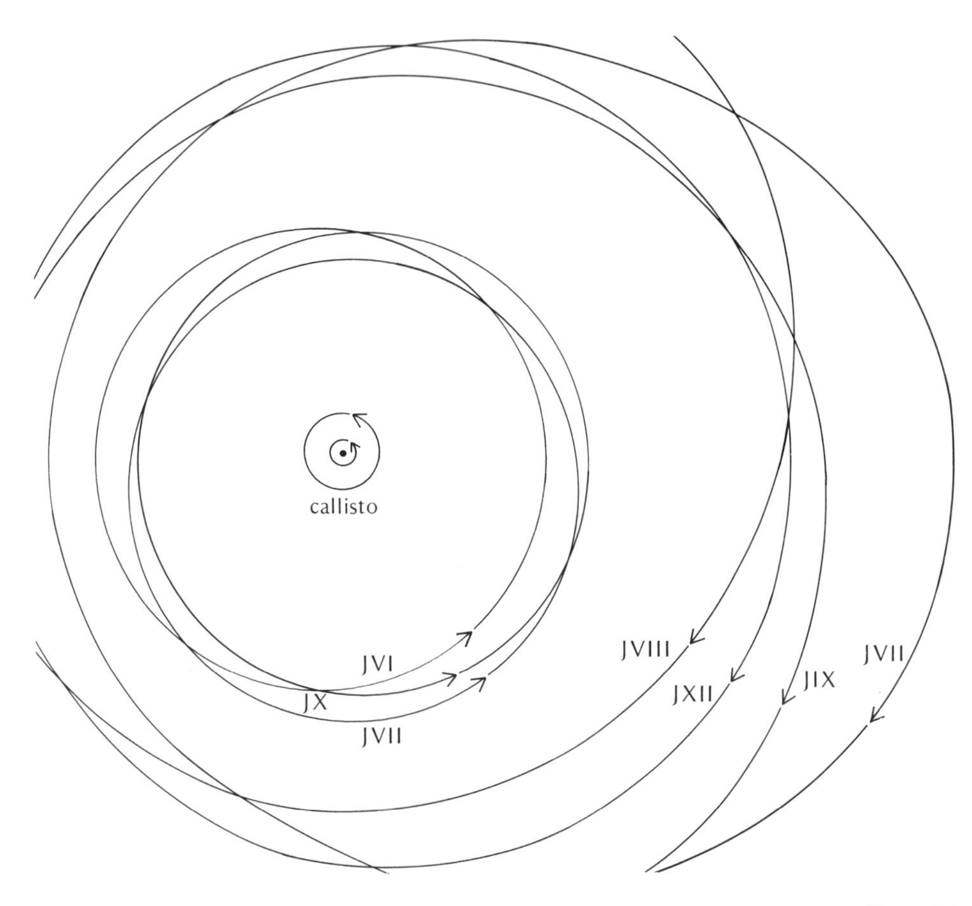

Figure 6-4

Schematic diagram of orbits of the outer Jupiter satellites. Callisto is the outermost Galilean satellite. Satellites beyond Callisto fall into two groups: an inner direct group with nearly coinciding orbits, and an outer retrograde group.

results from capture of asteroids at Jupiter's perihelion, while the outer retrograde group results from capture at aphelion.

Evidence of the complex history of satellites has recently come from photographs of the two Martian satellites, Phobos and Deimos, by the Mariner 9 spacecraft in orbit around Mars. These small satellites do not show the smooth, spherical surface that might be expected if they had simply grown from orderly accretion of innumberable tiny grains, with no further complications. Instead they are irregular in shape, being roughly 18 by 22 and 12 by 13 km in diameter, respectively. Both satellites—smaller than any other planetary bodies that have been photographed in detail—display heavily cratered surfaces, as shown in Figure 6-5. The irregular shapes and craters are testimony to a history that probably involved past fragmentation and collisions with other bodies. Possibly these satellites are examples of asteroid fragments captured gravitationally into Martian orbit, or possibly they are fragments of an original, larger Martian satellite that was struck and fragmented by an asteroid.

The unanswered question still facing theorists is whether *all* the satellites were originally interplanetary planetesimals that were captured, or whether some were direct products of the parent planet's formation. According to the first view, the solar system at the close of the formation of the planets was full of planetesimals that had not yet collided with the planets. Some of these were destroyed by collisions, some were ejected, and some impacted the planets,

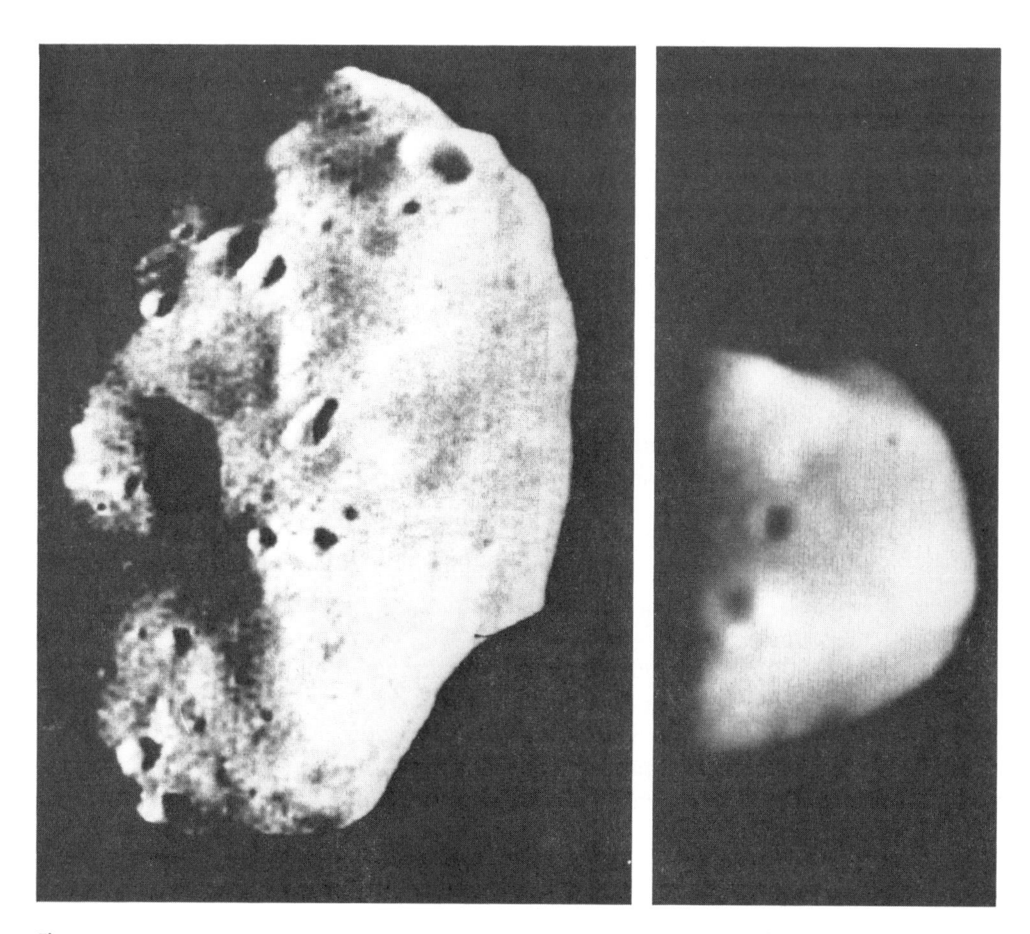

Figure 6-5

Mars' two satellites: Phobos (left) and Deimos (right). These photographs, by the Mariner 9 spacecraft, are among the first to show surface details and irregular shapes of the satellites. The craters give evidence of a complex collisional history. This print illustrates the relative sizes of the two bodies. (NASA.)

forming craters and influencing the obliquities as described above. Some were trapped temporarily at Lagrangian points, as is the case with Jupiter's family of Trojan asteroids. A few, however, were captured into satellite orbits; some of these may have escaped even again. Some retrograde satellites may have spiraled inside Roche's limit and disintegrated (the origin of Saturn's rings?), while others, influenced by tidal evolution, achieved stable orbits near the planet's equatorial plane.

Kuiper (1956b), in a detailed analysis, takes the second view. He divides the satellites into two groups, regular and irregular. Only the irregular satellites formed as above, while the regular satellites (such as Jupiter's Galilean family) are products of evolution of flattened nebulae around the Kuiper's massive protoplanets. According to the Kuiper model, regular satellite systems are analogous to the solar system itself. Individual satellites could have grown by accretion in disk-shaped clouds of particles surrounding primeval planets.

The Moon

The earth's moon is widely regarded as unique, partly because it has an unusually large mass *relative to its primary* (although several other satellites are larger in absolute terms). There have been three theories for the origin of the moon: (1) it accreted as a binary proto-planet together with the proto-earth; (2) it accreted entirely independently and was later captured by the earth; (3) it broke off the earth in a fission process (Wise, 1966). The objection to the first theory is that the moon is known to have a lower density and different composition from the earth, and thus is not likely to have evolved from the same proto-planet or the same part of the solar nebula. The objection to the capture theory is that the capture of the moon from another part of the solar system is extremely difficult in view of energy relations, whereas capture from a nearby location faces the same composition difficulty as the first theory. The objection to the fission theory is that geophysicists have been unable to affirm that fission is feasible.

After the first Apollo missions brought back lunar samples, it seemed at first that the fission theory gained some ground. The moon's composition was found to be quite like that of the earth's mantle, and the moon's low density was found to result from depletion of iron and the *siderophile elements*—the heavy elements that chemically tend to go with the iron during a planetary melting and differentiation process. The most straightforward explanation seemed to be that the moon had indeed been part of the earth's mantle *after* the earth's iron core formed. According to the fission hypothesis, the draining of the heavy iron to the earth's center was an impetus in speeding up the earth's rotation enough to cause fission, by which the moon was formed from mantle material.

More recent studies (Ganapathy et al., 1970) have shown, however, that abundance patterns of certain elements are different in the moon than in the earth. Further, the work of Larimer and Anders (1970) on abundances of these elements in meteorites suggests that they can be depleted by condensation processes before or as the planets form. This reopens the possibility that the moon and earth could have formed separately. The moon might thus have been captured, or have formed near the earth.

As Urey has remarked, the moon seems a highly unlikely object. Theoreticians have been led by frustration on more than one occasion to suggest facetiously that it does not exist.

SUMMARY

In this chapter we have traced the evolution of the solar nebula from the time the sun formed (about 4.7×10^9 years ago) to a point (about 3×10^8 years later) when the solar system took on much of its present-day appearance. In spite of many unanswered questions and gaps in our knowledge, we have been able to draw on a mixture of planetological and astrophysical evidence, and we can agree with a remark by the Soviet theoretician, Safronov (1967):

> Planetary cosmogony has been placed on the firm foundation of a variety of empirical data acquired from all related scientific fields.

We have found that the smaller bodies of the solar system—comets, asteroids, and meteorites—may be samples of the early planetesimals involved in forming the planets. Therefore we shall treat the small bodies in the next three chapters, showing that they provide abundant evidence about the earliest state of planetary bodies and about the interiors and chemistry of planets—a subject to which we shall proceed in later chapters.

References

Aitekeeva, Z. A. (1968) "Anomalous Satellites of Planets," *Solar System Res., 2,* 19.

Alfvén, H. (1968) "On the Structure of the Saturnian Rings," *Icarus, 8,* 75.

Anders, E. (1963) "Meteorite Ages," in *The Moon, Meteorites, and Comets,* B. M. Middlehurst and G. P. Kuiper, eds., p. 402 (Chicago: University of Chicago Press).

——— (1965) "Fragmentation History of Asteroids," *Icarus, 4,* 399.

Arnold, J. R. (1965) "The Origin of Meteorites as Small Bodies. II," *Astrophys. J., 141,* 1536.

Bailey, J. M. (1971) "Jupiter: Its Captured Satellites," *Science, 173,* 812.

Baldwin, R. B. (1963) *The Measure of the Moon* (Chicago: University of Chicago Press).

Blander, M., and M. Abdel-Gawad (1969) "The Origin of Meteorites and the Constrained Equilibrium Condensation Theory," *Geochim. Cosmochim. Acta, 33,* 701.

——— and J. L. Katz (1967) "Condensation of Primordial Dust," *Geochim. Cosmochim. Acta, 31,* 1025.

Bronshten, V. A. (1968) "Origin of Irregular Satellites of Jupiter," *Solar System Res., 2*, 23.

Cameron, A. G. W. (1962) "The Formation of the Sun and Planets," *Icarus, 1*, 13.

Chandrasekhar, S. (1946) "On a New Theory of Weizsäcker on the Origin of the Solar System," *Rev. Mod. Phys., 18*, 94.

Dermott, S. F. (1968a) "On the Origin of Commensurabilities in the Solar System, I. The Tidal Hypothesis," *Monthly Notices Roy. Astron. Soc., 141*, 349.

______ (1968b) "On the Origin of Commensurabilities in the Solar System, II. The Orbital Period Relation," *Monthly Notices Roy. Astron. Soc., 141*, 363.

Ganapathy, R., R. Keays, L. Laul, and E. Anders (1970) "Trace Elements in Apollo 11 Lunar Rocks: Implications for Meteorite Influx and Origin of Moon," in *Proc. Apollo 11 Lunar Sci. Conf.*, A. Levinson, ed. *2*, 1117 (Elmsford, N.Y.: Pergamon Press, Inc.; *Geochim. Cosmochim. Acta, Suppl. I*).

Hartmann, W. K. (1968a) "Growth of Asteroids and Planetesimals by Accretion," *Astrophys. J., 152*, 337.

______ (1968b) (with A. C. Hartmann) "Asteroid Collisions and Evolution of Asteroidal Mass Distribution and Meteoritic Flux," *Icarus, 8*, 361.

______ (1970) "Growth of Planetesimals in Nebulae Surrounding Young Stars," in *Evolution stellaire avant la sequence principale*, Proc. Liege Symp., June 30–July 2, 1969, Liege Collec. in −8°, 5th Ser., *19*, 215.

Herbig, G. H. (1970) "VY Canis Majoris II. Interpretation of the Energy Distribution," *Contrib. Lick. Observ. 317*.

Hoyle, F. (1960) "On the Origin of the Solar Nebula," *Quart. J. Roy. Astron. Soc., 1*, 28.

______ (1963) "Formation of the Planets," in *Origin of the Solar System*, R. Jastrow and A. Cameron, eds. (New York: Academic Press, Inc.).

Kuhi, L. V. (1966) "T Tauri Stars: A Short Review," *J. Roy. Astron. Soc. Can., 60*, 1.

Kuiper, G. P. (1951) "On the Origin of the Solar System," in *Astrophysics*, J. A. Hynek, ed., p. 404 (New York: McGraw-Hill, Inc.).

______ (1956a) "The Formation of the Planets, Parts I, II, III," *J. Roy. Astron. Soc. Can., 50*, 57, 105, 158.

______ (1956b) "On the Origin of the Satellites and the Trojans," *Vistas Astron., 2*, 1631.

______ (1957) "Further Studies on the Origin of Pluto," *Astrophys. J., 125*, 287.

______ D. P. Cruikshank, and U. Fink (1970) Letter in *Sky and Telescope, 39*, 80.

Larimer, J. W. (1967) "Chemical Fractionations in Meteorites—I. Condensation of the Elements," *Geochim. Cosmochim. Acta, 31*, 1215.

______ and E. Anders (1967) "Chemical Fractionations in Meteorites—II. Abundance Patterns and Their Interpretations," *Geochim. Cosmochim. Acta, 31*, 1239.

______ and E. Anders (1970) "Chemical Fractionations in Meteorites—III. Major Element Fractionations in Chondrites," *Geochim. Cosmochim. Acta, 34*, 367.

Lebofsky, L., T. V. Johnson, and T. B. McCord (1970) "Saturn's Rings: Spectral Reflectivity and Compositional Implications," *Icarus, 13*, 226.

Lord, H. C. (1965) "Molecular Equilibria and Condensation in a Solar Nebula and Cool Stellar Atmospheres," *Icarus, 4*, 279.

Lyttleton, R. A. (1936) "On the Possible Results of an Encounter of Pluto with the Neptunian System," *Monthly Notices Roy. Astron. Soc., 97*, 108.

Marcus, A. (1969) "Speculations on Mass Loss by Meteoroid Impact and Formation of the Planets," *Icarus, 11*, 76.

McCord, T. B. (1966) "The Dynamical Evolution of the Neptunian System," *Astron. J., 71*, 585.

_____ (1968) "The Loss of Retrograde Satellites in the Solar System," *J. Geophys. Res., 73*, 1497.

Öpik, E. J. (1958) "Meteor Impact on Solid Surface," *Irish Astron. J., 5*, 14.

_____ (1963) "Survival of Comet Nuclei and the Asteroids," *Adv. Astron. Astrophys., 2*, 219.

_____ (1966) "The Stray Bodies in the Solar System: Part II. The Cometary Origin of Meteorites," *Adv. Astron. Astrophys., 4*, 301.

Papanastassiou, D., and G. Wasserburg (1969) "Initial Strontium Isotope Abundances and the Resolution of Small Time Differences in the Formation of Planetary Objects," *Earth Planet. Sci. Lett., 5*, 361.

Podesek, F. (1970) "Dating of Meteorites by the High-Temperature Release of Iodine-Correlated Xe^{129}," *Geochim. Cosmochim. Acta, 34*, 341.

Poveda, A. (1965) "H-R Diagram of Young Clusters and the Formation of Planetary Systems," *Bol. Observ. Tonantz. Tacubaya, 4*, 15.

Reiffenstein, J. M. (1968) "On the Formation of the Rings of Saturn," *Planet. Space Sci., 16*, 1511.

Ringwood, A. E. (1966) "Genesis of Chondritic Meteorites," *Rev. Geophys., 4*, 113.

Safronov, V. S. (1966) "Sizes of the Largest Bodies Falling onto the Planets during Their Formation," *Sov. Astron.—AJ, 9*, 987.

_____ (1967) "The Protoplanetary Cloud and Its Evolution," *Sov. Astron.—AJ, 10*, 650.

Schwartz, K., and G. Schubert (1969) "The Early Despinning of the Sun," *Astrophys. Space Sci., 5*, 444.

Urey, H. C. (1952) *The Planets: Their Origin and Development* (New Haven, Conn.: Yale University Press).

_____ (1954) "On the Dissipation of Gas and Volatilized Elements of Protoplanets," *Astrophys. Suppl., 1*, No. 6, 147.

Watson, K., B. C. Murray, and H. Brown (1963) "The Stability of Volatiles in the Solar System," *Icarus, 1*, 317.

Whipple, F. L. (1964) "History of the Solar System," *Proc. Natl. Acad. Sci., 52*, 517.

Wickramasinghe, N. C. (1967) *Interstellar Grains* (London: Chapman & Hall Ltd.).

Wise, D. U. (1966) "Origin of the Moon by Fission," in *The Earth-Moon System*, B. Marsden and A. Cameron, eds. (New York: Plenum Publishing Corporation).

Wood, J. A. (1963) "On the Origin of Chondrules and Chondrites," *Icarus, 2*, 152.

_____ (1968) *Meteorites and the Origin of Planets* (New York: McGraw-Hill, Inc.).

COMETS

IN THE PRECEDING CHAPTERS we have followed the evolution of the solar system from the presolar evolution of a collapsing interstellar cloud, through the formation of the sun, and through the formation of planetary bodies. Among the major planets there must have been, in the early solar system, many small bodies. Some, at least those in the outer solar system, must have been ice-rich, with some admixture of stony material. In this chapter we shall see that comets are bodies that fit this description, to the best of our knowledge. Indeed, comets may be still-surviving representatives of the ancient building blocks of the planets. Therefore, it is appropriate to take up comets as our next topic.

The purpose of this chapter is to clarify first what comets look like, then what they are, and finally where they may come from. Although most persons have some conception of a comet, many persons have erroneous conceptions. For example, even in relatively scholarly works, one finds references to "a comet flashing across the sky." Such statements are totally misleading because comets are almost always visible for several days at a time and are so far away that their motion cannot be detected by the naked eye except by watching their progress among the stars for hours. Authors who speak of comets "flashing across the sky" probably confuse them with meteors, atmospheric phenomena that we shall discuss later (see also Chapter 9). By 1577 Tycho Brahe had shown that a comet he observed was more distant than the moon.

APPEARANCE AND NOMENCLATURE

Comets are named from the Greek word *komē,* or hair. The name describes the long wispy *tail* for which the comets are famous. Orientals knew comets by the equally descriptive term "broom stars." Comet tails generally extend in a direction opposite from that of the sun's, as shown in Figure 7-1. In addition to the tail, a comet has a *head,* composed of the *coma,* a bright, diffuse region immediately surrounding the star-like *nucleus*. The head is the source of most of the light. Examples are shown in Figure 7-2. Among the less spectacular comets, the inner coma, sometimes called the *central condensation,* is all that can be seen with the naked eye. With a large telescope, the true nucleus of even a spectacular comet appears as no more than a star-like point, attesting to the fact that cometary nuclei in physical reality must be very small, much smaller than a planet. Certain measurements indicate sizes on the order of a few kilometers (see below). Yet these small nuclei are able to produce diffuse, gaseous comas much bigger than entire planets, reaching up to 2 million km diameter, and tails that span an astronomical unit.

Comets are usually discovered when they are faint and approaching the sun at a distance of several astronomical units. In general, they brighten and their tails develop and lengthen as they approach the sun. However, there is a tendency for their comas to reach a maximum diameter at roughly 1.5 to 2.0 A.U. from the sun. Evidentally, this maximum occurs because no gases are released at greater distances, while the molecules that make up the coma are readily disassociated by solar ultraviolet radiation when the comet approaches closer to the sun. As the comet swings around the sun, it is very bright and has a well-developed tail, as shown in Figures 7-1 and 7-2. Often the comet is not visible from the earth at this stage, however, because it is lost in the glare of the nearby sun. As the comet moves away from the sun, it becomes once more visible and its coma and tail shrink again until the whole comet appears only as a faint, distant, star-like point even on the best photographs. Most comets then continue their outward journey reaching the extreme outer edges of the solar system, returning toward the sun only after many thousands of years.

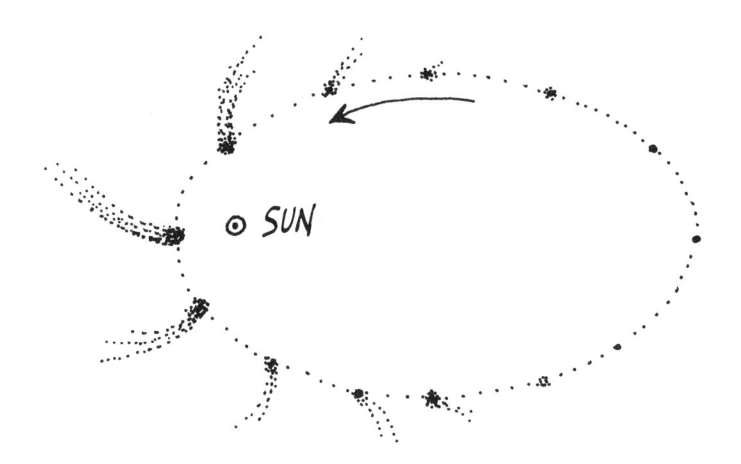

Figure 7-1

Orbit of a typical comet, showing development of the tail as the comet approaches and recedes from the sun.

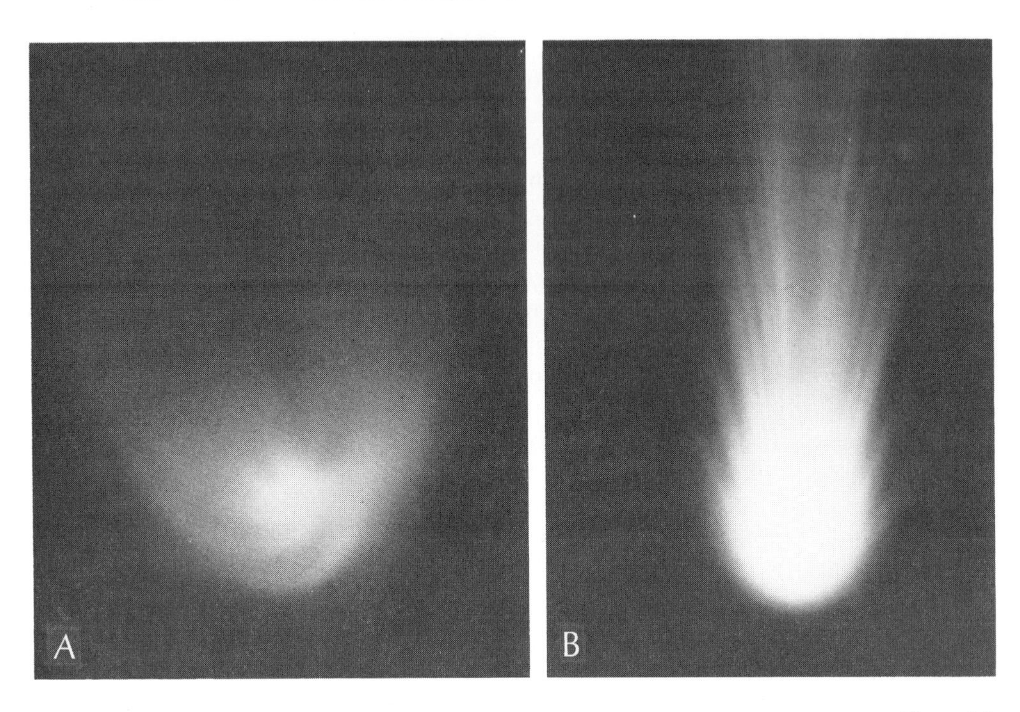

Figure 7-2
Heads of comets (in both photos the nucleus is too small to be resolved). A: "Pinwheel" pattern of streamers in the head of Comet Bennett, 1969i. (Lunar and Planetary Laboratory.) B: Linear streamers in the head of Comet Tago-Sato-Kosaka, 1969g. (University of Michigan Observatory/Cerro Tololo.)

COMET TAILS

There are two types of comet tails, Type I and Type II, shown in Figure 7-3. *Type I tails* are more or less straight, often structured by fine, linear streamers or "rays." Their spectra show emission lines which show that they are composed mostly of ionized molecules. As a mnemonic aid, it is helpful to think of the I in Type I: this numeral is linear and can stand for *I*onized. The principal light-emitting molecules are CO^+, $N_2{}^+$, and $CO_2{}^+$. Among other observed ionized and neutral molecules and atoms are OH^+, CH^+, and Na.

Type II tails are usually broad, diffuse, and gently curved. Their spectra show only the reflected spectrum of the sun. This indicates that instead of being composed of gas atoms and molecules that absorb or give off their own characteristic light, they are composed of grains of dust that merely reflect (scatter) sunlight.

Some comets have Type I tails, some have Type II, and some show both. Occasionally so-called Type III tails appear, diverging at a distinct angle from Type I tails.

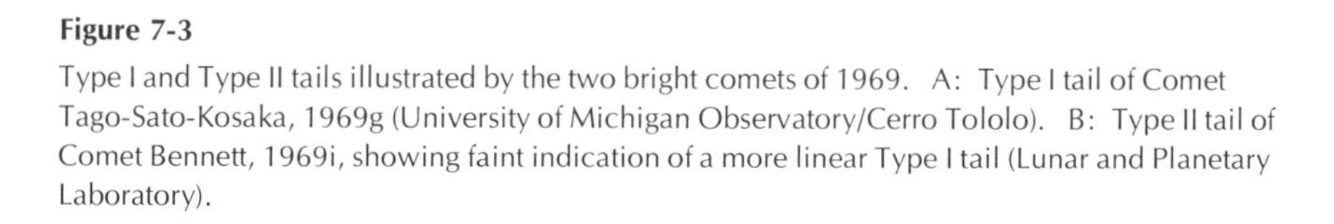

Figure 7-3

Type I and Type II tails illustrated by the two bright comets of 1969. A: Type I tail of Comet Tago-Sato-Kosaka, 1969g (University of Michigan Observatory/Cerro Tololo). B: Type II tail of Comet Bennett, 1969i, showing faint indication of a more linear Type I tail (Lunar and Planetary Laboratory).

Occasionally a comet displays a seemingly anomalous tail which *appears* to point toward the sun; comet Arend-Roland in 1957 was a widely publicized example. This is only a projection effect. As we look along the line from the earth to the comet, most of the tail may be on the antisolar side, but a portion of the curved tail may *appear in projection* to be on the sunward side. In the case of Arend-Roland, the "solar-directed tail" was part of a Type II tail seen in projection. It appeared as a narrow spike only as the earth passed through the comet's orbit plane. This indicated that the dusty Type II tails are planar and highly confined in the plane of the comet's orbit.

Comet tails, then, are composed of gas and dust emitted by the comet's nucleus and directed more or less away from the sun, with Type I being most nearly antisolar. Thus, during the part of its orbit when a comet moves out away from the sun, the tail leads; it is a false analogy to picture the tail trailing "behind" the comet like a girl's long hair as she speeds down the street on her bicycle.

Why does the tail point away from the sun? The tiny dust grains that compose the Type II tails are repelled from the sun by radiation pressure (see Chapter 3). Each particle follows an orbit determined by gravity and radiation. According to Kepler's second law, the particles farther from the sun revolve more slowly and so they lag behind, giving the Type II tail its characteristic curvature. But this does not explain all properties of Type II tails, not to mention Type I tails. Observations of comets showed that knots of material in the Type I tails were being accelerated away from the sun faster than could be explained by radiation pressure. This led Biermann (1951) to the concept of the solar wind, streaming out from the sun (Chapter 3). Magnetic interactions couple the ions of the Type I tails to the outward-expanding ionized gas, or *plasma,* of the solar wind. To correct our analogy with the bicyclist's hair, we would have to say that she is riding in a fierce wind, sometimes into it with her hair streaming back, and sometimes with it, her hair streaming out in front of her.

DISCOVERY OF COMETS

Roughly 10 comets are observed each year. About half of these are new discoveries and half are known comets on return trips around the sun, known as *"recovered comets."* Most of the recoveries are first detected on astronomical photographs taken specifically to locate them. Many new comets are found by amateur astronomers who make a hobby of scanning the skies with modest telescopes, checking any suspicious fuzzy objects against the catalogs of known nebulae (which can easily be mistaken for comets in small telescopes). In recent years, several comets have been discovered by airline pilots, who are favored by the very clear skies at high altitudes and by their alertness during hours when most people have no time to watch the sky. Problems of search and discovery are reviewed in detail by Roemer (1963).

When a newly discovered comet is confirmed, an announcement is sent to astronomers around the world by the International Astronomical Union, and the comet is assigned a designation consisting of the year and a letter in order of discovery or recovery: 1973a, 1973b, and so on. Comets are also popularly named after their discoverers: Comet Burnham, or Comet Ikeya-Seki (which had two independent discoverers). After a year or two, when observations of the comets have been collected and reliable orbits determined, they are assigned new, permanent, designations with Roman numerals for the order in which they passed their perihelion points: 1973 I, 1973 II, and so on.

ORBITS OF COMETS

Determination of the orbit of a comet is a complex process requiring the best fit of many observed positions to an elliptical orbit. Study of the computed orbits quickly shows that comets fall into two rather distinct classes: long-period or near-parabolic comets and short-period comets, illustrated in Figure 7-4. *Long-period comets* have periods greater than a few hundred years and move in extremely eccentric orbits. Aphelion distances may reach as much as 40,000 to 50,000 astronomical units, far beyond the outer planets. In these cases, intervals between solar approaches range from 100,000 to 1 million years. These aphelia are so far away that we can observe the comet in only a small portion of its orbit near the sun, where the comet is brightest. The record distance at which a comet has been seen is 11.5 A.U.; few are observed beyond 3 or 4 A.U.

Consider the orbits of long-period comets. If you were handed a 1-inch arc of a 1000-inch ellipse and of a similar parabola, you would have difficulty distinguishing them. This indeed is what happens when we try to derive orbits of long-period comets: We see only the small tip of the orbit near the sun, and it is very difficult to tell whether the orbit is a very elliptical orbit or parabolic or even a hyperbolic orbit. This is a critical decision, because an elliptical orbit would imply the comet was a member of the solar system while the last two cases would imply that it had come from among the stars.

★ PROBLEM ★

Justify the last statement.

Evidence suggests that most near-parabolic comets have elliptical orbits and thus are members of the solar system. Their eccentricities tend to cluster near 1.000 . . . , whereas interstellar comets would have a range of higher

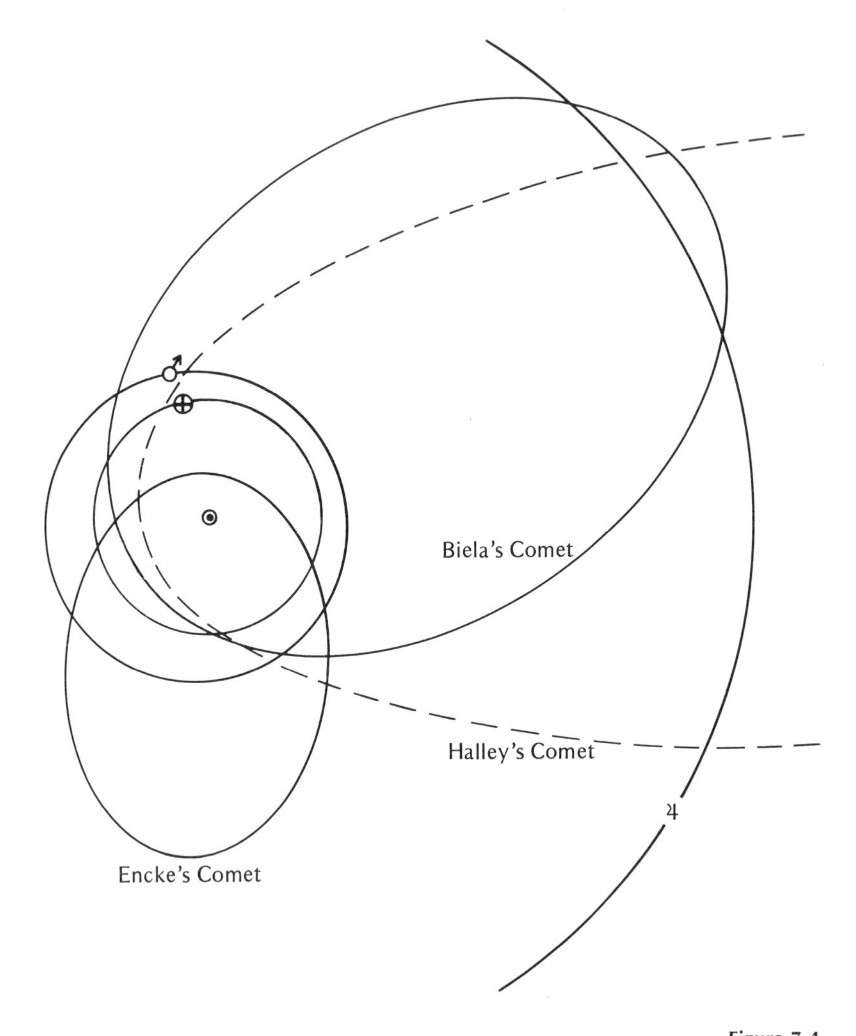

Figure 7-4

Orbits of typical comets, showing two short-period comets (Encke and Biela) and one long-period comet (Halley).

values (hyperbolic orbits). No definitive comet orbit has been observed to have an eccentricity higher than 1.004; all cataloged hyperbolic orbits may represent either observational error or pertubations of originally elliptical orbits.

The *short-period comets* have moderately eccentric elliptical orbits, more eccentric than the planetary orbits, but distinctly less eccentric than those of long-period comets. Typical aphelion distances are five to seven astronomical units.

A 1965 catalog of the 583 known comets gave the following breakdown of orbits:

56 short-period ($P < 200$ years), seen in a total of 343 apparitions
43 short-period, seen in only one apparition each
130 long-period elliptical orbits
284 long-period "parabolic orbits"
70 long-period "hyperbolic orbits"

Since hyperbolic orbits may result from observational error and since parabolic ($e = 1.000 \ldots$) are often adopted in order to make the calculations easier, we might regroup the breakdown in a more realistic fashion as follows:

99 short-period comets
484 long-period (near-parabolic) comets

A few comets may actually depart the solar system on hyperbolic orbits. Work pioneered by Stromgren (1914) has indicated that some comets, while approaching the sun on initially elliptical orbits, are perturbed by planets into hyperbolic orbits. Thereby we gain a clue as to the history of comets, as will be seen in the next section.

ORBITAL CLUES TO COMET HISTORIES—THE OORT CLOUD

Oort (1950, 1963) found that the distribution of semimajor axes of long-period comets indicated the existence of a swarm of comets surrounding the solar system out to about 50,000 astronomical units. This swarm is known as the *Oort cloud.* The inclinations of long-period comets are distributed almost at random, indicating that the Oort cloud of comets lies not in the plane of the solar system but surrounds it in the form of a spherical halo, and that about half the long-period comets have retrograde orbits.

Comets in the Oort cloud are so far from the sun that they move very slowly by cosmic standards (on the order of 0.1 km/sec relative to the sun) and are subject to perturbations by the nearer stars. It is such perturbations that occasionally send one into the inner solar system, where it may be discovered by terrestrial observers.

How many comets are in the Oort cloud? If we recall that roughly five new ones are discovered every year and that these may return only every 100,000 or more years, we could roughly estimate that at least 500,000 comets must be available in the cloud. Estimates many times higher than this (as much as 100 billion, or 10^{11} comets!) have been made, taking into account that many unknown comets probably have perihelia far beyond Jupiter, where comas and tails are not developed enough to be visible.

The short-period comets are highly concentrated to the plane of the solar system and about 93 percent have direct orbits. The majority of periods are 5 to 10 years. However, the shortest period among retrograde comets is 33 years. Apparently the short-period comets originated as bodies that originally came in from the Oort cloud and were eventually captured in short-period orbits as a result of interactions with giant planets. Fayet (1910) and Russell (1920) pointed out that Jupiter is the dominant perturbing influence on the incoming comets, and indeed most short-period comets have aphelia near Jupiter. Usually one of the nodes, or points where the comet's orbit crosses through the plane of the solar system, lies close to Jupiter's orbit. It must have been at this point that the comet passed very close to the planet, allowing a major perturbation and changing a long-period orbit into a short-period orbit. The difference between retrograde and direct short-period orbit statistics is understandable. During encounters with planets, retrograde comets move past the planet at relatively high velocity, whereas direct comets move "with" the planet, giving a longer opportunity for pertubation.

Oort (1963) and Öpik (1963, 1966) have discussed the dynamics of such interactions between Jupiter and the comets. Some comets coming in from the Oort cloud may have their aphelia distances increased and some reduced, typically to about 10,000 A.U. In a few cases the perturbations will so reduce the aphelion distance that a short-period orbit is created. According to Öpik (1963, p. 232) about 96 percent of these would ultimately be reejected from the solar system, after about 1 million to 100 million years, if they survived that long. The other 4 percent would ultimately collide with planets. Calculations of this type are a current topic in celestial mechanical research and must be carried out with large computers.

METEORS AND COMETARY METEOR SHOWERS

Meteors, or "shooting stars," are familiar to anyone who has watched the night sky for more than a few minutes. We will show that they are related to comets. These bright, fast-moving objects are tiny particles that heat up due to friction with the air molecules when they plunge into the atmosphere, just as a reentering spacecraft heats up and creates a spectacular fireball.

Velocities and Orbits

Velocities give one clue that meteors are related to comets. By computing orbits based on careful photographic observations of velocities and decelerations of meteors, it has been found that the meteors travel in orbits around the sun, similar to those of comets. Some of these orbits are directly related to individual comets.

★ QUESTION ★

What velocity do you think a meteor has?

Answer

From the observation that meteors flash across the sky in less than a second or two, and from the fact that the atmosphere is of the order of 100 km thick, we can at once surmize that the velocities of meteors must be tens of kilometers per second. This is correct; typical measured velocities are 20 to 30 km/sec (13 to 19 miles/sec) and range up to 72 km/sec.

Showers and Comets

Meteor showers are composed of large numbers of meteors that fall within an interval of a few hours or a few days and appear to come from the same direction in space. Showers are named for the constellation that lies in the direction from which they come. One of the most famous showers is the Perseid shower, which comes in mid-August each year. The *radiant,* or direction of its source, lies in the constellation Perseus. Some other well-known meteor showers are listed in Table 7-1. On an average night, you may see about eight

Table 7-1
Some meteor showers[a]

Shower name	Date of maximum activity
Lyrid	April 21, morning
Perseid	August 12, morning
Draconid	October 10, evening[b]
Orionid	October 21, morning
Taurid	November 7, midnight
Leonid	November 16, morning
Geminid	December 12, morning

[a] The showers last up to several days before and after the peak on the listed date. The best time to observe is on the listed date when the radiant constellation is high above the horizon. Often this is in the predawn hours; the sunrise side of the earth is on the leading side as the earth moves around the sun and hence the sunrise side is sweeping up meterors at the greatest rate. Observations show that the mean rate on a *non*shower night rises from about 3 meteors/hour after sunset to about 15 meteors/hour before dawn.

[b] The Draconid's orbit has been perturbed and the shower is weak at present. Perturbations may some day bring it back.

meteors per hour (one every 7 or 8 minutes if you watch the whole sky carefully) but during a shower the rate may rise to 60 per hour (typical of the Perseids) or more. The most spectacular shower of recent times occurred on November 17, 1966, when the observers counting Leonid meteors in the predawn hours in the western United States noted a dramatic rise in the rate which continued until they could no longer count fast enough to record them all. For about half an hour meteors fell like snowflakes in a blizzard and the estimated rate exceeded 2000/minute! A time exposure of part of this shower is shown in Figure 7-5. A similar shower of Leonid meteors fell in 1833.

Figure 7-5

The great Leonid meteor shower of November 17, 1966. The visible meteor infall rate was estimated to exceed 2000 per minute. This 43-second exposure shows meteors against the background of the constellation Ursa Major (the "big dipper"). (D. R. McLean, Lunar and Planetary Laboratory.)

How can we explain such observations? Obviously the earth is encountering a cluster of particles called a *meteoroid swarm.* In the case of the Leonid swarm, the duration of the shower indicated that the inner core was some 32,000 km (20,000 miles) across (McLean, 1967). Similarly, Davies and Lovell (1955) found a diameter 100,000 km (62,000 miles) for the inner core of the Draconid (Giacobinid) shower. In 1866 the Italian astronomer G. V. Schiaparelli (later more famous as the discoverer of the "canals" on Mars) showed that the Perseid meteoroids* were clustered along the orbit of Comet 1862 III. Within a few years it was found that several shower meteoroids lay in swarms that coincided with comet orbits. The inference was unmistakable: meteors and meteor swarms are associated with comets.

Their orbits are understandable in terms of celestial mechanics. Should a group of meteoroids become detached at low speed from a comet, they would initially lie in a swarm moving in the same orbit. But perturbation theory shows that the swarm would soon stretch out along the same orbit so that eventually the meteoroids would lie more or less smoothly distributed all the way around the ellipse. If the earth's orbit intersects the cometary orbit and intercepts a relatively young, undispersed cluster of meteoroids, a spectacular shower like that of November 17, 1966, ensues. Older swarms are more dispersed.

Physical Nature of Cometary Meteoroids

Carefully calibrated photographs of meteors show how they decelerate as they hit the atmosphere. From this we can calculate the drag due to atmospheric friction and hence estimate the density of the particles. Most cometary meteoroids are very low density, fragile objects. In fact, no known cometary *shower* meteor has been big enough or dense enough to reach the ground, but samples have been collected during rocket flights, as in Figure 7-6. An investigator of meteoroids has compared some of them to bits of burnt newspaper or the balls of fluff that collect under beds! Various studies (Jacchia, Verniani, and Briggs, 1967; Verniani, 1969) also show that different showers (i.e., different comets) produce meteoroids of different character. Most are very fragile and fragment upon hitting the atmosphere, but short-period comets, which have been exposed to the sun more often than long-period comets, tend to produce higher-density stronger particles. There are exceptions to this tendency; the lowest density meteors are the Draconids, associated with the periodic comet Giacobini-Zinner. The average densities of photographed meteors range from less than 0.01 g/cm^3 (Draconids) to 1.06 g/cm^3 (Geminids, comet association unknown but possibly associated with a prehistoric "burnt-out" comet). For comparison, the density of water or ice is 1.0 g/cm^3.

Denser particles are evidently also present among the shower microme-

* Meteoroid is the proper term for the particle in space. *Meteor* denotes only the luminous phenomenon during entry into the atmosphere.

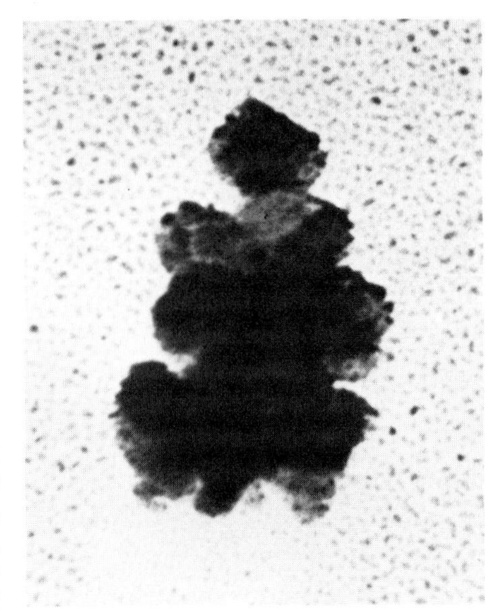

Figure 7-6
Typical low-density, irregular micrometeorite collected during flights at 75- to 150-km altitude. Particle diameter, 0.3 micron. Most of the collected particles are nonmetallic and many are frothy or "fluffy" in structure. (Dudley Observatory.)

teors. Among microscopic particles netted in space by spacecraft and rocket collectors were a possible Perseid which had a mass of 5×10^{-7} g, a density of 7.7 g/cm^3, was magnetic, and presumably mostly iron. Geminid micrometeoroids were found to have densities of 2.0 and 2.6, and no crystalline structure. These appeared to be glassy silicate particles.

Ejection Velocities from Comets

The upshot of this evidence is that comets, in the course of their evolution, disperse small solid particles. Davies and Turski (1962) found that Draconid meteors probably separated from their parent comet with velocities less than 10/sec (about 20 mph), which indicates that the particles are not ejected with any great force.

FRAGMENTATION AND OUTBURSTS OF COMETS

So far we have seen that comets lose gaseous material, which contributes to Type I tails, and dust particles. The small particles are driven away from the sun in Type II tails, and the larger ones spread along the orbit in meteoroid swarms.

More catastrophic mass loss can also occur. Some comets, for example, suddenly flare up in brightness, suggesting sudden releases of gas. Comet

Humason is known for its peculiar variability in brightness and unusual intensity of CO^+ emission lines in its spectrum. Such changes have been seen as much as 6 A.U. from the sun, considerably farther than the limiting distance for the activity of most comets (about 3 A.U.). Some comets return earlier and some later than predicted. As early as 1836, Bessel attributed this to accelerations produced by loss of mass from the nucleus. Whipple (1950) showed how the jet action of pockets of escaping gas in a rotating nucleus could either accelerate or decelerate a comet.

Most catastrophic of all is the fragmentation or complete disruption of comets. For example, Comet Wirtanen (1957 VI) broke into two pieces for no observable reason near the orbit of Jupiter.

Certain comets, known as the sun-grazers, pass only about 0.005 to 0.007 A.U. from the sun at perihelion and all have similar orbital elements. It is thought that these may all be pieces of a single parent comet that broke up after a close encounter with the sun. The sun-grazing comet 1882 II broke into several pieces during perihelion passage, and the 1965 sun-grazer Ikeya-Seki (see Figure 8-5) also broke into two pieces that separated at roughly 12 m/sec (about 25 mph) (Pohn, 1966). Another example was Comet Biela 1846 II, which split in 1846. When it returned in 1852, it was seen as a pair of comets, but during subsequent scheduled returns it was never seen again. Finally, several comets have disappeared entirely near their perihelia: Comet Ensor 1929 III is an example. That close solar passage promotes comet breakup gives us a clue about the nature of cometary nuclei.

When the sun-grazing comets pass perihelion, solar radiation excites the cometary gas and causes emission lines not normally present in most comets. Various metals, such as iron, nickel, and sodium, become prominent at this time and give us additional clues about comet composition.

PHYSICAL NATURE OF THE NUCLEUS

With the abundance of observational evidence on comets, what can we say about the actual cometary bodies, the nuclei? What causes the observed loss of material?

Masses and Dimensions

If a comet is observed just after fragmenting or as it passes near another celestial body such as a Jupiter satellite (Comet Brooks did this in 1886), we can attempt to observe the mutual perturbations of the bodies on one another and hence estimate the comet's total mass. A second method to determine a comet's mass is to determine the rate of loss of dust particles, which then indicates the total amount of dust present (if we can estimate the comet's age).

The results of these methods are very crude, but the best estimates suggest comets have masses of 10^{15} to 10^{19} g, comparable to the masses of small asteroids.

These methods are supported by independent estimates of the *dimensions* of comet nuclei based on their brightness and ability to reflect sunlight. Such measures indicate diameters of a few kilometers or possibly tens of kilometers in some cases, consistent with the mass estimates.

★ PROBLEM ★

Confirm this conversion from mass to diameter.

The "Dirty Iceberg" Model

In attempting to make a physical model of a comet nucleus, we are guided by the observations we have listed: Comets give off gases abundant in hydrogen, nitrogen, carbon, and oxygen. Within the nucleus the molecules formed from these elements must exist not as gas but in the forms of solid ices. This is true first because the temperatures at the distance of the Oort cloud are so low that the material would be frozen, and second because in the inner solar system any material solely in the form of gas would disperse, as evidenced by the comas and tails. Comets also give off fragile dust particles. Particles from older comets may be stronger. Comets are rather unstable, the nuclei tending to break up or perhaps disappear entirely if they are old or come too close to the sun.

Whipple (1950, 1963) combined these observations in what has become the most widely accepted model of a comet nucleus. Frequently, it is called the "dirty iceberg," or "dirty snowball," model. The nucleus is viewed as a several kilometer diameter frozen mass of ices—methane ice, ammonia ice, and familiar water ice. Mixed with the ices are dust particles—bits of silicates and possibly metals that were trapped when the comet formed. In the central regions of the nucleus these materials may be much more compacted than in the outer regions.

When the comet approaches within a few astronomical units of the sun, the surface temperature becomes too warm for the ices to be stable, and they sublime (go directly from the solid to the gas form), creating a gaseous coma around the nucleus. As they move out from the nucleus the gas molecules may become ionized and trapped in the solar wind, thus creating Type I tails. As the icy surface of the nucleus wears away the dust particles are exposed and loosened, drifting away to form the Type II tails and meteor swarms. The surface may be loosely consolidated, rather like a loosely packed snowball, and substantial pieces may break loose. Possible trapped pockets of gas might burst forth if the surface wears too thin.

The history of a comet, then, is a history of wearing-away, of sublimating the ices until nothing is left but a residue of dust particles. The inner part of the comet, with its compacted ices and dust particles, would be exposed in old, worn-down comets, accounting for the seeming association of the stronger meteoroids with older comets. The sublimation rate is greatest when the comet is closest to the sun. Watson, Murray, and Brown (1963) have shown that ices are unstable if closer to the sun than the asteroid belt. This explains why, after a certain number of passages around the sun, comets may "burn out" (lose all their volatile ices) and disappear from view while approaching or after passing the sun.

Sun-grazing comets, in addition to these eroding processes, suffer tremendous tidal forces when they pass perihelion. This helps account for their frequent fragmentation as they pass the sun.

The Sandbank Model

An alternative theoretical model of comet nuclei was the so-called "sandbank" model advocated by Lyttleton (1953) on the basis of observations of comet-related meteor swarms. This theory pictured the nucleus not as a single object, but as a cluster of dust and ice particles traveling together. A number of objections have been raised, and the sandbank model today is not widely accepted.

ORIGIN OF COMETS

What process in the history of the solar system could have given rise to a large number of kilometer-sized icy bodies on the outskirts of the solar system? Many exotic possibilities have been suggested, such as eruption from volcanoes on the Jupiter satellites or formation from interstellar material (Van Woerkom, 1948).

The most probable explanation stems from our ideas about the origin of the solar system. We have seen that the most abundant solids in the ancient solar nebula were methane, ammonia, and water ices with an admixture of silicate and metallic dust grains, and we have seen that such materials probably accreted into planetesimals which could eventually grow to the size of planets. The comets are very probably examples of these planetesimals, formed among the outer planets and then thrown into "deep freeze" in the Oort cloud by near-miss encounters with the outer planets, especially Jupiter.

Öpik (1963, 1966) has studied this process. Given a group of planetesimals, Öpik finds that only a small percentage ultimately collide with one of the planets. The others suffer a series of near-misses, being perturbed into successively more eccentric orbits, and ultimately ejected from the planetary system. At distances on the order of 20,000 A.U., the planetesimal finds itself moving very slowly, so far from the sun that stellar perturbations are effective. Therefore, it drifts around on a sort of "random-walk" orbit until one of the perturbing influences directs it back into the solar system. Thereupon, it falls into an elliptical orbit around the sun. If the orbit is highly eccentric, the comet loops close to the sun and is observed by us as a long-period comet. Should it happen to encounter one of the planets, especially Jupiter, in just the right way, it may be deflected into a short-period orbit and be observed by us on periodic passages. Eventually, if it remains near the sun long enough, its volatile ices become exhausted.

Because comet nuclei may be samples of the very first bodies in the solar system, there is great interest in space experiments involving them. Two types of experiments have been proposed. The first is to launch into orbit a quantity of ice, a "dirty iceberg," to see if cometary behavior can be simulated. (This is not as wasteful as it sounds, since boosters full of inert material are sometimes launched in rocket developmental tests.) The second type of experiment would be an attempt to send a space probe close to the nucleus of a comet. It has been suggested that this should be done when Halley's comet makes its scheduled return in 1986, or during the approach of Comet d'Arrest in 1976.

HAS A COMET STRUCK THE EARTH?

The most violent impact in modern history occurred in Siberia on June 30, 1908, when the arrival of an enormous meteorite set off an explosion with an energy of about 10^{21} to 10^{23} ergs (equivalent to a nuclear explosion of several hundred kilotons). Although smaller than the largest nuclear blasts (which in recent years have exceeded 20 megatons), the event knocked a man off a porch some 60 km (38 miles) away, was audible more than 1000 km (620 miles) away, and dumped millions of tons of dust into the air, causing a decrease in air transparency detected as far away as California.

At the time, Russia was preoccupied with events that produced a violent impact of another sort, and no substantial expedition to the site was organized until 1927, when the U.S.S.R. Academy of Sciences sent a scientific party. In this and succeeding expeditions, the impact site was identified as an area where trees had been seared and felled radially outward in an area of about 15 km (9 miles) radius, but no clear crater was found and no ordinary meteorite samples could be located. For some time there was a great mystery as to why there were no fragments from this enormous meteoritic body.

It is now widely believed that this object was cometary. The evidence is as follows:

1. The object evidently exploded in the air, since trees at "ground zero" stood upright but were stripped of branches. A loosely consolidated icy comet nucleus would be expected to volatilize and explode before it hit the ground.

2. The lack of meteorite fragments is consistent with our picture of a predominantly icy nucleus.

3. A 1961 expedition recovered soil samples that contained small spherules believed to be part of the object. The spherules would be consistent with the idea of an admixture of small grains of non-icy "dirt" in the dirty iceberg and their spherical shape could be the result of sudden melting during the explosion.

4. Observations of the motion of the object across the sky indicated that it was traveling toward the earth probably in retrograde motion at a very high velocity, perhaps 50 km/sec, which would be typical of a comet but not of ordinary meteorites (see Chapter 9). The observations and interpretations are described in detail by Krinov (1963).

5. For weeks afterward, the night sky in Europe and Russia was anomalously bright. This may have been due in part to atmospheric interaction with tail and coma material (although the comet was too small to have been noticed prior to the collision, being of the order 10^{10} to 10^{11} g in mass instead of about 10^{18} g, typical of observed comets).

In addition to the unique Siberian event of 1908, there is growing evidence for additional cometary material on earth. A certain rare type of meteorite, known as the Type I carbonaceous chondrite, has a volatile-rich composition similar to that predicted for the stony material that may be incorporated in comet nuclei. These meteorites penetrate through the earth's atmosphere, being much larger than the meteors we discussed earlier. Some researchers believe they may represent the more compacted siliceous material from the interiors of burnt-out comets (see Chapter 9).

References

Bessel, W. (1836) "Bemerkungen über mögliche Unzulenglichkeit der die Anziehung Allein Beruchsichtigenden Theorie der Kometen," *Astron. Nachr., 13,* 345.

Biermann, L. (1951) "Kometenschweife und solare Korpuskularstrahlung," *Z. Astrophys., 29,* 274.

Davies, J. G., and A. C. B. Lovell (1955) "The Giacobinid Meteor Stream," *Monthly Notices Roy. Astron. Soc., 115,* 23.

Davies, J. G., and W. Turski (1962) "The Formation of the Giacobinid Meteor Stream," *Monthly Notices Roy. Astron. Soc., 123,* 459.

Fayet, G. (1910) "Recherches concernant les excentricités des comètes," *Paris Mém., 26,* A.1 a A. 134.

Jacchia, L. G., F. Verniani, and R. E. Briggs (1967) "An Analysis of 413 Precisely Reduced Photographic Meteors," *Smithsonian Contrib. Astrophys., 10,* No. 1.

Krinov, E. L. (1963) "The Tunguska and Sikhote-Alin Meteorites," in *The Moon, Meteorites, and Comets,* B. M. Middlehurst and G. P. Kuiper, eds. (Chicago: University of Chicago Press).

Lyttleton, R. A. (1953) *The Comets and Their Origin* (New York: Cambridge University Press).

McLean, D. R. (1967) "The Leonid Meteor Shower of November 17, 1966," *Comm. Lunar Planet. Lab., 6,* 43.

Oort, J. H. (1950) "The Structure of the Cloud of Comets Surrounding the Solar System, and a Hypothesis Concerning Its Origin," *Bull. Astron. Inst. Neth., 11,* 91.

______ (1963) "Empirical Data on the Origin of Comets," in *The Moon, Meteorites, and Comets,* B. M. Middlehurst and G. P. Kuiper, eds. (Chicago: University of Chicago Press).

Öpik, E. J. (1963) "The Stray Bodies in the Solar System. Part 1. Survival of Cometary Nuclei and the Asteroids," *Adv. Astron. Astrophys., 2,* 219.

______ (1966) "The Stray Bodies in the Solar System. Part 2. The Cometary Origin of Meteorites," *Adv. Astron. Astrophys., 4,* 301.

Pohn, H. A. (1966) "Observations of a Double Nucleus in Comet Ikeya-Seki," *Sky and Telescope, 31,* 376.

Roemer, E. (1963) "Comets: Discovery, Orbits, Astrometric Observations," in *The Moon, Meteorites, and Comets,* B. M. Middlehurst and G. P. Kuiper, eds. (Chicago: University of Chicago Press).

Russell, H. N. (1920) "On the Origin of Periodic Comets," *Astron. J., 33,* 49.

Schiaparelli, G. B. (1866) *Bull. Meteorol. Observ. Coll. Romano, 5,* 10.

Stromgren, E. (1914) "Über den Ursprung der Kometen," *Publ. Copenhagen Observ., 19,* 62.

Van Woerkom, A. J. J. (1948) "On the Origin of Comets," *Bull. Astron. Inst. Neth., 10,* 445.

Verniani, F. (1969) "Structure and Fragmentation of Meteroids," *Space Sci. Rev., 10,* 230.

Watson, K., B. C. Murray, and H. Brown (1963) "The Stability of Volatiles in the Solar System," *Icarus, 1,* 317.

Whipple, F. (1950) "A Comet Model. I. The Acceleration of Comet Encke," *Astrophys. J., 111,* 375.

______ (1963) "On the Structure of the Cometary Nucleus," in *The Moon, Meteorites, and Comets,* B. M. Middlehurst and G. P. Kuiper, eds. (Chicago: University of Chicago Press).

ASTEROIDS

ASTEROIDS* ARE SMALL, noncometary members of the solar system. Most asteroids lie between Mars's and Jupiter's orbits, but some are known to follow orbits in other parts of the solar system. Asteroids that can be observed from earth range in size from a few kilometers to a few hundred kilometers across. Their orbits and their sizes distinguish them semantically from the principal planets, and also from the abundant small meteoroids strewn through interplanetary space.

DISCOVERY OF ASTEROIDS

As mentioned earlier, the popularization of Bode's law about 1772, followed by its confirmation through the discovery of Uranus in 1781, led to widespread belief that there ought to be a planet at 2.8 A.U. In 1800 six German astronomers, under the leadership of Johann Schröter, determined to search out the missing planet. Before these "celestial police" (as they were nicknamed) could succeed, the announcement came that the Italian astronomer Giuseppe Piazzi had discovered an unknown body while making routine stellar observations at the observatory at Palermo, Sicily. The new object moved with respect to the stars and Piazzi called it a new planet, Ceres. By the fall of that year, the famous astronomer Gauss had derived the first general method for determining orbits from observations of celestial bodies, and he computed Ceres's orbit. Ceres lay at 2.8 A.U., just as predicted by Bode's law.

Since Ceres seemed too small to be the sought-for planet, the German astronomers continued their survey program. In March 1802 one of them, Olbers, discovered a second asteroid, which he named Pallas. Like Ceres, Pallas was also small and faint. The addition of a second small asteroid raised the possibility that one normal planet had once occupied the predicted position but had somehow fragmented, producing many small bodies. On the

* Asteroids are often called "minor planets," especially by astronomers who specialize in their dynamical properties. They are also occasionally called planetoids, although this usage is dying out.

basis of this hypothesis, the search continued. Juno was discovered in 1804 and Vesta in 1807.* No more had been found by 1815 and the search stopped until a Prussian amateur named Hencke began a one-man program in 1830 and eventually discovered Astraea. Other discoveries by various workers followed. By 1890, the total known was 300. In 1891 the German astronomer Max Wolf began the first photographic patrol, detecting asteroids by

Table 8-1
Characteristics of selected asteroids

No.	Name	Semi-major axis, a (A.U.)	Eccen-tricity, e	Inclina-tion, i	Diameter (km)	Albedo (est. visual Bond albedo)	Rotation period (hr)	Axis inclina-tion to ecliptic
Part 1: Belt asteroids								
1	Ceres	2.77	0.08	10°.6	800[a]	0.04[a]	9^h1	
2	Pallas	2.77	0.23	34.8	490[a]	0.05[a]	10?	
3	Juno	2.67	0.26	13.0	250[a]	0.1[a]	7.2	41°?
4	Vesta	2.36	0.09	7.1	490[a]	0.2[a]	5.3	25°
5	Astraea	2.56	0.19	5.3	130?		16.8	
6	Hebe	2.43	0.20	14.8	230?		7.3	75°?
7	Iris	2.38	0.23	5.5	220?	0.1?	7.1	75°?
14	Irene	2.59	0.16	9.1	170?		11?	
Part 2: Trojan asteroids								
588	Achilles	5.21	0.15	10.3	70?			
624	Hektor	5.12	0.02	18.3	90?		6.9	80° ± 2°
911	Agamemnon	5.15	0.07	21.9	80?		7?	
1437	Diomedes	5.08	0.05	20.6	70?		18	
Part 3: Apollo asteroids and Apollo-like asteroids								
—	Apollo	1.47	0.56	6				
433	Eros	1.46	0.22	10.8	35 × 16[a]		5.3	70°?
1566	Icarus	1.08	0.82	23.0	1.1[a]	0.1		
1620	Geographos	1.24	0.34	13.3	3?			
Part 4: Other asteroids								
944	Hidalgo	5.82	0.66	42.5	20?			

[a] Based on direct measures; see Dollfus (1970). Other values are based on a formula by Anders (1965) and reports at the International Astronomical Union Colloquim (Vol. 12), 1971. Probable error on measured diameters ±20%; on albedos, 40%. Additional references: *Ephemerides of Minor Planets* and T. Gehrels (1970, private communication), and J. Veverka (1971, private communication).

* Ceres, Pallas, Juno, and Vesta are normally considered to be the largest asteroids. However, recent observations suggest that (324) Bamberga has a very low albedo and hence is large, possibly exceeding Pallas in diameter.

their trails on long-exposure plates guided on the stars. This mechanization greatly increased the rate of discovery. Keeping track of the many asteroids—no mean task—was an undertaking performed by the Rechen-Institut in Heidelberg until World War II, when it was assumed by Russian astronomers.

The annual Soviet catalog of asteroid orbits, *Ephemerides of Minor Planets,* currently lists more than 1750 asteroids. Many more have been observed once or twice but lost before enough observations could be made to determine a good orbit.

Asteroids are designated by number when an accurate orbit has been found and the asteroid recovered at a second opposition. The discoverer is also given the right to give a name to a newly numbered asteroid, and the names range the fields of human interest from mythology to universities and political figures to wives and girl friends.

Table 8-1 lists a number of asteroids and selected information about them.

ORBITS OF ASTEROIDS

The Asteroid Belt

The great majority of asteroids are confined to a swarm known as the *asteroid belt* lying between Mars and Jupiter. Mars is at 1.5 and Jupiter at 5.2 A.U.; the mean distances of most asteroids are between 2.1 and 3.3 A.U. from the sun, but due to their eccentricities, their perihelia, and aphelia occupy a wider zone.

Figure 8-1 illustrates some typical asteroid orbits. Few are as nearly circular as the orbits of the major planets. Planetary eccentricities are mostly less than 0.1, but in the asteroid belt eccentricities are more typically 0.1 to 0.3. The belt asteroids also have more inclined orbits than do the planets; typical asteroid inclinations range up to 30° while the planets are mostly inclined less than 3°. On the other hand, the orbits of the asteroids are far more regular than the orbits of long-period comets, which are highly eccentric and almost randomly inclined. Asteroids and short-period comets have some orbital similarities, which we shall discuss later. All the asteroids in the belt have prograde revolutions, as do the planets.

Kirkwood Gaps

We have already seen in Chapter 7 that Jupiter, because of its large mass, is an important perturbing influence in the solar system; gravitationally, it dominates a large region. Thus, we might expect that it would have a strong perturbative influence on the asteroids, which are its nearest neighbors. This comes about in the following way.

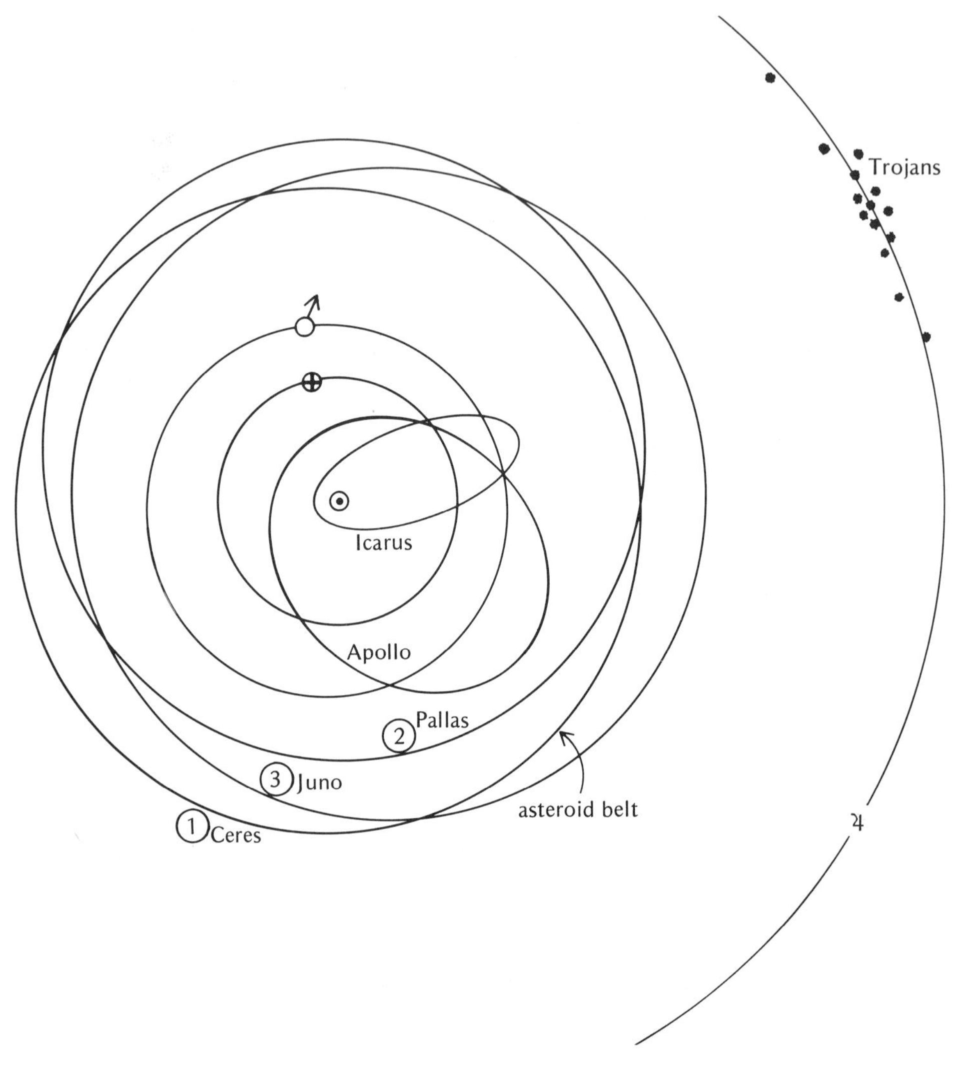

Figure 8-1
Orbits of selected asteroids.

Since the asteroid belt is inside Jupiter's orbit, the asteroids have shorter periods than Jupiter, a consequence of Kepler's third law. Figure 8-2 illustrates that most of the belt occupies a region of orbits with periods ranging from about $\frac{1}{4}$ to about $\frac{1}{2}$ Jupiter's period. Suppose an asteroid had a period exactly half that of Jupiter. If we start by picturing it at its closest approach to Jupiter, we shall find that this configuration is repeated after every two revolutions. Jupiter will be moving around the sun slower than the asteroid, but the two will move almost side by side for some time, allowing Jupiter to exert a substantial perturbative force. The important point is that as long as the asteroid stays in any such

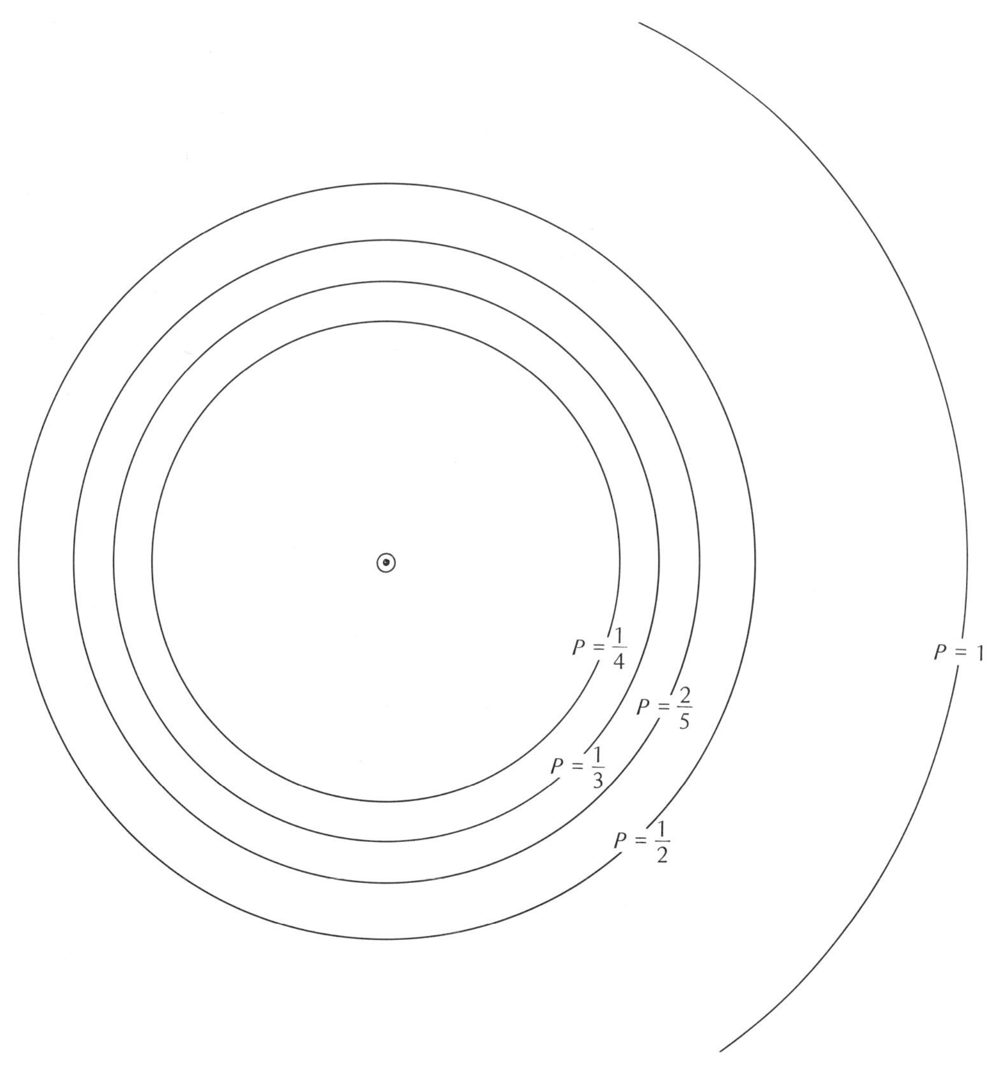

Figure 8-2
Commensurabilities in asteroid orbits (see the text).

orbit, the perturbation will be repeated regularly in the same way. This is called *resonance*, or *commensurability*. Asteroids in other orbits would also suffer perturbations but not repeated almost identically at such regular intervals; perturbations on them would thus not resonate.

The situation is rather like a child swinging on a swing. If you give him a push at just the right moment on each cycle (or every other cycle, as in our example), the swing oscillates more widely because of the resonance that occurs. But if you push at random intervals, you may sometimes accelerate, sometimes decelerate the swing, and have little net effect.

The asteroid in a *resonant,* or *commensurate,* orbit suffers a repeated perturbation which is very effective in changing the orbit. Therefore, any asteroid with a period equal to one half of Jupiter's will soon be forced into a different orbit. The same would seem at first glance to be true for orbits of $\frac{1}{3}$, $\frac{1}{4}$, and $\frac{2}{5}$ the period of Jupiter, and other periods in proportion to Jupiter's as the ratio of two small integers, but the matter is not quite so simple. Some commensurate orbits, instead of being quickly changed, are actually stabilized by resonance effects. This is true, for example, among the "Hilda group" of asteroids, which have a period $\frac{2}{3}$ that of Jupiter. Similar effects are known among Jupiter's and Saturn's satellites. A full theoretical explanation of such resonance phenomena in terms of celestial mechanics unfortunately does not yet exist, and this remains a fruitful area for further dynamical research.

Jupiter and the Trojan Asteroids

We noted in Chapter 3 while discussing the Lagrangian points that at points L_4 and L_5, 60° behind and ahead of Jupiter, small groupings of asteroids have collected. Members of these two groups have been named after the Greek and Trojan heroes of the Homeric epics (although names from the two warring sides are unfortunately mixed—there is a Trojan among the Greeks and a Greek spy in the Trojan camp!). These asteroids are thus called the Trojan asteroids. Fourteen are now cataloged, but a recent survey of faint asteroids by van Houten, van Houten-Groeneveld, and Gehrels (1970) revealed 45 more near the L_5 point alone. Based on their measures of the mass distribution of the Trojans, which is the same as that for the normal asteroids, these authors anticipate a total of about 700 observable Trojans at each Lagrangian point. Because of perturbations by the planets, the Trojans oscillate rather widely, ranging up to 20° along Jupiter's orbit from the actual Lagrangian points. It is thus possible that some Trojans may someday escape entirely, or new ones be captured, although celestial mechanicians believe that the present-day pattern is fairly stable.

Mars-Crossing Asteroids

Not all asteroids are confined to the space between Mars's and Jupiter's orbits. A few on the inner fringe of the asteroid belt cross inside Mars's orbit. The *Mars-crossing asteroids* are asteroids that cross the orbit of Mars but not that of the earth.

Thirty-four such asteroids are known. Calculations by Anders (unpublished) indicate that their half-life* against collision with Mars or serious

* *Half-life* is here defined as the time for half of the existing asteroids to be destroyed by the mechanisms mentioned. The half-life does not depend on the number of asteroids present and is analogous to the concept of half-life defined for radioactive decay of atoms.

perturbation by Mars is of the order of 3 billion (3×10^9) years, somewhat less than the age of the solar system. This suggests that in the past there were more, but not a lot more, large Mars-crossers.

Anders (1964) suggests that the Mars-crossing asteroids may be the source of most meteorites that fall on earth (see Chapter 9). The combination of collisional breakup plus perturbative changes in the orbits may send asteroidal fragments from the inner fringe of the asteroid belt onto Mars-crossing orbits. After billions of years, such fragments may be perturbed *by Mars* into an earth-crossing (Apollo-type) orbit, where it is "immediately" (within about 10^8 years) swept up by the earth or possibly Venus or Mars (Öpik, 1963).

The Apollo Group

Not all asteroids follow orbits beyond or near Mars. A few have elliptical orbits that bring them into the inner solar system. Those with orbits crossing the orbit of the earth are called Apollo asteroids (the prototype being named Apollo). Only ten Apollo asteroids—all of diameter about 1 km or somewhat greater—are cataloged. Öpik (1963) estimates on the basis of discovery rates that the total number must be considerably more than forty, and Whipple (1967a) estimates on statistical grounds a total of about 50 large enough to be observed. There must be still many more smaller bodies, too small to be discovered except during unusually close approach to the earth. Probably all sizes exist, from a few kilometers down to the size of meteorites. Table 8-2 lists the Apollo asteroids and their orbital characteristics.

It is perhaps sobering to note that Apollo asteroids have a possible potential for striking the earth. For an actual collision to occur, the asteroid's inclined orbit must cross through the plane of the ecliptic at the earth's orbit. According to Öpik's calculations, most Apollo asteroids will indeed end their days by colliding with the earth. The effects of a collision with a 1-km-diameter asteroid would be truly catastrophic by human standards. The crater alone would be about 13 km (8 miles) in diameter, and the devastated area around the crater would reach out perhaps 50 km (31 mi), covering 7900 km^2 (3000 mi^2). The shock wave would be fatal even farther away, and the dust ejected into the high atmosphere would color sunsets around the world for years. The famous Siberian "meteorite" (comet nucleus?) that fell in 1908 was probably only about 70 m in diameter, less than one tenth the size of a small Apollo asteroid, yet it devastated 3000 km^2 and its explosive impact was heard more than 1000 km from the impact site. It is fortunate for us that the half-life of a typical Apollo asteroid (i.e., the expected time in orbit until it collides with the earth) is of the order 10 million to 100 million years (Öpik, 1963; Arnold, 1965)!

A famous example of an Apollo asteroid is Icarus, which not only can come close to earth but also passes only 0.19 A.U. from the sun at perihelion. On June 14, 1968, Icarus passed—as predicted—only about 6 million kilome-

Table 8-2
Orbits of Apollo asteroids and near earth-crossing asteroids[a]

Asteroid designation: Number (if cataloged)[b]	Asteroid designation: Name	Perihelion distance, q (A.U.)	Semimajor axis, a (A.U.)	Eccentricity, e	Inclination i (°)	Aphelion distance, Q (A.U.)
Part I: Apollo asteroids						
1566	Icarus	0.19	1.08	0.825	23.0	1.97
	Adonis (1936A)	0.44	1.87	0.76	1	3.30
	1971FA	0.56	1.46	0.615	22.1	1.81
	Hermes (1937UB)	0.62	1.64	0.62	5	2.66
	Apollo (1932HA)	0.65	1.47	0.56	6	2.30
1685	Toro	0.77	1.37	0.443	9.6	1.96
	(1959LM)	0.83	1.34	0.38	3	1.85
1620	Geographos	0.83	1.24	0.336	13.3	1.66
	(1950DA)	0.84	1.68	0.506	12.2	2.53
	(1948EA)	0.89	2.26	0.605	18.4	3.63
Part II: Near earth-crossing asteroids						
	(1953EA)	1.03	2.44	0.58	20.3	3.86
	(1960UA)	1.05	2.24	0.53	4	3.43
	(1969AA)	1.06	2.15	0.50	24	3.22
	(1950LA)	1.08	1.68	0.36	26	1.60
1221	Amor	1.08	1.92	0.44	11.9	2.76
1627	Ivar	1.12	1.86	0.40	8.4	2.60
1580	Betulia	1.12	2.20	0.49	52.0	3.27
433	Eros	1.14	1.46	0.22	10.8	1.78
887	Alinda	1.15	2.52	0.54	9.1	3.88

[a] Based on E. Roemer (1970, private communication), and International Astronomical Union Minor Planet Circulars, which contains occasional updates of these orbits.
[b] A number of Apollo asteroids, including Apollo, were lost before good orbits could be determined—hence they have no catalog numbers.

ters (4 million miles) from the earth. This incident touched off a number of "scare" articles in the tabloids by "reporters" who warned of collision dangers and allegedly did not trust the assurances of astronomers that the earth would be spared.

The fact that the half-life of Apollo asteroids is only 1 percent of the age of the solar system tells us something important about them: Either they are the last remnants of a tremendous population of early Apollo asteroids, or they are being replenished. The first possibility is ruled out by Öpik (1963), who shows that an absurdly high initial mass of asteroids would be required (100 times the mass of the sun!). The question, then, is: If Apollo asteroids are being replenished, where do they come from?

A widely held opinion is that they are asteroids that came from the fringe of the asteroid belt and were once on orbits that crossed the orbit of Mars. Mars then perturbed them into earth-crossing orbits. This view is expressed in a lengthy review by Anders (1964). One piece of evidence is the theoretical work by Arnold (1965), who showed with computer-simulated orbits that Mars could throw Mars-crossing asteroids into orbits like those of the Apollo group.

In this connection it is interesting to note that of the 17 Apollo or near earth-crossing asteroids in Table 8-1, 65 percent have aphelia just inside or just outside the normal limits of the asteroid belt (2.1 to 3.3 A.U.). Only 24 percent have aphelia well inside the belt. This accords with the view that Apollo objects are being perturbed from the fringes of the belt and also gives a good indication that the spatial extent of the asteroid belt is constrained by dynamical processes.

A second view of the Apollo asteroids is taken by Öpik (1963, 1966), Wetherill (1967), Marsden (1970), and others, who argue that not enough asteroids could come from the Mars-crossing group. Öpik therefore suggested that the Apollo "asteroids" are not really asteroids at all, but burnt-out short-period comet nuclei (see Chapter 7). The Apollo "asteroids," in this view, might be physically dissimilar to the ring asteroids. This would accord with the fact that Apollo asteroid orbits bear a resemblance to some short-period comet orbits in terms of semimajor axis and eccentricity, and that some comets with tenuous comas are difficult to distinguish observationally from asteroids. Marsden (1970) has described the kinds of perturbations that may send comets into asteroid-like short-period orbits. Asteroid (944) Hidalgo has an orbit more like a comet than an asteroid, and has probably been subjected to nongravitational forces such as the gas-jetting suspected in comets. Among all cataloged asteroids, Hidalgo is the prime suspect as a burnt-out comet nucleus. Several others are also suspects; indeed, some observers have reported a faint halo around Irene. Finally, Icarus has a near-spherical shape that might be expected in a burnt-out comet nucleus but not from an asteroid fragment (Gehrels et al., 1970).

Anders and Arnold (1965) synthesized these two conflicting views of Apollo asteroid origins by dividing the Apollo objects into two groups: high velocity and low velocity. The high-velocity, or "cometary," group, had more comet-like orbits and contained Icarus, which was later found to have the spherical shape that could result from slow wasting of a dirty-iceberg comet nucleus. The low-velocity, or "asteroidal," group had orbits more like those which could be attained by fragments perturbed out of the asteroid belt, and contained Geographos, which was later shown to have a highly irregular shape, as might be expected to result from the smash-up of two stony asteroids.

Hirayama Families

In 1918 the Japanese astronomer K. Hirayama published the first of a series of papers pointing out that various asteroids clustered into groups with similar orbital elements. The groupings are called *Hirayama families.*

It is widely supposed that each Hirayama family contains the fragments of an asteroid that suffered a collision and fragmented long ago. The pieces are spread out along the orbit of the original body in much the way meteor streams spread along comet orbits. When the original asteroid collided with another (smaller) asteroid, it was blown apart and the pieces separated with ejection velocities. Brouwer (1951) shows that the differences in velocity among family members are about 0.1 to 1.0 km/sec (220 to 2200 mph)—values that must approximate the original ejection velocities. Such speeds are only a small fraction of orbital velocity; hence all the fragments remain in approximately the initial orbit. The speeds are also consistent with average ejection velocities of fragments from hypervelocity impacts in the laboratory (Gault, Shoemaker, and Moore, 1963). Chapman et al. (1971) found that of 11 asteroids studied, the only two with identical spectrocolorimetry were members of one of the Hirayama families, supporting the contention that Hirayama families are fragmented parent bodies. However, Chapman (in press) has more recently found asteroids associated with single families but with differing spectrophotometry.

Brouwer (1951) describes nine families and 19 additional "groups" identified by similar orbits. The number of asteroids in the families ranges from 9 to 62, and in the groups, 4 to 11. Arnold (1970) reviewed Brouwer's and Hirayama's data, confirmed the existence of families 1 to 9, and added a number of new families. Arnold also confirmed the existence of *jet streams,** families formed either by narrow jets of ejecta shot out from specific collisions or by dynamical clustering of orbits.

Of the 34 Mars-crossing asteroids mentioned above, 20 can be identified with four Hirayama families. These 20 contain 98 percent of the mass of the Mars-crossers (Anders, 1964). Possibly the eccentric orbits of the Mars-crossers increases their probability of disruption by collision, accounting for the large percentage in the form of Hirayama families. Nonetheless, in the crowded central part of the asteroid belt, the percentage of Hirayama families may be higher than we now recognize, and simply harder to measure because of the large number of orbits.

It is intriguing to think that in the Hirayama families we see the actual

* A term introduced by 1970 Nobel prize winner H. Alfvén.

drifting debris from the titanic, catastrophic collisions that have sporadically marked solar system history.

ROTATION, SURFACE CHARACTERISTICS, AND COMPOSITION OF ASTEROIDS

In almost all cases, asteroids are too small to be resolved by earth-based telescopes*; our knowledge of their sizes, shapes, and surface properties must be indirect. Photometry (measurement of brightness of light), colorimetry (measurement of color, or brightness dependence on wavelength), and polarimetry (measurement of the extent of polarization of light) are important aids. Presence or absence of spectral absorption bands characteristic of certain minerals is indicative of composition.

The simplest photometric observation is simply a record of the brightness of an asteroid, called a *light curve* (brightness versus time, as in Figure 8-3). Asteroid brightnesses are not constant, but vary regularly. This can have two possible causes:

1. The asteroid is elongated in shape and is tumbling end over end, causing a periodic change in exposed surface area and hence brightness (this is the reason many artificial earth satellites vary in brightness as they move across the sky). In this case the true period would include two maxima in the light curve.
2. The asteroid has dark and light markings on its surface, with one hemisphere averaging darker than the other (the moon has this characteristic). In this case the true period would include only one complete cycle in the light curve.

A choice can be made between these possibilities by looking for asymmetries in the light curve. A tumbling object would show repeated symmetry in the variations of its light curve; surface markings would generally produce asymmetric curves. Most of the smaller asteroids show indications of elongated or irregular shape, consistent with the idea that they are fragments of larger original bodies; they may also show indication of surface markings. Complications arise in the combined case, when an irregular rotating asteroid has light and dark surface markings.

Analyses of asteroid rotations, based on the work of Gehrels and others, suggest that the larger asteroids retain ancient, original rotations imparted to them during the formation of planetary bodies (Alfvén, 1964; Fish, 1967; Hart-

* An exception is the asteroid Eros. When it approached within 23 million km in 1931, astronomers with large telescopes were able to detect its elongated shape and watch it rotate with a period 5.27 hours.

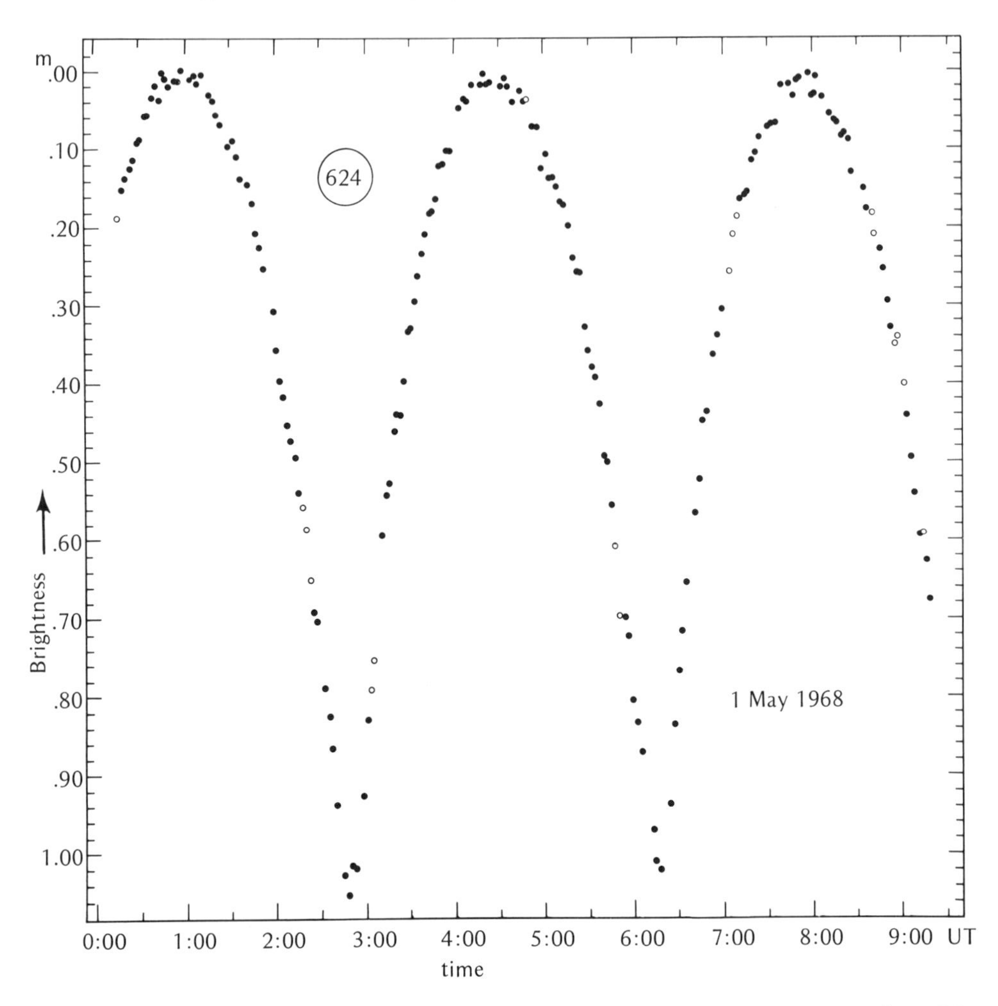

Figure 8-3
Light curve of the Trojan asteroid 624 Hektor. (T. Gehrels, Lunar and Planetary Laboratory.)

mann, 1969a; Kopal, 1970). This supports the thesis that they are preserved examples of the original planetesimals.

Photometry not only indicates rotation, nonspherical shapes, and markings, but also indicates the obliquity of asteroids. To see how this can be, suppose an irregular asteroid can be seen pole-on from the earth during one apparition; some months later we would see it at some quite different aspect angle, because of our relative motion. When seen pole-on, its light curve would have zero amplitude; but later the light variations would become appreciable. By studying variations in the light curve with aspect angle, we can derive the position of the asteroid's axis of rotation. Gehrels and Owings (1962, 1967) and others have performed several such analyses, but agreement among various workers is not good. Table 8-1 includes obliquities for a few of the best-observed asteroids.

What would the surfaces of such ancient bodies be like? Colorimetry allows us to study the reflective properties of asteroids at different wavelengths. These properties depend on the mineralogical and physical nature of the surface material and allow rough description of the surface. Broadly speaking, the asteroids exhibit properties similar to the moon. They are generally redder than unreflected sunlight, as is the moon, and have similar light variations with phase angle (sun–asteroid–earth angle). This suggests that their surfaces are similar to the moon: possibly cratered and overlain with rocky debris. The closeup photography of Mars' small satellite Phobos—shown in Figure 6-5—probably gives the best idea of how an asteroid may appear at close range.

Polarimetric studies show further similarities between asteroids and the moon. The interpretation is that asteroidal rocks have intricate microfracturing or are at least partly mantled by fine dust-like particles, as is the moon. (Veverka, 1971.)

McCord (1970), who is pursuing a program of infrared planetary spectroscopy designed to identify minerals in planetary rocks, reported in his first observations of asteroids that the large asteroid Vesta showed absorption bands caused by ferrous iron minerals, and has a composition very similar to that of certain basaltic stone meteorites known as *basaltic achondrites*. The large asteroids Ceres and Pallas and several others did not show the iron mineral absorptions, suggesting that different asteroids may contain different characteristic surface rocks. Chapman et al. (1971) confirm these differences among asteroid spectra and colorimetry.

Feasibility of Irregular Shapes

If the earth, moon, and other planets are almost exactly spherical, how can most asteroids be highly irregular in shape? The asteroids are in a sense large chunks of rock (or stony-iron material); obviously rocks can take a variety of shapes. Perhaps, then, the question should be rephrased: *Why are the larger planets round?*

It is a question of strength of material. The larger the asteroid or planet, the greater the pressure at the center. If the central pressure exceeds the strength of the rocky material, the material will be able to deform, either by plastic flow or by fracture, as a result of failure of the normal elastic properties of the solid rock. Higher internal temperatures also favor deformation toward an equipotential shape. In the case of an ideal nonrotating body, this shape will be perfectly spherical.

For ordinary rocky materials, such as those in stony meteorites, the critical diameter at which the transition from nonspherical to spherical occurs can be estimated to be several hundred kilometers. Larger asteroids and planets have centers deformed to match equilibrium shapes. Some loosely consolidated meteorites—possibly representative of primitive planetesimal material—have much lower strengths. Bodies composed of such material may have critical diameters as low as 120 km. The quoted diameters are only estimates, because even solid rock can deform plastically, given enough time (see Chapter 10), and the theoretical treatment of long-term deformation of such material is not well developed.

Nonetheless, observations support these estimates. Gehrels (1967) has shown that the large asteroid Vesta (diameter ca. 490 km) is nearly spherical, while most smaller asteroids are markedly nonspherical, suggesting their fragmentary nature. Small, spherical asteroids, such as Icarus, may be original accretions that were near-spherical from the start and never suffered major collisions.

★ MATHEMATICAL THEORY ★

The pressure inside a planet can be computed from the hydrostatic equation (see Chapter 10). For the central pressure, we find

$$P_c = \frac{2\pi G R^2 \rho^2}{3},$$

where R = asteroid radius and ρ = asteroid mean density. At sufficiently high pressure, depending on the temperature, deformation can occur. A crude estimate of the onset of this deformation can be made by equating the central pressure to the crushing strength of the rock. Such a procedure is not entirely defensible, since most reported crushing strengths are measured under laboratory temperatures and pressures and in an unconstrained condition. In a highly confined planetary interior, rock would have a higher resistance to crushing than can be measured in the laboratory (thus opposing deformation), yet with the higher temperatures and long time scales available, deformation is favored. To calculate exactly the deformation in a planetary body we would have to know the exact fluid mechanical or rheological equations characterizing the ability of material inside the planet to deform over long time-periods (see also Chapter 10). Unfortunately, these are unknown. Hence the uncertainty in the estimates. Wood (1963) lists the following crushing strengths for meteorites, which may represent asteroidal material: $S = 2 \times 10^9$ dynes/cm^2 for most stones, as low as 6×10^7 dynes/cm^2 for loosely consolidated stones, and 3.7×10^9

dynes/cm^2 for iron meteorites. The figures quoted for the critical diameters come from equating the central pressure to the first two values of S using mean density $\rho = 3.5$ g/cm^3 as found for stony meteorites.

Centrifugal Force versus Gravity

Because the asteroids are spinning rapidly, with periods as short as a few hours, the question has often been asked whether they spin so fast that loose material would be thrown off their equators by centrifugal force. The answer is no. We can show that an asteroid would have to rotate in less than about 1.8 hours in order for centrifugal force to exceed gravity. Even the fastest rotating asteroid, Icarus, has a period 2.25 hours, so loose debris could exist on its surface.

★ MATHEMATICAL THEORY ★

Equating centrifugal force with gravity, we have

$$\frac{mv^2}{R} \equiv \frac{4\pi^2 mR}{P_c^{\,2}} = \frac{GMm}{R^2}.$$

Thus we can solve for the critical period at which centrifugal force equals gravity:

$$P_c = \sqrt{\frac{3\pi}{G\rho}}.$$

We note that this varies only as the square root of the mean density variations among planets, and thus has a value of a few hours for all planets. It is indeed a limit on the possible rotation periods for planets forming from initially loose material.

DIAMETERS, ALBEDOS, AND ABSOLUTE MAGNITUDES

Several methods have been used to estimate the diameters of asteroids. Among direct observations are micrometer measurements of angular size, interferometer* measurements of angular size, measurements of occulations of stars, and comparison of infrared thermal radiation with visual reflected sunlight. Such measures, combined with the known distance (which comes from the known orbit), give the linear diameter. The methods have been

* The interferometer is a device used in astronomy to measure extremely small angles and is based on interference between light waves.

applied only to the largest asteroids and those that pass closest to the earth.

Once the diameter of an asteroid has been determined, its albedo (the percentage of sunlight reflected) can be determined from measures of its apparent brightness. Diameters and albedos are listed for some asteroids in Table 8-1. Ceres, Pallas, and Juno have low albedos, somewhat higher than lunar values. The albedo of Vesta, although less certain, is higher, possibly indicating different composition.

For most asteroids whose diameters are unknown, the property simplest to measure is the brightness. Since the brightness depends on distance from both the sun and earth, a standardized way of expressing it has been defined as follows. The *absolute magnitude* of a solar system body, such as an asteroid, is the brightness at zero phase angle (sun–object–observer) when the object is 1 astronomical unit from the sun and 1 astronomical unit from the observer. The absolute magnitude is usually labeled g, and is listed in *Ephemerides of Minor Planets,* mentioned earlier.

The measurements of albedo of lunar surfaces give typical values of about 0.07. If we know the absolute magnitude g and assume this albedo for most asteroids, we can calculate estimated diameter for them. Anders (1965) in this way derived the following formula for estimating diameter D in kilometers:

$$\log D_{\text{km}} = 3.686 - 0.200g.$$

NUMBER AND MASS OF ASTEROIDS

Nearly 2000 asteroids with well-determined orbits have been named and cataloged. Many others have been observed in asteroid search programs or on routine astronomical photographs but too poorly to determine an orbit. To estimate the "total number," we must specify what minimum size we are considering, since the smaller the size, the more asteroids there are. Only 3 asteroids are bigger than 300 km in diameter, but about 220 are bigger than 100 km across. The relation between number and size has been found to be approximately a power law. If we represent the size by estimated mass (since that is the most fundamental planetary parameter), then the total number N of small observable asteroids larger than mass m has the form

$$N \propto m^{-0.65},$$

according to the data of Kiang (1962) and van Houten et al. (1970). Although the astronomer Baade (1934) made a widely quoted estimate that there were 44,000 asteroids brighter than photographic magnitude 20 (close to the faintest objects that can be well photographed with large telescopes), the data analyzed by Kiang (1962) indicate a total of some 74,000.

MASS DISTRIBUTION AS A CLUE TO ASTEROID HISTORY

The equation cited directly above shows the form of the mass distribution for the smaller asteroids. The distribution is illustrated in Figure 8-4. This mass distribution—or its equivalent diameter distribution—had been used as a clue to asteroid origin. In 1958 Kuiper and his coworkers published an important survey of the asteroids and pointed out, among other findings, that a "*hump*" appeared in the distribution at about diameter 60 km, and that the *three largest asteroids* are too large to fit on the smooth curve defined by the rest. Kuiper et al. speculated that the "hump" divided the large, original asteroid from smaller, fragmented asteroids. As Hawkins (1960), Hartmann (1969b), and others have shown, the power-law mass spectrum of the small asteroids matches that produced when rocks are broken apart. Anders (1965) concluded that the "hump" is a remnant of the original size distribution of planetesimals in the asteroid belt at the time when the accretion process of planet formation stopped; this original distribution peaked at a diameter of about 60 km (38 miles), which is interpreted as a mean size of the original planetesimal asteroids. The apparent excess growth of the three largest asteroids was interpreted by Hartmann (1968a) to result from their having reached the size range where the *gravitational* cross section exceeded the *geometric* cross section; therefore, they were able to sweep up other particles much faster than were their smaller counterparts (see Chapter 6). The larger asteroids apparently were on their way to becoming planets when the growth process stopped.

COLLISIONS AMONG ASTEROIDS

Collisions among asteroids are probably not uncommon. Such collisions have been studied in considerable detail, one motivation being the belief that stony and iron meteorites may be the smaller fragments from such collisions.

Piotrowski (1953) found that in the present-day asteroid belt, collision velocities are typically several km/sec, sufficient to produce complete fragmentation when similar-sized bodies collide. More recently Wetherill (1967) reviewed the collision theory in detail and predicted rates of collision among objects with different sizes and different orbits. Hartmann (1968b) used a simplified theory to show that if the original distribution of asteroid masses had been the "hump"-shaped distribution hypothesized by Anders, later collisions would have produced the mass distribution now observed.

Briefly, the evidence in favor of widespread collisions among asteroids is: (1) theoretical prediction of numerous collisions; (2) agreement of the observed mass distribution of small asteroids with that found for rock fragments; (3) exis-

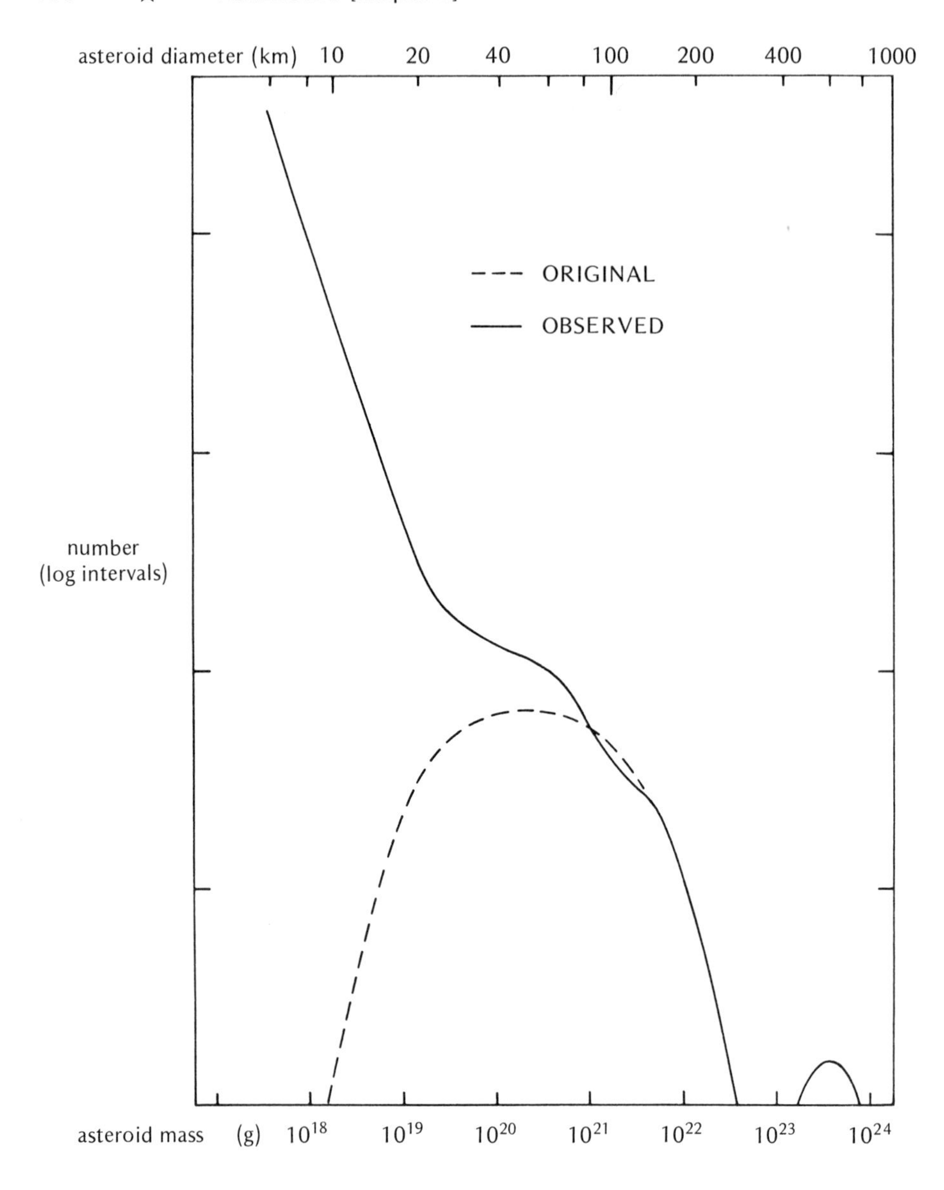

Figure 8-4

Schematic mass distribution among the asteroids. Solid line shows the observed distribution at present; dashed line shows a hypothetical reconstruction of the initial mass distribution (see the text).

tence of Hirayama families, which may be undispersed clusters of fragments; (4) high obliquities among the small asteroids, suggesting that their rotations were not original but produced in random manner by the fragmentation events; and (5) irregular shapes among most smaller asteroids.

ORIGIN AND EVOLUTION OF ASTEROIDS

The miscellaneous observations and theories described above can now be put together to give the following hypothetical outline of asteroid history. Four and a half aeons ago between Mars and Jupiter, planetesimals were growing and the process of planet formation was underway. The largest planetesimals had exceeded the critical radius at which gravitation dominates the accretion process, and they had begun the rapid growth toward planetary dimension. If the process had continued, one of them, probably Ceres, would have entirely dominated and swept up the other asteroids.

Instead of going to completion, however, the growth process stopped. This may have happened because Jupiter perturbed asteroid orbits significantly and increased the collision velocities, or because the solar nebula dissipated. At that time the planetesimals displayed Anders' "hump"-shaped distribution of sizes, with about 60 km (38 miles) being the most common diameter. Many of the larger of these have survived to the present day and some may exhibit their original rotations.

As collision velocities increased, *destruction* of asteroids began to dominate the growth process. A number of 60- to 100-km asteroids, which were the most common, collided and fragmented, producing most of the innumerable smaller fragments that we see today. Many fragments may have been refragmented by further collisions. Thus the smaller asteroids are mostly fragments, while the largest ones are original planetesimals (Kuiper, 1950).

Collisions perturbed orbits and produced Hirayama families. Some asteroids reached Mars-crossing orbits from which eventually they either collided with Mars or had Mars encounters which sent them into earth-crossing orbits. Such bodies account for at least some of the Apollo asteroids (some being possible burnt-out comet nuclei). Many of them have collided with the earth. The smallest ones are so numerous that they hit the earth frequently, falling to the ground as meteorites.

This is a tentative reconstruction of asteroid history; other suggestions have been made. A suggestion by Alfvén is that accretion can occur in the asteroid belt, especially in the Hirayama families or jet streams, where collision velocities are low. Another hypothesis suggests that asteroids are much more intimately associated with comets than indicated above. It is widely agreed, however, that asteroids are some sort of unique survivors from early solar system history. A spacecraft mission to an asteroid would thus be of great interest. Such a mission is being informally studied by NASA planners and might occur in the late 1970s.

ZODIACAL LIGHT AND INTERPLANETARY DUST

Asteroids and comets are only the largest of many interplanetary bodies. Interplanetary space, particularly in the inner solar system contains innumerable dust particles, molecules, and atoms of gas. Millman (1967) defines a

> . . . *meteoritic complex* . . . to include the small solid particles in interplanetary space, ranging in size from dust grains bigger than the largest molecules up to objects of the smallest asteroid class, appreciable fractions of a kilometer in diameter.

We shall not devote much space to the interplanetary gas, because the gas is not directly related to the planetary material. It spreads outward from the sun, creating the solar wind, which we mentioned in Chapters 3 and 7. Nevertheless, the interplanetary gas is a research topic of much current interest. Research on the gas and its interactions with planets and with magnetic fields is often called *space science.*

You can satisfy yourself that abundant dust exists in interplanetary space by observing it with the naked eye. It can be seen in the form of the *zodiacal light,* a faint reflection of sunlight off dust grains concentrated in the ecliptical plane. The zodiacal light is visible shortly after sunset and before sunrise (Figure 8-5). In the evening, just after the stars have come out, the zodiacal light is a faint band of light, widest at the western horizon, and tapering upward to a width of only 5 or 10°. Seen from mid-northern latitudes, it slants upward to the left, following the ecliptic. In other words, it lies along the plane of the solar system. It is about as bright as the Milky Way, and under good conditions can be traced at least 45° from the sunset point.

Zodiacal light is seen only from sites with dark skies. One must be several miles from dense population and several hundred meters from street lights. Averted vision and a technique of scanning—swinging the head back and forth—may be necessary at first to make the zodiacal light noticeable.

Measurements of the brightness, distribution, color, and polarization of the zodiacal light have helped to clarify the nature of the interplanetary dust. The particles responsible for most of the zodiacal light are tiny dust grains of the order of 1 micron (10^{-4} cm) in diameter. Early arguments that electrons were responsible for zodiacal light have been discounted. Whipple (1967b) estimates that the total mass of the interplanetary dust is about 2.5×10^{19} g, a mass less than that of a single moderate-sized asteroid. Most of this mass is concentrated among the smallest particles.

Figure 8-5

Zodiacal light. The diffuse bright patch is zodiacal light; to the right the tail of comet Ikeya-Seki extends above the horizon. The photograph was made with a 35-mm camera, unguided, with a 25-second exposure at f/2 on Tri-X film. (D. Milon.)

The shape of the zodiacal dust system can be partly inferred from the narrow width of the zodiacal light; our cross-section view shows that the system is highly flattened toward the plane of the solar system. The radial distribution in the plane of the solar system is a more difficult datum to derive from the observations. Some of the dust particles associated with the zodiacal light exist quite close to the sun. The outer luminous atmosphere of the sun, called the corona, grades into the interplanetary medium and has two components, the K-corona and the F-corona. The *K-corona* is the gaseous portion and the *F-corona* is the dust, mixed with the gas. This dust can be traced to within 4 solar radii of the sun by its infrared radiation.

Measurements of the concentration of dust in the region from four solar radii out to about 0.5 A.U. are difficult, but there is some evidence of a smooth decrease in the dust density outward from the sun (Blackwell and Ingham, 1967). Powell et al. (1967) find that the number density of particles from about 0.5 A.U. to the orbit of Mars is roughly constant and may decline beyond Mars. The density of zodiacal dust is in the range 10^{-24} to 10^{-26} g/cm^3 near the earth's orbit.

Studies of the dynamics of zodiacal dust show that the particles are transient. The Poynting–Robertson effect and trapping in the solar wind (see Chapter 3) drive the smallest particles either into the sun or out of the solar system. In addition, collision phenomena—ranging in scale from atomic impacts to collisions with planets—wear down or eliminate dust particles. Whipple (1967b) shows that particles larger than a micron have a mean lifetime of only about 10^5 (100,000) years. When particles, after sputtering and abrasion, reach a size less than about 1 micron, they are rapidly removed from the solar system in a period of about a year (Biermann, 1967). Whipple concludes that the dust cloud is completely replenished on a time scale of about 1.7×10^5 years.

This continual replenishment of the interplanetary dust raises the question of its origin. Where does it come from? Replenishment requires about 2.5×10^{14} g/yr, or 8 tons/sec to be added constantly to the meteoritic complex in the form of particles ranging from microscopic dust grains to metric dimensions in order to maintain the zodiacal light.

As we pointed out in Chapter 7, observations indicate that the small meteoric particles are mostly cometary. It is believed that most of the zodiacal light particles also came from this source (i.e., they are strewn out by comets as the cometary nuclei wear away). An additional source of dust in the solar system is the collision and grinding down of the asteroids. Whipple speaks of loosely consolidated "half-baked" asteroids—asteroids that underwent minimal melting (see Chapter 9)—as possible sources of some of the dust. Bits of such asteroids could be brought from the asteroid belt into the inner solar system by the Poynting–Robertson effect.

The total amount of this material falling on the earth is estimated to be roughly 10^{12} g/yr (about 3000 tons/day), in the form of small dust particles. This quantity is difficult to determine, and the estimate has an uncertainty of an order of magnitude. These particles are slowed and trapped by the earth's atmosphere and settle out slowly. Dust-particle collections in unpolluted areas such as arctic wastes, sea floors, and the upper atmosphere are used to measure the number and nature of interplanetary particles striking the earth.

METEORITE EVIDENCE

In Chapter 9 we shall study evidence that at least some meteorites are samples of asteroid fragments and that they support the hypothetical history sketched above. Important assertions relating to asteroids and discussed further in Chapter 9 are: (1) Meteorites show pressure effects and cooling rates which indicate that they formed inside bodies of asteroidal dimension; (2) in the two cases where orbits could be determined, the meteorites had aphelia near the asteroid belt and may have originated there; (3) petrologic studies show that many meteorites have been subjected to sudden shocks as would occur in asteroidal collisions; (4) dating techniques show that most meteorites solidified about 4.5 to 4.7 billion years ago but that many were subsequently fragmented in discrete events; (5) various evidence suggests that the meteorites came from a relatively small number of parent bodies; and (6) at least one asteroid (Vesta) is observed to have photometric surface properties matching those of a type of meteorite. This evidence is discussed further in Chapter 9.

References

Alfvén, H. (1964) "On the Origin of the Asteroids," *Icarus, 3,* 52.

Anders, E. (1964) "Origin, Age, and Composition of Meteorites," *Space Sci. Rev., 4,* 583.

——— (1965) "Fragmentation History of Asteroids," *Icarus, 4,* 399.

——— and J. R. Arnold (1965) "Age of Craters on Mars," *Science, 149,* 1494.

Arnold, J. R. (1964) "The Origin of Meteorites as Small Bodies. II. The Model," *Astrophys. J., 141,* 1536.

——— (1965) "The Origin of Meteorites as Small Bodies," *Astrophys. J., 141,* 1536.

——— (1970) "Asteroid Families and Jet Streams," *Astron. J., 74,* 1235.

Baade, W. (1934) "On the Number of the Asteroids Brighter than Photographic Magnitude 19.0," *Publ. Astron. Soc. Pacific, 46,* 54.

Biermann, L. (1967) "Theoretical Considerations of Small Particles in Interplanetary Space," in *The Zodiacal Light and the Interplanetary Medium,* J. Weinberg, ed., p. 279 (Washington, D.C.: *NASA SP-150*).

Blackwell, D. E., and M. F. Ingham (1967) "Toward a Unification of Eclipse and Zodiacal-Light Data," in *The Zodiacal Light and the Interplanetary Medium,* J. Weinberg, ed., p. 17 (Washington, D.C.: *NASA SP-150*).

Brouwer, D. (1951) "Secular Variations of the Orbital Elements of Minor Planets," *Astron. J., 56,* 9.

Chapman, C. R., T. V. Johnson, and T. P. McCord (1971) "Spectrophotometry of Asteroids," in *Physical Studies of Minor Planets,* T. Gehrels, ed. (Washington, D.C.: *NASA SP-267*).

Dollfus, A., ed. (1970) *Surfaces and Interiors of Planets and Satellites* (New York: Academic Press, Inc.).

Fish, F. F. (1967) "Angular Momenta of the Planets," *Icarus, 7,* 251.

Gault, D. E., E. M. Shoemaker, and H. J. Moore (1963) "Spray Ejected from the Lunar Surface by Meteoroid Impact," *NASA Tech. Note Dd 767.*

——— and D. Owings (1962) "Photometric Studies of Asteroids, IX. Additional Light Curves," *Astron. J., 135,* 906.

Gehrels, T. (1967) "Minor Planets. I. The Rotation of Vesta," *Astron. J., 72,* 929.

——— E. Roemer, R. Taylor, and B. Zellner (1970) "Minor Planets and Related Objects. IV. Asteroid (1566) Icarus," *Astron. J., 75,* 186.

Hartmann, W. K. (1968a) "Growth of Asteroids and Planetesimals by Accretion," *Astrophys. J., 152,* 337.

——— (1969a) "Angular Momentum of Icarus," *Icarus, 10,* 445.

——— (1969b) "Terrestrial, Lunar and Interplanetary Rock Fragmentation," *Icarus, 10,* 201.

——— (1968b) (with A. C. Hartmann) "Asteroid Collisions and Evolution of Asteroidal Mass Distribution and Meteoritic Flux," *Icarus, 8,* 361.

Hawkins, G. S. (1960) "Asteroid Fragments," *Astron. J., 65,* 318.

Hirayama, K. (1918) "Groups of Asteroids Probably of a Common Origin," *Tokyo Ann. App., 6.*

Kiang, T. (1962) "Asteroid Counts and Their Reduction," *Monthly Notices Roy. Astron. Soc., 123,* 509.

Kopal, Z. (1970) "The Axial Rotation of Asteroids," *Astrophys. Space Sci., 6,* 33.

Kuiper, G. P. (1950) "On the Origin of Asteroids," *Astron. J., 55,* 164.

——— Y. Fujita, T. Gehrels, I. Groeneveld, J. Kent, G. van Biesbroeck, and C. van Houten (1958) "Survey of Asteroids," *Astrophys. J. Suppl., 3,* 289.

Marsden, B. G. (1970) "On the Relationship between Comets and Minor Planets," *Astron. J., 75,* 206.

McCord, T. B. (1970) "Asteroid Vesta-Spectral Reflectivity and Compositional Implications," *Science, 168,* 1445.

Millman, P. M. (1967) "Observational Evidence of the Meteorite Complex," in *The Zodiacal Light and the Interplanetary Medium,* J. L. Weinberg, ed., p. 399 (Washington, D.C.: *NASA SP-150*).

Öpik, E. J. (1963) "The Stray Bodies in the Solar System, I. Survival of Cometary Nuclei and the Asteroids," *Adv. Astron. Astrophys., 2,* 219.

——— (1966) "The Stray Bodies in the Solar System, II. The Cometary Origin of Meteorites," *Adv. Astron. Astrophys., 4,* 301.

Piotrowski, S. (1953) "The Collisions of Asteroids," *Acta Astron., A5,* 115.

Powell, R. S., et al. (1967) "Analysis of All Available Zodiacal Light Observations," in *The Zodiacal Light and the Interplanetary Medium,* J. L. Weinberg, ed., p. 225 (Washington, D.C.: *NASA SP-150*).

van Houten, C. J., I. van Houten-Groeneveld, and T. Gehrels (1970) "Minor Planets and Related Objects, V. The Density of Trojans near the Preceding Lagrangian Point," *Astron. J., 75,* 659.

Veverka, J. (1971) "The Polarization Curve and the Absolute Diameter of Vesta," *Icarus, 15,* 11.

Wetherill, G. W. (1967) "Collisions in the Asteroid Belt," *J. Geophys. Res., 72,* 2429.

Whipple, F. L. (1967a) "The Meteoritic Environment of the Moon," *Proc. Roy. Soc. (London), A296,* 304.

——— (1967b) "On Maintaining the Meteoritic Complex," in *The Zodiacal Light and the Interplanetary Medium,* J. L. Weinberg, ed., p. 409 (Washington, D.C.: *NASA SP-150*).

Wood, J. A. (1963) "Physics and Chemistry of Meteorites," in *The Moon, Meteorites, and Comets,* B. M. Middlehurst and G. P. Kuiper, eds. (Chicago: University of Chicago Press).

METEORITES AND METEORITE IMPACT CRATERS

METEORITES ARE DEBRIS that collide with the earth after floating in space for many years. The goal of *meteoritics,* the study of meteorites, is to determine where meteorites come from, to understand how they relate to either the primordial or the present-day planetary bodies, and to understand and utilize the impact craters they create.

In the previous chapters we have traced the development of planetary bodies from tiny grains condensed out of the solar nebula through the planetesimal stage, which are illuminated by the cometary and asteroidal bodies of the last two chapters. Viewed in a chronological light, the meteorites give us information about the next stage of planetary evolution. Although the planetesimals probably accumulated from "cold" solid bodies, there is evidence that they subsequently were heated and at least partially melted. Possible contributing sources of heat include the energy of collision during accretion, external heating, and internal heating due to decay or radioactive elements, especially short-lived radioactive elements created at the beginning of the solar system. Among various lines of evidence for heating are the still-molten interior of the earth, the presence of abundant lava flows on the moon, and many evidences from meteorites, which we shall discuss in this chapter. The meteorites are thus viewed as our only indicators of the interior conditions in the early planetary bodies, and our only physical samples of planetary interiors. This chapter will discuss the meteoritic evidence and Chapter 10 will extend the discussion to the interiors of the earth and other planets. Certain definitions must be given first.

DEFINITIONS

Meteoroid

A meteoroid is any extraterrestrial solid body, floating in space.

Meteor

A meteor* is a heated, glowing meteoroid in transit through the atmosphere, having not yet hit the ground. A meteoroid may enter the atmosphere at 10 to 70 km/sec, and friction between it and the atmosphere heats its surface and ionizes atmospheric molecules. Airless planets such as the moon would have no meteors, for an atmosphere is needed to heat the meteoroids and make them visible.

Meteorite

Meteorites are those meteoroids or their fragments which strike the ground. There are three broad classes: stones, irons, and stony-irons. We cannot assume that the *meteors* we see on an average night are physically related to the *meteorites* we see in a museum. Indeed, there is contrary evidence which we shall consider.

Fireballs

Fireballs are the very brightest, infrequent meteors. They can be much brighter than the full moon, and in rare cases rival the sun.

Falls

Meteorites witnessed during descent (through light or sound phenomena) are called falls.

Finds

Meteorites found lying on or buried in the ground are called finds. Many finds may have originally fallen thousands of years ago. The people in locales of some ancient falls recount legends or place names associated with the sky, as at Campo del Cielo, Argentina. Such legends may have persisted since the time of the fall.

Names

Meteorites are generally named after a city or geographical landmark near where they fell or were found.

* The word stems from a Greek root referring to "things high in the air." Until the 1900s it was used to refer to rainbows, lightning, snowfalls, etc., as in "the meteor of the ocean air shall sweep the clouds no more" (Oliver Wendell Holmes). Hence the discipline "meteorology" refers to the atmosphere—not meteors.

Parent Bodies

Meteorites are probably fragments of larger objects known as parent bodies. The interiors of these hypothetical parent bodies sustained pressures and temperatures suitable for forming the minerals found in meteorites.

HISTORY OF METEORITE STUDIES

Early Chinese, Greek, and Roman writings describe the fall of "stones from the sky." Our ancestors were apt to revere such allegedly celestial stones. The famous black stone of Ka'ba in Mecca, sacred stone of the Muslims, is apparently a meteorite that was enshrined about 600 A.D. or perhaps considerably earlier. Around 600 A.D., a group of Pueblo Indians near Winona, Arizona, buried a much smaller stone meteorite in a stone-slab crypt (Heineman and Brady, 1929) (Figure 9-1). Other examples of meteorite veneration from Mexico, India, and elsewhere could be cited.

During the Middle Ages in Europe, fallen meteorites were likely to be brought by incredulous peasants to their local priests or city magistrates. Thus a number of meteorites, such as the Ensisheim stone, which fell in 1492, were preserved in European village churches.

In 1751 a widely witnessed fall occurred in Zagreb. The Bishop of Zagreb carefully collected and protected pieces of the iron meteorites and sent them, along with a tabulation of witness testimony, to the Austrian emperor. This was passed on to the Vienna museum but was not treated as important.

By the 1790s, philosophers and scientists were aware of many allegations that stones had fallen from the sky, but the most eminent scientists were skeptical. The first great advance came in 1794, when a German lawyer and physicist, E. F. F. Chladni, published a study of some alleged meteorites, one of which had been found after a fireball had been sighted. Chladni accepted the evidence that these meteorites had fallen from the sky and correctly inferred that they were extraterrestrial objects that were heated by falling through the earth's atmosphere. Chladni even postulated that they might be fragments of a broken planet—an idea that set the stage for early theories about asteroids,* the first of which was discovered seven years later. Chladni's ideas were widely rejected, not because they were ill conceived, for he had been able to collect good evidence, but because his contemporaries simply were loathe to accept the idea that extraterrestrial stones could fall from the sky. Gradually the evidence became more compelling. A 1796 paper accepted the celestial origin but argued that the stones had been swept up by tornadoes, only to fall again to the earth (J. A. Wood, 1968).

* This accounts for the early idea that asteroids were fragments of a single planet that exploded for unknown reasons.

Figure 9-1

The burial of the Winona stone meteorite in a crypt constructed by prehistoric Indians. (Museum of Northern Arizona, Anthropological Collections; photo by the author.)

The last holdouts were members of the austere French Academy, who were supposed to be the scientific leaders of the day. These gentlemen pooh-poohed the idea of stones falling out of the sky and argued that the whole business was nonsense because the peasant-witnesses could (supposedly) not give reliable observations. In retrospect, it appears that part of the Academy's reluctance was generated by subtle social pressures. In the years following the French revolution, there was a philosophic wave of antipathy to religious authorities. The Academy frowned on samples that had come to reside (through historicosocial mechanisms) in village churches or had been collected by country priests.

As if in answer, a shower of fragments from a large meteorite pelted the French town of L'Aigle on April 26, 1803. The Academy sent the noted physicist Biot to investigate the matter. His exhaustive report finally convinced the scientific world that stones did fall from the sky and that Chladni had been right.*

For decades, geologists and chemists tried to make sense of the extraterrestrial stones. It was soon realized that they had an exceedingly complex history, not a simple, single-event origin. By the middle of this century, for reasons we shall discuss, researchers were in general agreement that meteorites are fragments of several *parent bodies,* or planetesimals with dimensions of several hundred kilometers.

IMPACT RATES

A number of studies have been made to estimate the amount of material striking the earth per year. The number of particles varies greatly with size. Microscopic particles, called *micrometeorites,* are abundant and constantly present in the atmosphere, whereas objects large enough to hit the ground and form substantial craters come only every year or so.

The smallest particles have been studied by detectors on spacecraft. The impact rate of larger meteorites has been estimated from statistics of witnessed falls, and the rate of the largest, rare, crater-forming meteorites has been estimated by counting the number of craters in datable undisturbed geologic provinces such as the eastern United States and Canada. An excellent review of these data is by Vedder (1966).

Has the rate of meteorite infall been constant in the past history of the earth? The dating of ancient surfaces on the earth and moon has allowed attempts to reconstruct the past history of the meteorite influx. The evidence indicates that the rate has remained more or less constant (within an order of magnitude) for the last 3 billion years but was much higher during the first billion years of solar system history (Baldwin, 1963; Hartmann, 1970). The initial high rate may reflect the rapid sweeping up of the last planetesimals left over after a planet's formation.

The flux of meteoroids in other parts of the solar system is different from that measured near the earth. Measurements by the Mariner IV spacecraft in 1965, for example, indicated that the flux of micrometeoroids near Mars's orbit was about 5 times higher than that near the earth (Alexander et al., 1965), presumably due to contributions from the nearby asteroid belt.

* The knowledge only slowly filtered to the fledgling country, America. The politician and naturalist President Thomas Jefferson was said to have remarked upon hearing two Yale professors report on a meteorite fall at Weston, Connecticut, "It is easier to believe that Yankee professors would lie, than that stones would fall from heaven."

★ MATHEMATICAL THEORY ★

Averaging over the many estimates, Vedder derives a rough law that gives the rate of fall N for meteorites larger than any given mass m; in the mass range 10^{-13} to 10^{13},

$$\log N = -17 - \log m,$$

where N = number of meteorites (mass larger than m)/cm^2 sec
m = mass (g)

★ PROBLEM ★

It takes a meteorite of about 10^{11} g mass to form a 1-km-diameter crater. Compute the average frequency of such a catastrophe on the land area of the United States. The area of the United States is 9.4×10^6 km^2, or 9.4×10^{16} cm^2.

Solution

Substituting for the mass, we find that the log of the impact rate is -28. The impact rate is thus 10^{-28} meteorite/cm^2 sec, or

$$\frac{10^{-28}\ \text{impact}}{\text{cm}^2\ \text{sec}} \times 9.4 \times 10^{16}\ \text{cm}^2 \times \frac{\pi \times 10^7\ \text{sec}}{\text{yr}} = 30 \times 10^{-5}\,\frac{\text{impact}}{\text{yr}}$$

on the earth. Here we have used the convenient-to-remember approximation that 1 year $\simeq \pi \times 10^7$ seconds. Inverting this result, we have

$$3 \times 10^3\,\frac{\text{years}}{\text{impact}}.$$

Thus Vedder's rough law shows that the time scale between such catastrophes would be on the order of several thousand years. A more accurate prediction can be made by studying in detail the meteorite mass range for the impact in question. Vedder's more detailed work indicates an impact of this size every 10,000 to 100,000 years, which incidentally is the approximate age of famous Arizona meteorite crater—a crater somewhat over 1 km across.

PHENOMENA OF METEORITE FALLS

Sound

The largest meteorites produce energetic shock waves when they enter the atmosphere at supersonic velocities. These "sonic booms" have long been known, being described in terms of contemporary culture: "like distant guns at sea" (England, 1795); "like a horse and carriage clattering over a bridge" (America, 1913). The Tunguska fall of 1908 in Siberia was a mass of roughly 10^{10} to 10^{11} tons (see Chapter 7); its explosions were heard over 1000 km (600 miles) from its flight path (Krinov, 1963).

Brightness

The larger the object, the brighter it will appear. Figure 9-2 shows the relation between observed brightness and size of meteorite. The Sikhote–Alin fall in Siberia, shown in Figure 9-2, was a 70-ton iron meteorite; it dazzled the eyes of observers and cast shadows even though the sun was in a clear sky (Krinov, 1963).

Train

Many large meteorites, such as Sikhote–Alin (Figure 9-2), leave a dusty train made up of debris blown off during atmospheric passage. Many break apart, spreading pieces in a line along the flight path (Figure 9-3) or detonate with a burst of light. Meteors and meteorites seen at night may leave a faintly lumi-

Figure 9-2
Dust train left in the morning sky by the fall of the Sikhote-Alin meteorite, Siberia, in 1948. (*Sky and Telescope.*)

Figure 9-3
The bright fireball of April 25, 1966. This object passed from south to north over the east coast of the United States at 7:14 P.M. and broke into the string of fragments shown here. (Associated Press.)

nous train visible from a few seconds to several minutes. This is caused by photochemical reactions among high-altitude air molecules disturbed by the meteor's passage.

Temperature

Passage through the atmosphere is so swift that only the outer surface of a meteorite is heated. There is no time to transport significant heat to the meteorite's interior, which remains cold. The widespread idea that a meteorite may remain too hot to touch for some hours after its fall is a fallacy, because the outer hot layer, about 1 mm thick, rapidly cools. A thin *fusion crust,* usually black, is produced on meteorite surfaces by atmospheric heating. Krinov (1960) and Mason (1962) have discussed fusion crusts in some detail.

Velocities

A typical meteorite strikes the atmosphere at about 16 km/sec (36,000 mph). Those smaller than 1000 tons lose a substantial part of their initial velocity due to atmospheric drag, but larger meteorites can strike the ground with most of their initial velocity. Meteoroids with the highest velocities (up to 70 km/sec) are most likely to fragment in the air.

METEORITE CRATERS

Crater Formation

Craters are formed by interplanetary collisions of bodies of different size, whether a planetesimal and a planetesimal fragment, or the earth and a meteorite. When meteorites strike the surface of a planet at velocities of many kilometers per second, their kinetic energy of motion is converted into mechanical, thermal, and acoustic energy. The explosive energy release creates a crater surrounded by a raised rim of upthrust and ejected material. Such a crater superficially resembles an explosion crater, as seen by comparing Figures 9-4

Figure 9-4

The Arizona meteorite crater. The scale can be judged by museum buildings on the far rim. The arrow at the right marks a massive boulder ejected from the crater.

and 9-5, which show the Arizona meteorite crater and a nuclear explosion crater. Chapter 11 contains illustrations of volcanic craters, some of which also resemble impact craters.

The bigger the meteorite, the bigger the crater. Figure 9-6 shows the relation between the mass of the impacting body and the crater diameter. The figure shows that at higher impact velocities, larger craters are formed. Since the gravitational attraction of different planets varies, and since the meteorite approach orbits toward different planets vary, the same meteorite could have different impact velocities on different planets and could thus form different-sized craters.

Shoemaker (1963) has discussed the sequence of events during the impact. As the meteorite enters the ground, tremendous pressures build up and the ground deforms in fluid fashion. Previously flat-lying layers are pushed up and out like the petals of an opening flower. The final crater shows evidence of folding, shocking, and fracturing of the rocks, and the rim is steeply turned up. These phenomena illustrate that normally brittle rock units can, under appropriate conditions such as sudden high pressure, deform plastically.

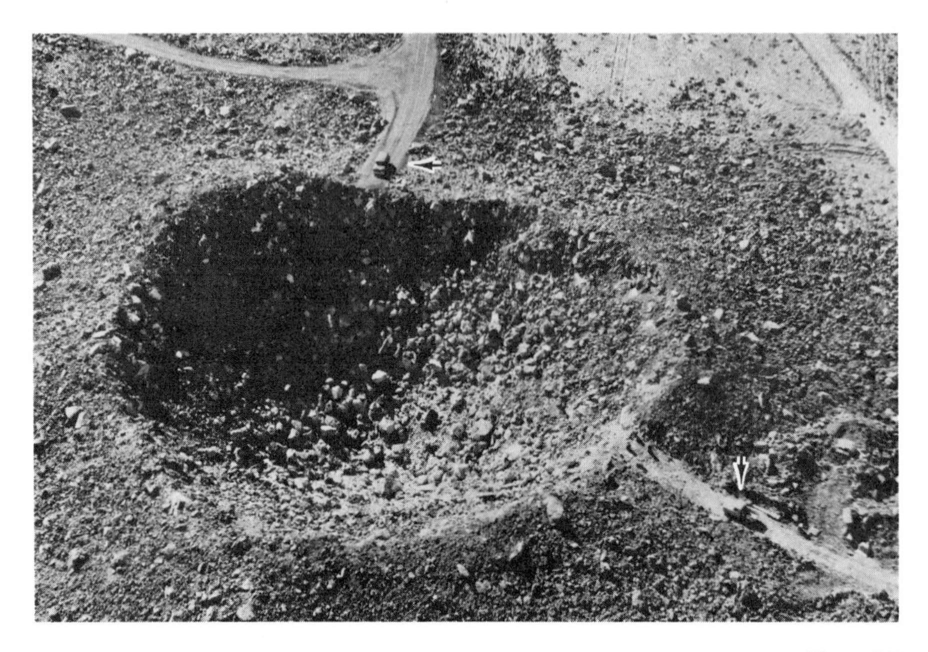

Figure 9-5

Explosion crater from detonation in basaltic rock. General structure is similar to that of high-velocity meteorite impact craters. Note ejected boulders as large as trucks (arrows), which indicate scale. (AEC.)

Figure 9-6

Diameters of craters formed by meteorites with various masses and with various impact velocities.

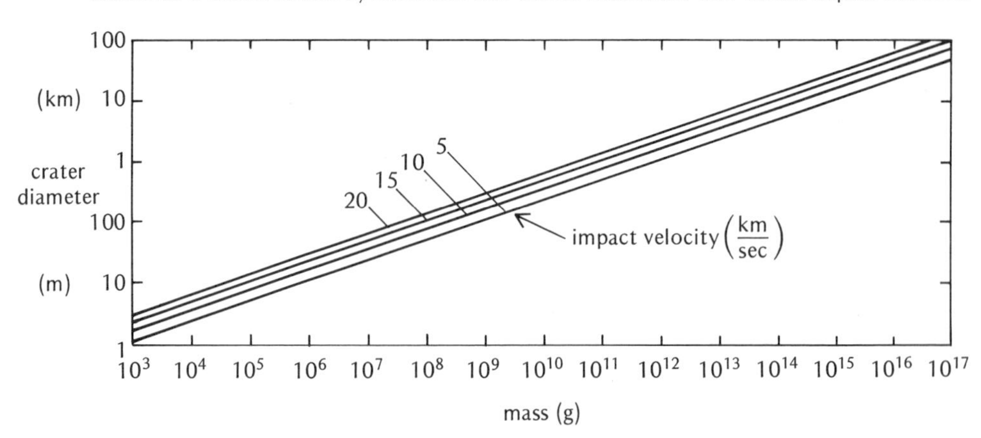

Craters on Two Planets and Three Satellites

Craters on the moon are familiar to most readers (Figure 9-7; see also Chapter 11). Craters on the earth are less common only because the earth's surface is repeatedly eroded and churned by mountain-building activity. Ancient terrestrial craters have thus been lost. Perhaps the best-preserved substantial crater is Meteor Crater, in northern Arizona (Figure 9-4). (More properly, it should have been named Meteorite Crater.) It was formed by a large iron meteorite, pieces of which are still strewn on the desert nearby.

When the first photographs of Mars showed prominent craters that looked

Figure 9-7
Cratered terrain on the moon, photographed by Apollo astronauts. (NASA.)

like those on the moon, there was some surprise. In retrospect there should have been no surprise, since Mars is close to the asteroid belt and thus very prone to bombardment by asteroidal fragments. Indeed two astronomers, Öpik and Tombaugh, had predicted Martian craters in little-noticed papers as early as 1950!

Mariner 9 photographs show craters on the Martian satellites Phobos and Deimos (see Figure 6-5). These craters outnumber their comparable-size counterparts on the nearby Martian surface, and thus serve as an indicator that the Martian surface has undergone some erosion.

Central Peaks and Related Phenomena in Craters

Many lunar craters of presumed impact origin show central mountain peaks. Mechanisms that have been proposed to explain these include (1) volcanic extrusions triggered by the impact; (2) slumping of materials from crater rim areas into the center; (3) slow isostatic rebound of material from underneath the crater, pushing up and favoring volcanism in the central area; and (4) rebound immediately after the impact, analogous to rebound of a liquid drop falling into a pool. Recent evidence suggests that a resistent rock layer beneath a forming crater encourages a rebound (hypothesis 4), although other characteristics of central peaks support hypothesis 3 (see Baldwin, 1963).

C. A. Wood (1968) has shown that on the moon, central peaks are most common in the youngest and largest craters, e.g., about 60 to 100 km (40 to 60 miles) in diameter. Central peaks occur in Martian craters of about 10 to 30 km, and still smaller terrestrial craters show evidences of central uplift. Present data suggest that central peaks are related to other phenomena, such as concentric mountain rings and faults inside and around craters. These phenomena appear at smaller scales on larger planets. Study of these "planetary scaling" differences from planet to planet may give a better understanding of the role of gravity in large-scale geologic deformation processes.

Astroblemes

In the 1950s and 1960s, geologists recognized a large number of very ancient terrestrial meteorite craters, in some cases eroded almost beyond recognition. For these, Dietz (1963) coined the term *astrobleme,* or star wound. Frequently the original crater and surface layers have been planed off by erosion; hence we often see only the exposed circular root structures. A large number of astroblemes have been recognized in the ancient, stable geological province surrounding Hudson's Bay, known as the Canadian shield. A number of astroblemes have been discussed in detail by Dietz (1963), Baldwin (1963), and Beals and Halliday (1967) and in the *Journal of Geophysical Research, 76,* for August 10, 1971. Many other circular structures are of uncertain origin and are called cryptovolcanic structures. The discovery of nearby meteorites, evidence of high-energy explosions, shocked rock, or evidence that the disturbance disappears at depth is usually regarded as proof that the structure is a true astrobleme.

Table 9-1 gives a list of selected known terrestrial craters and astroblemes.

Secondary Impacts

Craters produced by meteorites falling upon a planetary surface from space are called *primary impact craters.* In such an event fragments may produce

Table 9-1

Selected meteorite craters and probable astroblemes

Name	Location	Estimated original crater diam.[a] (km)	Estimated approx. age[b] (years)
Manicouagan Lake	Canada	65	$2.10 \pm 0.04 \times 10^8$
Vredefort Ring	South Africa	40	
Clearwater Lakes	Canada	32 (2 craters)	$2.90 \pm 0.30 \times 10^8$
Rieskessel	Germany	24	$1.44 \pm 0.7 \times 10^7$
Deep Bay	Canada	12	6×10^8?
Ashanti	Ghana	11	10^6
Wells Creek	Tennessee	10 (4)	10^8?
Crooked Creek	Missouri	6.5	3×10^8?
Sierra Madera	Texas	4.9	2×10^6
Brent	Canada	3.5	$5.8 \pm 0.4 \times 10^8$
New Quebec	Canada	3.4	$>10^4$
Holleford	Canada	2.3	7×10^8?
Meteor Crater	Arizona	1.2	$<10^5$
Wolf Creek	Australia	0.85	
Monturaqui	Chile	0.47	
Aouelloul	Mauritania	0.26	
Henbury	Australia	0.22 (several)	
Boxhole	Australia	0.18	
Odessa	Texas	0.17 (several)	$>1 \times 10^4$
Kaali Järv	Estonia, USSR	0.11 (many)	4×10^3?
Wabar	Arabia	0.10 (2)	10^3–10^4?
Campo del Cielo	Argentina	0.08 (many)	
Pamir	Tadzhik, USSR	0.079 (many)	
Sikhote-Alin	Siberia, USSR	0.026 (many)	25 (fall, 1947)
Dalgarange	Australia	0.021	2×10^4
Haviland	Kansas	0.017	

[a] When more than one crater is present, the diameter of the largest is given and the number is indicated in parentheses.

[b] Various sources including Baldwin (1963), Beals and Halliday (1967), Faul (1966), and more recent radiogenic dates.

Figure 9-8
Secondary impact crater produced by rocky debris from a volcanic explosion, Kilauea, Hawaii. The crater is about 1 m across and was 44 years old when this photo was taken.

secondary impact craters. Secondary impact craters can be formed by fragments from volcanic eruptions as well as from meteorite craters, as shown in Figure 9-8.

The preceding discussion suggests the variety of possible crater types. A planetary surface may thus consist of a confusing mixture of primary impact craters, volcanic and collapse craters, and secondary impact craters from both meteorite and other secondary ejecta (see Chapter 11).

Chronology Based on Meteorite Craters

Meteorite craters can be used to estimate the ages of planetary surfaces. On a given planet the number of primary impact craters in a geologically inactive province increases with the length of time the surface is exposed. The number of primary impact craters should thus be a measure of the *relative age* of the surface. If we know the rate of crater formation, the rate can be divided into the crater density to give the *absolute age*.

The ages so determined are *crater retention ages*. On very inactive planets like the moon, craters may date back to the time of formation of the planet itself, but on active planets such as the earth or Mars, crater retention ages indicate only the interval during which erosion has not yet destroyed craters of whatever size is counted. Thus crater retention ages can be a function of crater size, since the largest craters may last longer than small craters. The incidence of primary impact craters is thus an important index of the amount of geological activity on a planet—i.e., erosion and deposition.

Crater counts have been made for both the moon and Mars and relative ages have been found for various provinces, as shown in Figure 9-9. Even

Figure 9-9

Crater counts for three planets. Most Martian regions fall between the ancient lunar terra surfaces and the younger lunar maria. The earth has only scarce, recent craters, older craters having been destroyed by erosion.

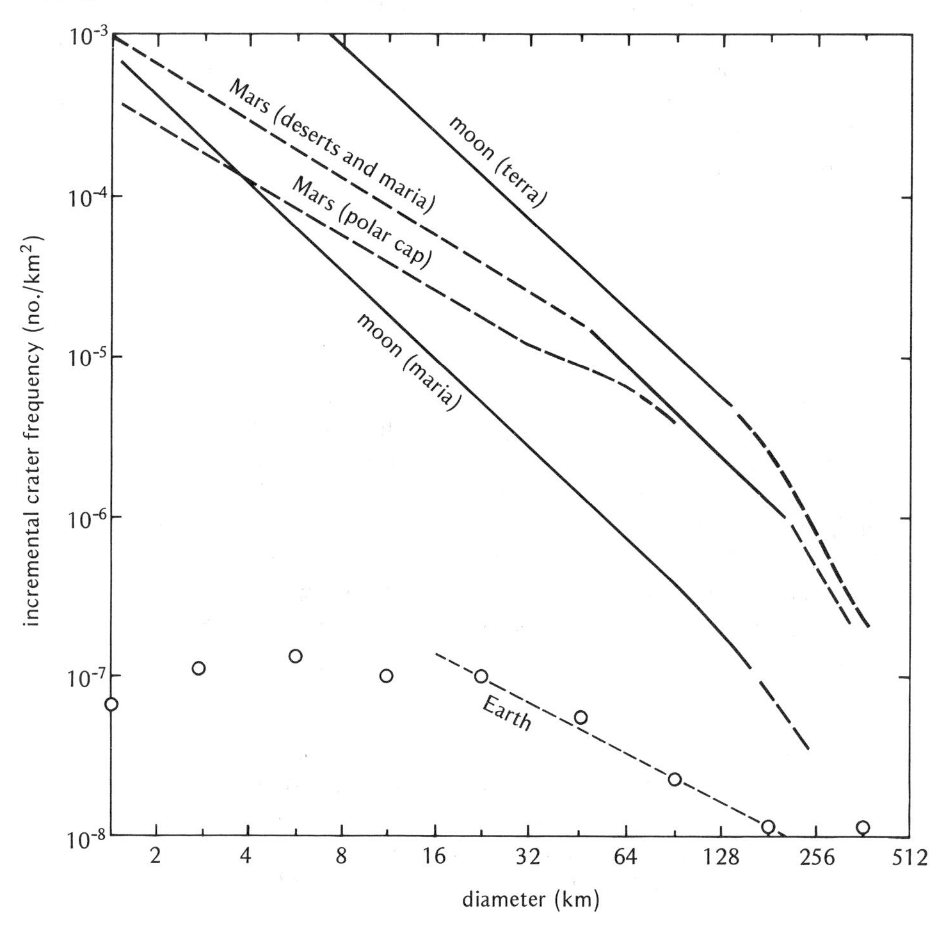

absolute ages can be inferred with accuracy probably better than an order of magnitude. For example, the well-known bright lunar ray crater Tycho has a very low crater density in its interior and is thought to have an age around $\frac{1}{2}$ to 1 billion years. Absolute dates cannot be estimated if the surface is so old that it is *saturated* with craters, for then it is impossible to determine the total number of impacts that have occurred. This applies to parts of the moon, which probably date back to the beginning of the solar system. Another difficulty with the method is that craters smaller than about 2 km on the moon (and possibly Mars) include not only primary impact craters, but also secondary impact and volcanic craters. The accuracy of crater retention ages is thus limited by our ability to discriminate small primary impact craters. If crater counts have to be limited to large craters of known origin, craters are farther apart and only large areas can be dated.

Crater-count dating is a useful technique as long as we have only photographs of other planets. Once samples from many provinces of the moon and planets are abundantly available, crater-count dating can be replaced by much more accurate radioisotope dating (discussed below in this chapter).

Once various cratered provinces have been dated accurately by radioisotopic techniques, crater counts in provinces of various ages can be used to reconstruct the prehistoric cratering rates experienced by the planet. Dates obtained from Apollo samples have allowed this technique to be used on the moon, giving the results described above under the heading "Impact Rates."

ORBITS OF METEORITES

Observed Orbits

Orbits of meteorites (prior to collision with the earth) are measured by means similar to those used to study orbits of meteors (see Chapter 7). However, meteorites are sufficiently rare that only two well-determined orbits are available. The first is the Pribram fall (Czechoslovakia, April 7, 1959) and the second, Lost City, Oklahoma (January 3, 1970). Figure 9-10 diagrams their orbits. In each case the aphelion lies near the asteroid belt, and the meteorite may originally have been in an asteroidal orbit before being perturbed into an earth-crossing orbit that led to collision with the earth (see also Table 9-2).

Incidence of Fall

The hourly incidence of observed falls is thought to peak in the afternoon and is lowest before dawn (Wood, 1961; Mason, 1962). This datum suggests an important clue on meteorite origins, as illustrated in Figure 9-11. As the figure shows, objects that catch up to the sunset side of the earth are preferentially moving in direct orbits with aphelia outside the earth's orbit. This suggests

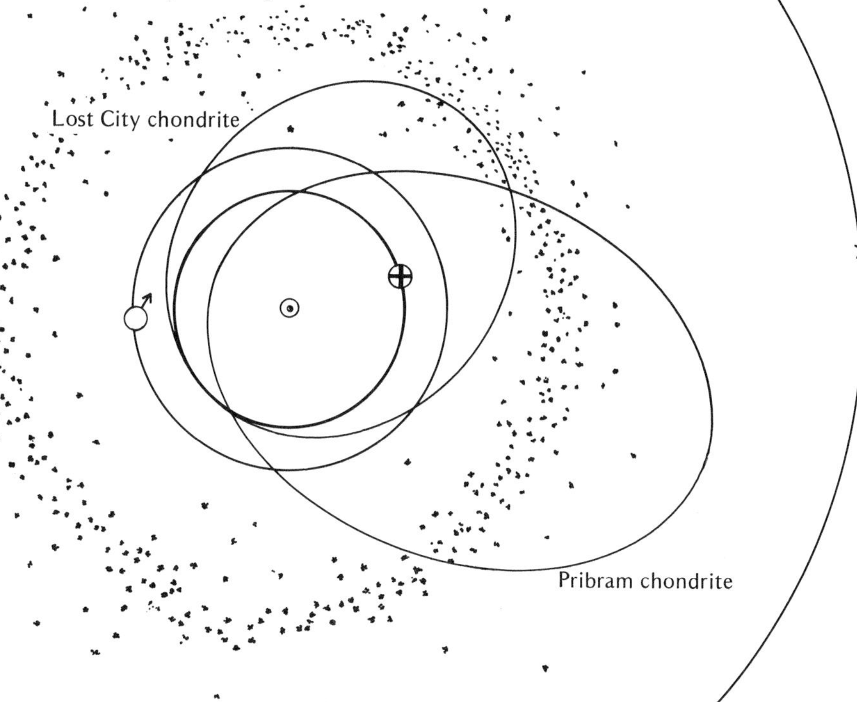

Figure 9-10

The only two well-determined orbits for meteorites, showing relative sizes, aphelion–perihelion distances, and relation to asteroid belt.

particles with asteroid-type orbits but is inconsistent with most comet orbits, which may be either direct or retrograde. Some analysts have questioned this datum, because it is difficult to correct for sociological and other selection effects; e.g., many more potential witnesses may be out in the evening than in the early morning hours.

DISTINCTIONS BETWEEN METEORITES AND METEORS

There has been continued controversy over the possible relationships between meteorites and meteors. The most acceptable idea, which is supported by

Table 9-2

Orbits of meteorites, Apollo asteroids, and meteors

Object	Perihelion distance, q (A.U.)	Semimajor axis, a (A.U.)	Eccentricity, e	Inclination, i	Aphelion distance, q' (A.U.)
Meteorites					
Pribram chondrite[a]	0.790	2.46	0.678	10°4	4.13
Lost City chondrite[b]	0.97	1.66	0.42	12.0	2.35
Apollo asteroids (see Table 8-1)					
Apollo (1932HA)	0.65	1.47	0.56	6	2.30
Hermes (1937UB)	0.62	1.64	0.62	5	2.66
Peculiar Apollo asteroid (see Table 8-1)					
Icarus	0.19	1.08	0.825	23.0	1.97
Low-velocity fireballs[c]					
Feb 13, 1965	0.69	2.30	0.69	13.0	3.91
May 31, 1966	0.60	2.97	0.79	9.3	5.34
Oct 27, 1966	0.98	2.12	0.53	26.3	3.26
Dec 11, 1966	0.61	2.13	0.71	0.8	3.66
High-velocity fireballs[c]					
Nov 30, 1965	0.27	∞	1.00	131.5	∞
Jul 31, 1966	0.44	32.30	0.98	41.8	64.2
Shower meteors and fireballs[c]					
1965 Draconid	0.99	3.46	0.71	32.1	5.94
1965 Leonid	0.98	52.6	0.98	161.8	104
1965 Taurid	0.42	2.20	0.80	2.4	3.98
1966 Perseid	0.96	40.20	0.97	114.4	79.4

[a] Ceplecha, Rajche, and Sehnal (1959).
[b] Smithsonian Center (1970).
[c] 100 fireball orbits are given by McCrosky (1967).

Figure 9-11

Preferential evening and morning arrival times for direct and retrograde meteoroids.

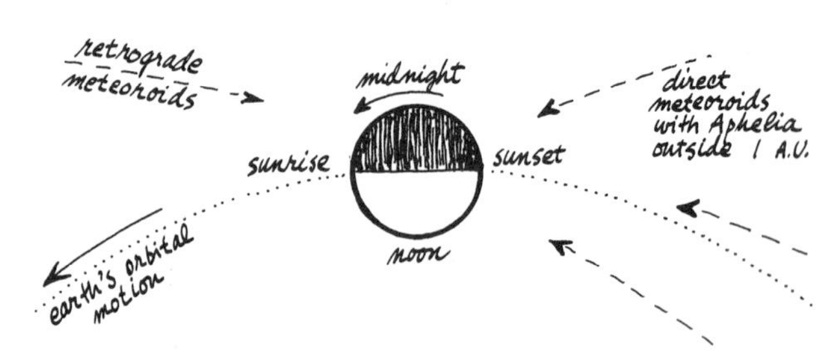

recent evidence, is that *most* meteors, especially faster meteors, are low-density material related to comets, while *most* meteorites are high-density material related to asteroids. (Compare Chapters 7 and 8.) Nonetheless, there is a nagging suspicion that some meteorites may come from elsewhere besides asteroids. Urey (1967) for some years championed the idea that many meteorites came from the moon. This was discounted after the first lunar landings. Öpik (1966) championed the more viable proposition that many meteorites may come from comets or "burnt-out comets."

One approach to this problem has been through study of orbits. Table 9-2 shows a sampling of available data. The two well-determined meteorite orbits resemble those of Apollo asteroids (see Chapter 8). The Smithsonian Astrophysical Observatory has catalogued a number of meteor and fireball orbits (McCrosky, 1967). Two groups, the low- and the high-velocity groups, form a transition between Apollo asteroids, which have moderate eccentricities and aphelia in the asteroid belt, and the shower meteors, which have high eccentricities and are known to be associated with comets (Anders and Arnold, 1965).

The question is: Where on Table 9-2, if at all, is the division between asteroidal material and cometary material? The most widely accepted answer is that most meteorites and most Apollo asteroids are probably fragments from the asteroid belt, but Öpik's theory is that many of these are burnt-out comets. Icarus, as mentioned in Chapter 8, has been shown to be spherical, as might be expected if it is an old comet nucleus residue but not if it is a fragment of a broken asteroid. Most observers agree that the percentage of cometary objects must be quite high among the shower meteors and fireballs, which have more mixed and more highly eccentric orbits than meteorites and Apollo asteroids.

A second approach to the relation between meteorites and meteors is through study of composition. McCrosky (1967) found that among 100 meteoric fireballs—none of which was substantial enough to hit the ground—densities averaged around 0.4 g/cm^3 and not greater than 1.2 g/cm^3. Such densities typify expectations for cometary nuclei composed of ice and dust but are much less than densities of stony and iron meteorites. Also, certain rare meteorites known as carbonaceous chondrites (see below) have high volatile contents, as might be expected of stones embedded in icy comet nuclei. It is tempting to conclude that comets produce low-density meteors and fireballs and a few stony meteorites with high volatile content.

A possible solution to the question of meteorites versus meteors and asteroids versus comets follows a certain middle ground. Probably both asteroids and comets are examples of primeval planetesimals, and both may contribute to the objects striking the earth. Inner asteroids, heated by the sun, may have lost or never retained volatiles and may produce stony and iron meteorites. Comets, formed in the outer solar system, retained ices and produce fragile, loosely cemented dust grains. Thus there may be no sharp distinc-

Table 9-3

Classification of meteorites[a]

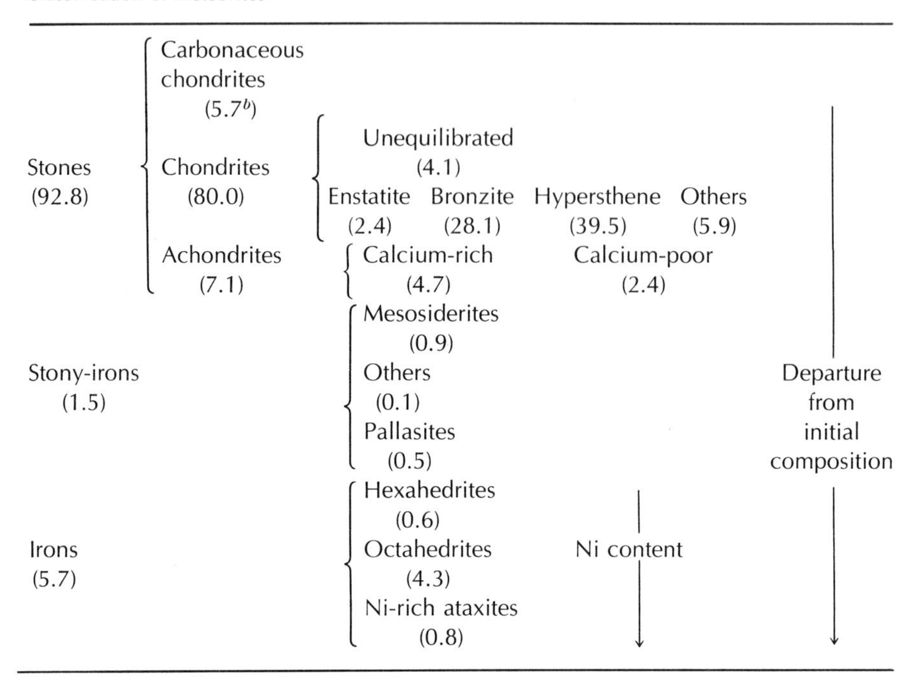

Class	Subclass	Group				
Stones (92.8)	Carbonaceous chondrites (5.7[b])					Departure from initial composition ↓
	Chondrites (80.0)	Unequilibrated (4.1)				
		Enstatite (2.4)	Bronzite (28.1)	Hypersthene (39.5)	Others (5.9)	
	Achondrites (7.1)	Calcium-rich (4.7)	Calcium-poor (2.4)			
Stony-irons (1.5)		Mesosiderites (0.9)				
		Others (0.1)				
		Pallasites (0.5)				
Irons (5.7)		Hexahedrites (0.6)	Ni content ↓			
		Octahedrites (4.3)				
		Ni-rich ataxites (0.8)				

[a] Numbers in parentheses give measured percentage of falls in each category.
[b] This number may eventually be increased at the expense of the other percentages (see the text).

tion—only a smooth transition—between the asteroid/meteorite materials and the comet/meteor materials.

Taking a purely petrological view and applying results of decades of research on meteorites, geochemists have developed a classification scheme based on the degree of chemical processing of the meteorite material (Table 9-3). We shall follow this theme of chemical evolution of the meteorite parent bodies in the following sections.

Basically, the idea is this: If early planetesimals were made of primordial accreted condensates, the interior parts of these planetesimals would have been heated by the short-lived isotopes or other heat sources, as was the moon. The outer parts would have been least affected.* Near the surfaces of various planetesimals, the primordial material could have been preserved, while in the

* An alternative idea, recently advocated, is that the early planetesimals were heated from the *outside* by events during the sun's T-Tauri stage. Their surfaces supposedly melted while the interiors stayed cool. This

inner parts it could have been completely melted. During complete melting, chemical *differentiation* occurs; the heavy iron and the elements with a chemical affinity for iron (called *siderophile elements*) drained toward the planetesimal centers while the lighter, silica-associated elements (called lithophile elements) were displaced upward. Similarly, in the case of iron making, a silicate-rich slag floats on the molten iron.

An initial look at meteorites, then, leads us to suspect that stones represent less heated parts of planetesimals; irons, the melted cores; and stony-irons, the interfaces. To the first approximation, this idea is supported by detailed meteorite studies. An excellent nontechnical review of the problem is given in a short book by J. A. Wood (1968). Mason (1962) gives a more technical review.

CHONDRULES

Chondrites are stony meteorites that (in almost all cases) contain BB-sized, round spherules. These small round bodies are called *chondrules,* after the Greek term for seedgrains. The chondrules are composed principally of the minerals olivine and orthopyroxene,* and they show by their structure that they were formed by rapid cooling of droplets of molten material.

The question of how such droplets formed is still hotly debated. There has hardly been much advance since the English mineralogist Sorby in 1877 invented the process of making thin sections of rocks for geological studies through the microscope and then first described the chondrules. Sorby inferred from his studies that the chondrules formed from "Melted . . . glassy globules, like drops of a fiery rain." A number of chemical indicators, summarized by Anders (1964), suggest that the initial high-temperature cooling of the chondrules occurred very fast, with a possible drop from 1500°K to solidification within minutes. One bit of evidence is the glassy structure of some chondrules. Free metal particles in the chondrules suggest that the final stages of cooling to a few hundred degrees was much slower, requiring 10^4 (10,000) to 10^7 (10 million) years (J. A. Wood, 1968).

The chondrites seem to be the least differentiated of meteorites, and thus it appears that chondrules must have formed at about the same time as planetesimal growth. What process in the earliest history of the solar system could produce "drops of a fiery rain"?

Many researchers have argued that the chondrules must have formed from

* Minerals and rock types will be discussed in more detail in Chapter 11.

---

view is not widely accepted but has attracted wide interest. In either case, observations require that some regions melted while others were preserved in nearly the original state.

surface materials of young planetesimals. An early idea was that they might result from volcanic eruptions of magma, producing rapidly cooling droplets of splashed lava. Ringwood (1966) and Urey (1967) proposed that chondrules must have been produced as secondary droplets sprayed out from impact events that occurred on the surfaces of planetesimals the size of large asteroids. The final slow cooling processes suggest that the chondrules might have been incorporated into insulating debris while they were cooling. This would occur if they were shot out and then fell back to a planetary surface after an impact event (harkening back to the 1877 theory of "drops of fiery rain").

An alternative interpretation has been favored by a growing number of researchers. Suess (1949), Levin (1958), and especially Wood (1958, 1963, 1968) have all discussed the possibility that chondrules condensed directly out of the cooling nebular gas. Wood pointed out that high densities would be required and suggested that these occurred when shock waves spread through the nebula during events in the unstable T Tauri stage of the sun. Whipple (1966) and Cameron (1966) suggested that turbulence in the nebula led to electrical charge separation, which in turn led to lightning strokes that produced chondrules from preexisting solid accreted particles. Blander and Abdel-Gawad (1969) suggested that the chondrules may have condensed in the nebula very suddenly from a supercooled vapor, thus accounting for their initial rapid solidification.

Whatever their mechanism of formation, it appears that chondrules were ubiquitous by the time of completion of the planetesimals and were incorporated into them.

CARBONACEOUS CHONDRITES

Carbonaceous chondrites are the material that comes closest to representing the unaltered, ancient planetesimal material. (The material in the solid planetesimals differed from the solar nebula's composition due to the condensation processes discussed in Chapter 6.) There are several indications that they have not been highly heated, compressed, or otherwise altered: (1) They have unusually high abundances of volatile compounds such as water; (2) they have low densities; (3) they are rich in organic compounds that would condense in the solar nebula but would be lost during any subsequent heating (Anders, 1963a); and (4) they contain the heavier elements in nearly their original cosmic abundance pattern.

Carbonaceous chondrites were divided into three types by Wiik (1956) on the basis of chemical analysis. The Type I carbonaceous chondrites are most extreme in the "primordial characteristics" listed above. They have some 20 percent water content, a density of only 2.2 g/cm^3 compared with 3.6 g/cm^3 for other kinds of chondrites, minerals stable only at low temperatures, and the highest volatile content. Thus it appears that Type I carbonaceous chondrites have been subjected neither to high heat nor high pressure (i.e., they have not been buried deep inside planet-sized bodies). Ringwood (1966) has shown that they have a composition suitable for the primeval parent material and that the other chondrites could be derived from them or from similar material by chemical processes that removed certain elements preferentially.

Type I carbonaceous chondrites have no chondrules—an internal inconsistency of the nomenclature! It is theorized that they represent the primeval agglomerate from which planetesimals formed prior to the chondrule-producing events or processes. Type II carbonaceous chondrites do have chondrules, but the stony matrix between the chondrules resembles the Type I material and has had only minimal chemical alteration (J. A. Wood, 1968).

DuFresne and Anders (1963) showed that the carbonaceous chondrites were exposed to the action of liquid water at some point in their history. This, plus the high volatile content and low density, suggests that the parent bodies of the carbonaceous chondrites were in some ways similar to the "dirty icebergs" postulated for *comet nuclei*. These may have been more ice-rich than the interiors of the stony asteroids. The characteristics of some recent falls of carbonaceous chondrites have suggested that they may be much more common than previously supposed and that their orbits may be cometary. Evidence from Apollo lunar samples suggests that most meteoritic material striking the moon resembles carbonaceous chondrites in composition (Ganapathy et al., 1970). Thus the entry of only 5.7 percent in Table 9-3 as the frequency of carbonaceous chondrite falls may have to be revised upward. The quick terrestrial weathering of the crumbly carbonaceous chondrites may explain why they have not hitherto been recognized as common, and also why materials from cometary fireballs have never been recovered. All carbonaceous chondrites are falls; none are finds. A relation to comets would support their identification as primordial planetary material, since comets may have been volatile-rich planetesimals ejected from the solar system and added to the surrounding Oort cometary cloud. (See Chapters 6 and 7, and comments above on "Distinctions between Meteorites and Meteors.")

Carbonaceous chondrites get their name from their high abundance of carbon. This carbon is not in the form of free carbon, but as early as 1864 was shown to exist as complex organic matter. There has been controversy, particularly in the early 1960s, over whether some of these organic compounds may have been of biogenic origin. Some theorists suggested that the molecular precursors of life were not produced on earth but in the planetesimals, which later crashed on the planets and started the evolution of life on earth. Some analysts have reported "organized elements," or microfossils, in meteorites, but this evidence for extraterrestrial precursors or products of life is not unambiguous. Better evidence is the discovery of apparently extraterrestrial amino acids in the Murchison meteorite, which fell in Australia in 1969 (see Chapter 13).

CHONDRITES

Problems of Composition

Analysis of chondrites prior to 1880 indicated that in their most common elements they are all rather alike, and that they have not been subjected to enough chemical processing to have changed their composition radically. The thrust of much modern work is to compare the chondrite abundances with the cosmic abundance pattern, usually represented by the sun. Elemental abundances of rock-forming elements in chondrites are much more similar to the original cosmic abundances than are abundances in the earth's crust or in other meteorites—except for carbonaceous chondrites (Ringwood, 1966; J. A. Wood, 1968, p. 14). This indicates that they have been subjected to relatively modest modification since the original parent material was created and incorporated into planetesimals.

This finding is illustrated in Figure 9-12. The graph compares the abundances of rock-forming elements in various meteorites with the abundances in the supposed parent material, the carbonaceous chondrites. The earth's crust is strongly enriched in lithophile (silicon-affinity) elements, while the iron and stony-iron meteorites are primarily composed of siderophile (iron-affinity) elements. The ordinary chondrites (dashed line), however, are only slightly modified from the composition of their parent material.

The fact that the chondrites all have approximately the same composition means that chemical alterations have been so slight that *major* constituents have not been removed or added; i.e., chondrites have not been strongly differentiated. Nonetheless, some chemical processes have occurred among the chondritic materials. Iron is one of the best indicators of this. It may either be strongly bound to oxygen atoms in rock-forming minerals (*oxidized*) or it may exist as pure metal (*reduced*).

Prior (1920) proposed several rules (known as *Prior's rules*) which recognized this oxidation–reduction relation: Prior assumed that the total iron content in ordinary chondrites was constant and that as the amount of oxidized iron decreased, the amount of reduced iron correspondingly increased. The oxidation–reduction state was thus pictured as a measure of the chemical alteration that occurred in the meteorite, while the supposedly constant iron content implied that none of the chondrites had been so highly altered as to lose any initial iron.

In 1953 Urey and Craig changed this finding. In a now-classic review of all the earlier analyses of chondrites, they rejected two thirds of the earlier studies because of errors and uncertainties (see Mason, 1962; J. A. Wood, 1968). Among the 94 remaining chondrites, Urey and Craig showed that there were a number of subgroupings. First and most basic was a distinction into different groups according to the total iron content. Urey and Craig initially listed two groups, the high-iron *H group*, and a low-iron *L group*. Sub-

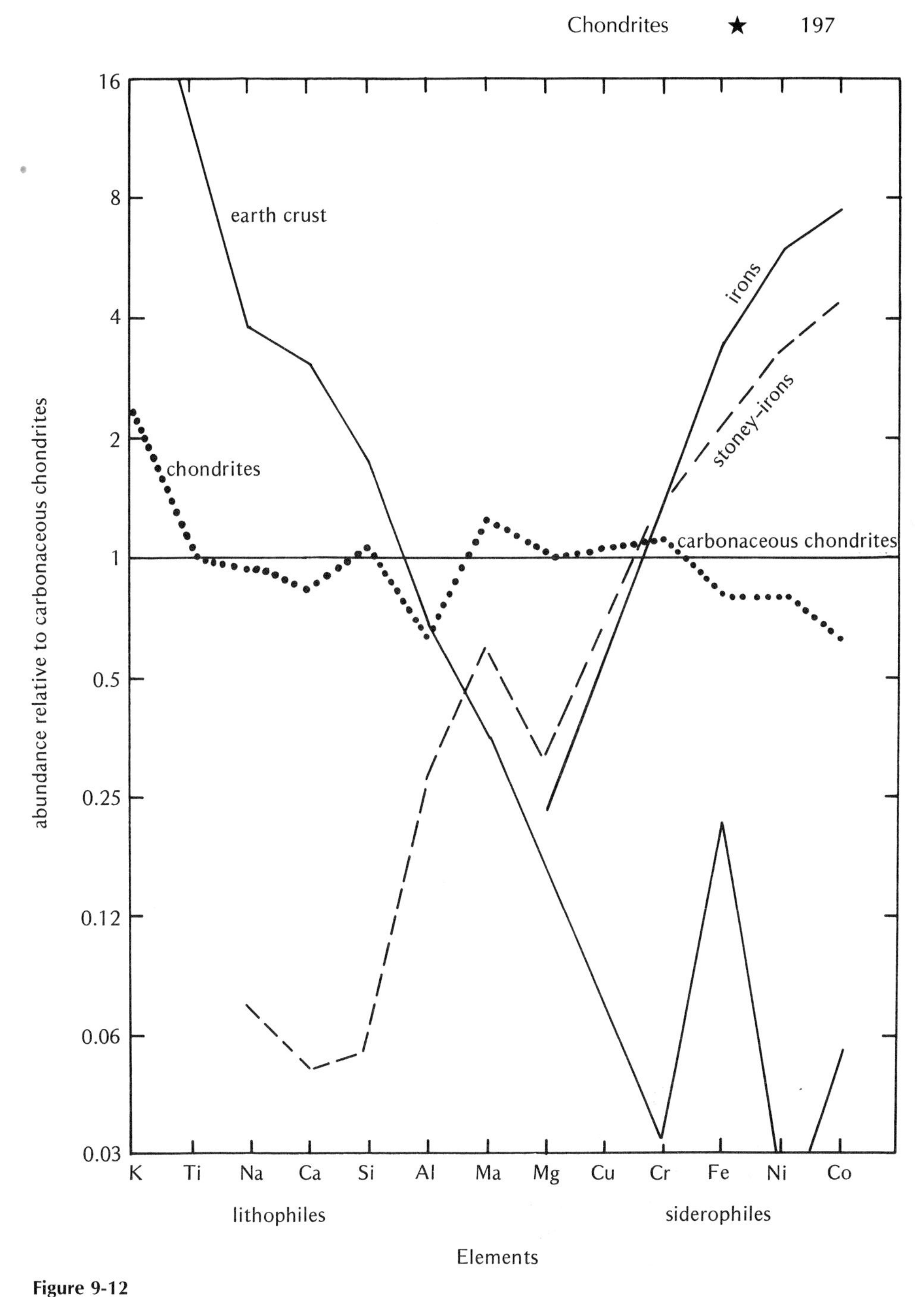

Figure 9-12

Elemental abundances in various meteorites and the earth's crust, relative to abundances in carbonaceous chondrites (horizontal line). Planetary crusts in general are enriched in lithophile elements, while interiors are enriched in siderophiles.

sequently, finer subdivisions have been proposed. The finding of a discontinuity in total iron abundance is evidence that more than one parent body may be involved in producing the meteorites.

The amount of chemical processing, the state of iron, the abundances of trace elements, and other properties are used in subdividing the chondrites into the groups shown in Table 9-3.

The unequilibrated chondrites, as well as the carbonaceous chondrites, are rich in oxidized iron and contain compounds that would not be stable together had they ever been heated to high temperatures. Of the noncarbonaceous chondrites, they have been subjected to the least complex chemical processing. One might speculate that they represent the materials from the outermost layers of the parent planetesimals, where they were subjected to little internal heat.

The enstatite, H-group (bronzite), L-group (hypersthene), and other chondrites are distinguished by their particular mineral and chemical states. Of these, the enstatite chondrites are the most extreme in that they contain almost completely reduced iron. The categories are discussed in much more detail by Mason (1962), Wood (1963), and Anders (1964).

Cooling Rates and Size of Parent Body

Most of the chondrites were heated to unknown temperatures, presumably while they were part of substantially large planetesimals, where radioactive heating or some other form of heating occurred. After this heating, they cooled. The chondrites contain metal particles from which the rate of cooling can be calculated by measuring the properties of the individual crystals in the metal (see J. A. Wood, 1968, pp. 29–39, for a detailed discussion). These calculations show that the chondritic material cooled about 1 to 10°K per million years. This implies that chondritic material was insulated by a layer of overlying material at least 10 to 100 km thick. Such dimensions are thus lower limits to the radii of the meteorite parent bodies.

ACHONDRITES

Achondrites are stony meteorites without chondrules. The crystal structure of achondrites is more coarse than in chondrites. Coarse crystal structure is usually an indication of slow cooling in insulated surroundings—for example, it is found in underground intrusions of solidified magma. Of all the meteorites, achondrites are closest to terrestrial rocks. In particular they are similar to igneous rocks (rocks produced from molten material).

These characteristics indicate that the achondrites were produced when some kind of parent material melted and differentiated. The question naturally arises whether chondrite-like material might be the parent material which produces achondrites simply by melting. The answer appears to be a qualified yes.

In the first place, chondrules would be destroyed if chondrites were melted. Indeed, the chondrites contain a range of chondrule states, from perfectly preserved glassy examples to nearly obliterated, recrystallized examples (see J. A. Wood, 1968, p. 43), grading into achondrites. Furthermore, the iron content has been severely reduced in many chondrites, as we would expect if the iron had drained away from the silicate phase during the melting and differentiation. Another argument that achondrites come from chondritic material comes from the physical chemistry of the recrystallization process, which indicates that molten chondritic material could crystallize to produce mineral groups similar to those observed among the achondrites (Mason, 1962).

The suggestion that achondrites were produced by melting chondrites is qualified, because the parent material of the achondrites was probably not precisely the same as the chondrites. Compared to cosmic and chondrite compositions, achondrites show some differences, for example in certain isotope ratios. Recent evidence suggests that the location in the solar nebula may be an additional factor in separating achondrites from chondrites.

In summary, the origin of achondrite material could be as follows. *More-or-less* chondritic parent bodies partially melted. Initially chondritic material melted, destroying the chondrules. Molten metal drained toward the centers of these bodies, leaving "inner mantles" of achondritic material and "outer mantles" of highly modified chondritic material. Parent bodies formed near the sun, with higher early temperatures, may have been more achondritic than distant parent bodies.

STONY-IRONS

These are divided into two main classes, depending on the minerals in the silicate stony part. The *mesosiderite* silicates are mainly composed of plagioclase and pyroxene minerals. The *pallasite* silicates are principally olivine.

The stony-irons are generally assumed to be material from the interfaces between the silicate and metal portions of highly differentiated interiors of meteorite parent bodies. The iron may have been concentrated in a central core, or it may have separated into local pools throughout the parent body. Thus the stony-irons might either come from a single interface between an iron core and silicate mantle or from different iron-silicate interfaces, or from a combination of both situations in various parent bodies.

IRONS

Iron meteorites are the most useful in telling us about the size, number, and history of the meteorite parent bodies.

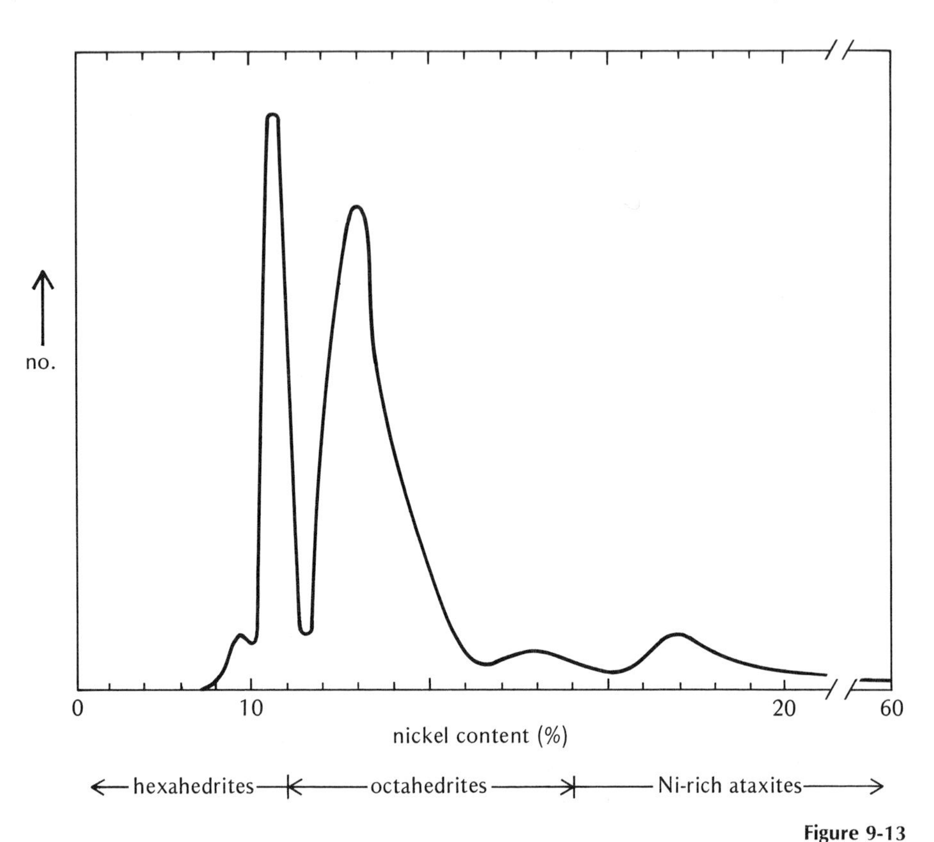

Figure 9-13
Classification scheme for iron meteorites, according to nickel content among irons.

Composition and Classification

Iron meteorites are composed almost entirely of metal. They are classified according to nickel content (Figure 9-13), and their names reflect the crystal structure.

The *hexahedrites* are low in nickel and consist of a single nickel-iron mineral, *kamacite.* A discontinuity in nickel content divides them from the *octahedrites,* which contain three nickel-iron minerals, kamacite, *taenite,* and *plessite. Nickel-rich ataxites* have the most nickel and consist mostly of plessite.

Widmanstätten Pattern: Size of Parent Bodies

In octahedrites the two major minerals, kamacite and taenite, occur in crystals that interlock in an interesting geometry called the Widmanstätten* pattern,

* Named after Count Alois von Widmanstätten, director of the Imperial Porcelain works in Vienna and discoverer of the pattern in a meteorite in 1808.

shown in Figure 9-14. This pattern is prominent only after the meteorite is cut, polished smooth, and then etched with acid. Hexahedrites and ataxites have no Widmanstätten pattern.

For a time it was thought that very large pressures and very long cooling times (10^9 years) were required to allow the Widmanstätten pattern to develop. It was said that the parent body must have been the size of the moon ($R = 1738$ km) in order to provide these conditions (Uhlig, 1954). However, more recent studies show that the pattern could develop in much smaller bodies. Goldstein and Short (1967) and Wood (1964) measured the fine structure of the metal crystals and were able to deduce the cooling rate experienced by the octahedrites when they were at temperatures in the range 900 to 700°K. Two thirds of the octahedrites had cooling rates of 1 to 10°K/million years. From this, one can calculate the size of the parent body required to give the necessary insulation and observed cooling rate in a central iron core. The results are

Figure 9-14

Examples of iron meteorites. The upper two specimens are pieces of the Canyon Diablo iron which produced the Arizona meteorite crater, showing their appearance as picked up in the field (length, approx. 3 cm). The lower specimen is a piece of the Arispe iron, cut, polished, and etched with acid to show the Widmanstätten pattern and Neumann bands (fine parallel light streaks).

that the octahedrites must have been inside parent bodies of radius 70 to 200 km. Goldstein and Short found that the Widmanstätten pattern could arise under conditions found inside bodies of such size.

In addition to Widmanstätten patterns, another early reason for believing that meteorites came from lunar-sized bodies was the discovery of diamonds in both irons and stones. Formation of diamonds requires pressures higher than exist in bodies as small as those mentioned above. However, a number of researchers have argued that high transitory pressures produced during the impacts of meteorites may have produced the diamonds—see the reviews by Mason (1962, p. 57) and J. A. Wood (1968, p. 53)—thus obviating this argument for parent bodies as large as the moon.

Neumann Bands: Evidence for Shock

Neumann bands are fine linear striations visible in a polished and etched surface of most hexahedrites and some other irons. They are caused by an alteration of the crystal structure within the kamacite crystals, known as *twinning* (not to be confused with the interlocking Widmanstätten arrangement of the crystals themselves).

Uhlig (1955) showed that the Neumann bands were the result of strong shocks applied to the meteorites when they were at low temperatures, not above 900°K and probably below 600°K. The interpretation of this finding is that after the parent bodies had cooled until their interiors were solid,* they were subjected to violent impacts. These impact events were probably collisions with other parent bodies. The collisions fragmented the planetesimals and dispersed the fragments—irons, stony-irons, and achondrites—from the inner parts of the larger bodies. Chondrites were dispersed from the less-heated regions of various parent bodies.

Gallium–Germanium Groups—Evidence for Different Parent Bodies

Trace elements, such as gallium and germanium, are often useful for subdividing meteorites and other rocks. In the case of the iron meteorites, these trace elements reveal intriguing relations. When gallium content is plotted against germanium content (or nickel or other trace elements), clusters of datum points at once become apparent. Four main iron-meteorite groups were thus identified by Goldberg et al. (1951) and by Lovering et al. (1957), and named I through IV. Subsequently, some subdivisions, IIIA and IIIB, IVA and IVB, have been proposed. The groups are further discussed by Anders (1964) and Wasson (1970).

* It requires 40 million years for a body of radius 100 km to cool from 1000° to 300°K (Anders, 1963b).

The gallium–germanium groups are distinguishable not only by composition, but also by crystal structure and history. Goldberg et al. (1951) and others have shown that the crystal patterns differ from group to group, indicating differences in thermal histories. Goldstein and Short (1967) found different cooling times for the groups, indicating different sizes of parent bodies. For example, Groups I and IIIB can be interpreted as samples of cores of parent bodies of radius 200 km, while groups IIIA and IVA have diverse cooling rates and might represent different pockets of iron in bodies of radius 150 to 200 km.

The most important implications of the discrete gallium–germanium groups is that *the iron meteorites have come from different, discrete environments.* Possible interpretations are that the groups represent either (1) some 4 to 11 different parent bodies, or (2) different regions within 4 to 11 parent bodies.

BRECCIATED METEORITES

Breccias are rocks composed of angular fragments of other, crushed rocks. The fragments were cemented together by high pressure to form a new rock. Many stones and some irons are brecciated. Brecciated meteorites are divided into two categories: (1) *monomict breccias,* in which all the fragments are of the same meteorite type, and (2) *polymict breccias,* where fragments of different meteorite types appear.

Monomict breccias were probably formed during violent events on the parent planet—either stresses generated during interior melting or collisions between parent planets.

Polymict breccias give some of the strongest evidence for *collisions* among different parent planets and fragments of parent planets, for this seems to be the only way intimately to mix broken pieces of meteorites formed in totally different environments (Wahl, 1952).

TEKTITES

Tektites are small, unique, and rare glass objects rich in silica (Figure 9-15). The largest have a mass of only 200 to 300 g (about 1/2 lb). Their origin is very mysterious, and as we will see, there has been a great controversy over whether they came from the moon or the earth (O'Keefe, 1963; Faul, 1966).

The high silica content of tektites (about 73 percent SiO_2) is an argument that they come from the earth, where silica-rich sediments abound, and not from the moon, whose igneous rocks are not silica-rich.

On the other hand, Chapman and Larson (1963) showed in wind-tunnel experiments that the tektites had either entered or reentered the atmosphere at

Figure 9-15
A typical southeast Asian tektite, showing bubbles, striations, and ablated tip. (Length, approx. 4 cm.)

hypersonic speeds, causing melting and reshaping. Chapman and Larson's finding was a final proof that tektites were "stones that fell from the sky." This supported the theory of extraterrestrial origin.

Tektites are found only in restricted areas, and each group has a different age. By far the largest and richest tektite field is in Australia and Southeast Asia. The Australasian tektites were formed about 7×10^5 (700,000) years ago. Another group of tektites comes from Africa's Ivory Coast Republic and is about 1.5×10^6 (1.5 million) years old. A group of Czechoslovakian tektites called moldavites (after the Moldau River) are about 1.48×10^7 (14.8 million) years old. The oldest tektites are American, coming from fields in east-central Texas and south-central Georgia, and dating from about 3.4×10^7 (34 million) years ago. A single tektite was also found at Martha's Vineyard, Massachusetts.

If tektites solidified in space, entered the earth's atmosphere, and fell in distinct groups, where did they originate?

The evidence cited above does not prove a genetic relation between tektites and meteorites, even if the tektites were at one time in space. Radioisotopic dating techniques (see the next section) show that tektites differ from meteorites. The most important difference is that tektites have not been in space nearly as long as normal meteorites. Most iron meteorites have been exposed in space for some 10^8 (100 million) years, and most stone meteorites for some 10^7 (10 million) years. These ages give the time since the collision that dispersed the meteorite fragments. But tektite exposure times have been less than 300 years—perhaps only minutes—which is much too short for asteroidal or cometary origin. This restricts their place or origin to the earth–moon neighborhood.

Opinion in the mid-1960s favored the view that the tektites were debris from major impacts on the moon. The debris were supposed to be shot off the moon with such high speed that they escaped. Some large pieces were temporarily caught in orbit around the earth and as they encountered atmospheric drag and spiraled into the earth's atmosphere they dropped fragments in restricted regions (O'Keefe, 1963). This theory has been largely abandoned since tektites have not been found on the moon.

The more successful theory, which has gained more advocates in recent years, emphasizes that several tektite fields are not too far from ancient astroblemes which have about the same age as the tektites. According to this theory, tektites represent jets of fused silica debris shot out through the earth's atmosphere during the astrobleme-forming impacts, only to reenter the atmosphere and fall back in restricted locations (Faul, 1966). A recent review of chemical data indicates that this theory is the most satisfactory explanation of tektites (Taylor and Kaye, 1969). Tektites add evidence that cosmic objects have played a role in complicating the earth's crustal geology.

METEORITE AGES

Age Determinations

Meteorite ages can be determined by study of radioactive isotopes. If a given isotope P (called the parent) is unstable, it decays into a different isotope D (called the daughter). The time for half the P atoms to decay is called the *half-life* for isotope P. If we start out with 1000 atoms of P, after one half-life there will be 500 atoms; after two half-lives, 250; after three half-lives, 125; and so on. If we know that a given mineral crystal originally contained 1000 atoms of element P with a half-life of 1 billion years, and if we measure only 250 atoms in the crystal now, its age would be 2 billion years.

In simplest outline this is the theory of radiometric age determinations. In practice there are many complications. A difficult figure to get is the original number of P atoms. This may be estimated by measuring the abundance of some other stable isotope X whose initial abundance ratio to P is known through theory. Thus the measures are usually not expressed as absolute numbers of atoms, but rather as ratios, P/X and D/X.

The most important concept about radiometric ages is that a number of *different kinds of time intervals* can be measured. If you are told the "age of a rock," you must inquire what kind of age is meant: the time since the rock crystallized, or the time since the actual sample you are looking at was broken off, or the time since the sample was last shocked, or chemically altered, etc. Table 9-4 lists the various age relations that can be determined. They are discussed in more detail below.

Table 9-4
Types of meteorite ages

Type of age and interval measured	Isotopes involved (examples)	Meteorite type	Measured years (years)
Formation interval (nucleosynthesis to gas retention)	^{129}I–^{129}Xe	Stones and irons	6×10^7 to 8×10^8
Solidification age (solidification to present)	^{87}Rb–^{87}Sr ^{235}U–^{207}Pb ^{238}U–^{206}Pb ^{232}Th–^{208}Pb	Stones and irons	4.4×10^9 to 4.7×10^9
Gas retention age (onset of gas retention to present)	^{40}K–^{40}Ar	Stones (irons contain little K)	4×10^8 to 4.7×10^9
Cosmic-ray exposure age (onset of CR exposure to present)	He^3 produced by cosmic rays	Stones	10^5 to 10^8
		Irons	10^6 to 10^9
Date of fall (end of CR exposure to present)	^{14}C	Stones	0.0 to 10^4
	^{26}Al	Irons	0.0 to 10^4 or more

Formation Interval

The formation interval is the interval of time between the creation of certain short-lived parent isotopes and their incorporation into solid bodies that could retain gaseous daughter isotopes (see Chapter 6). *The formation interval is an approximate measure of the time interval from formation of the sun* (or from certain presolar events) *to formation of planetesimals.* The range of variation among measured values of formation intervals indicates that growth of planets required about 2×10^7 (20 million) years (see Chapter 6).

Solidification Age

This is the interval from the solidification of the rock to the present; it is what is usually meant by "age of the meteorites." Solidification ages are used to measure the "age of the solar system" and average 4.5×10^9 (4.5 billion) years. Particularly well known are the *lead–lead* method and the *rubidium–strontium* method.

Gas Retention Age

This is the interval from the most recent degassing of the rocks to the present. The method utilizes gaseous daughter elements, such as argon, which collect in the crystal-lattice interstices. An event such as strong shocking of the rock during a meteoroid collision can cause tiny fractures and heating, which allow the gas to escape. New gas then starts to collect, and the measuring of this gas dates the *shock* event. If the rock was never shocked strongly enough to cause heating, the gas retention age may be the same as the solidification age, although simple leaking of the gas by diffusion, without shock events, may cause the measured gas retention age to be less.

By studying clustering of gas retention ages Heymann (1967) found that a group of black hypersthene chondrites had been shocked and reheated roughly 5×10^8 (500 million) years ago due to an impact on their parent body. Taylor and Heymann (1969) found that all chondrites with short gas retention ages (<2 billion years) show evidence of strong shock and reheating, presumably due to collisions and impacts among the various parent bodies and fragments of parent bodies. Thus the gas retention ages seem to refer in many cases to major, shock-producing, high-energy collisions that fragmented parent bodies and occurred every few 10^8 (hundred million) years during solar-system history.

Cosmic-Ray Exposure Age

Sometimes called a CRE age, the cosmic-ray exposure age measures the duration of the meteorite's exposure to space. Generally this is the interval from the meteorite's first exposure to space (or when it was first within a meter or less of the surface of a larger body) until its impact on the earth. Some meteorites had a multiphase exposure history, first being just under the surface of the parent body and then being exposed directly to space when the parent body fragmented.

Exposure ages may be measured either (1) by utilizing radioactive and stable isotopes created by nuclear reactions with cosmic rays, which penetrate roughly 1 m into the parent body, or (2) by counting microscopic *particle tracks* left in the material by passage of cosmic rays through it. The former method is widely applied to meteorites. The latter method is used for the tektite exposure ages cited earlier. Both methods have also been used on lunar samples.

An important result of studying cosmic-ray exposure ages is that some of them are clustered in time, even more clearly than the gas retention ages. Certain groups of meteorites can thus be identified with certain collision-fragmentation events that occurred in space. These events may represent either the breakup of original parent bodies or the subfragmentation of individual pieces of the original parent bodies. Dates of some of the events are given in Table 9-5.

A puzzle associated with CRE ages is that stones have much lower ages than irons. If they come from the same general source, why have they been exposed for different periods? Eberhardt and Hess (1960) suggested that meteorites erode due to their mutual collisions and collision with micrometeoroids.

Table 9-5
Dates of meteoroid collisions[a]

Meteorite type	Collision dates determined by clustered ages (years before present)	Types of age determination
Bronzite chondrites	5×10^6	Cosmic-ray exposure age
	2×10^7	Cosmic-ray exposure age
Hypersthene chondrites	2×10^7	Cosmic-ray exposure age
	7×10^6(?)	Cosmic-ray exposure age
	5×10^8	Gas retention age
Aubrites	4×10^7	Cosmic-ray exposure age
Hexahedrites	2×10^8(?)	Cosmic-ray exposure age
Medium octahedrites	6×10^8	Cosmic-ray exposure age

[a] See Anders (1963b, 1964) and J. A. Wood (1968) for discussions of the data. The collisions and suspected collisions greater than 10^8 years ago were probably major collisions that fragmented large objects, possibly original parent bodies. More recent collisions may have fragmented only pieces of parent bodies. Taylor and Heymann (1969) have shown that many meteorites give evidence of other major collisions scattered in time between 0.6 and 4.5 billion years ago.

If stones fragment more often than the stronger irons, their outer layers would be exposed more recently, on the average (Dohnanyi, 1970). Since cosmic rays can penetrate only in these outer layers (especially in the outer 20 cm), stones would usually appear to be younger than irons. The history of a meteorite in space might be likened to that of a rock struck by intermittent sledgehammer blows superimposed on a continuous sandblasting.

Date of Fall

This measures the interval from when the meteorite ceased being shielded from cosmic rays (its arrival on earth) to the present. Stones in particular are not identifiable much longer than about 10,000 years on the earth's surface because they are subject to weathering and erosion.

ORIGIN AND HISTORY OF METEORITES

We shall now summarize the diverse and complex evidence discussed above and present some additional evidence. The reader should be able to cite support for most of these statements.

Accretion and chemical fractionation processes in the solar nebula—as discussed in Chapter 6—produced grain-like planetesimals with compositions different from the nebular composition and varying with distance from the sun. Such planetesimals may have ranged in composition from the nearly primordial Type I carbonaceous chondrites to slightly fractionated chondrites. This material was incorporated into *parent bodies,* or planetesimals of asteroidal dimension, up to at least 400 km in diameter and probably larger. Because of both orbital data on meteorites and the chemical data on their parent bodies, most investigators identify the asteroids as the source of most, if not all, meteorites—with the possible exception of carbonaceous chondrites.

Figure 9-16

Collision of two meteorite parent bodies, showing mixed, fragmental nature of resulting debris. One of the parent bodies has a mantle-core structure and the other a "plum-pudding" structure.

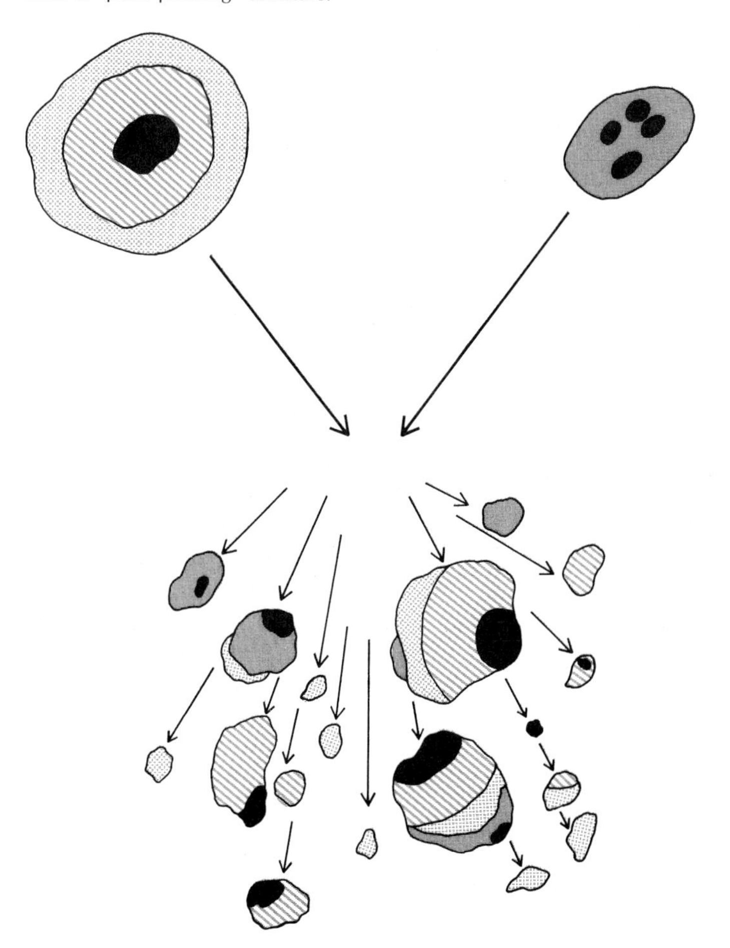

The parent bodies were internally heated. In their outer layers, the chondritic materials were metamorphosed. In their interiors the chondritic materials were destroyed by melting and differentiated into achondritic and iron portions (Figure 9-16). The source of the heat that caused the melting may have been a short-lived radioactive isotope with a half-life of about 7×10^5 years to 10^8 years (Fish, Goles, and Anders, 1960). This isotope may have been created during the formation of the sun. Theorists in the early 1960s considered ^{26}Al as a prime candidate, but it has been ruled out by recent geochemical studies of meteorites and lunar samples. Currently other heating processes are being more seriously appraised.

Anders (1965), Wetherill (1967), and others have shown that the early asteroids would have suffered repeated collisions that would have (1) produced the mass distributions presently observed among meteorites and asteroids, (2) produced the sorts of age distributions observed for the collision events, and (3) produced Hirayama families (see Chapter 8). These collisions must have fragmented the parent bodies and dispersed chondrites, achondrites, and irons into space. Carbonaceous chondrites might have been added from the surface layers or from smaller (cometary?) bodies that never melted. There is some evidence that the number of large (>400 km in diameter) parent bodies fragmented may have been rather small, perhaps as low as half a dozen.

★ QUESTION ★

Discuss gallium–germanium groups in this context.

Figure 9-17 summarizes the meteorite histories according to the above data and views.

Where do meteorites come from? Do they originate in the asteroid belt? Anders (1964), Dohnanyi (1970), and others have favorably discussed this hypothesis. Its advantages are obvious from the above. The only serious objection is that it is difficult to see how the meteorites would get from the asteroid belt to the earth. The collisions among asteroids do not impart enough energy to knock the pieces out of the asteroid belt. Instead, the meteorites may come from asteroids or asteroid fragments that were on Mars-crossing orbits and then were perturbed by Mars. Mars is massive enough to perturb objects out of the inner fringe of the belt and onto earth-crossing, Apollo asteroid orbits.

Öpik (1966) is on the forefront of those who have argued that most meteorites do not come from the asteroid belt. His argument hinges on a reconstruction of the orbital history of meteorites which leads him to the conclusion that "burnt-out" comets produce what we call Apollo asteroids, and that they are the true sources of meteorites (see Chapter 8).

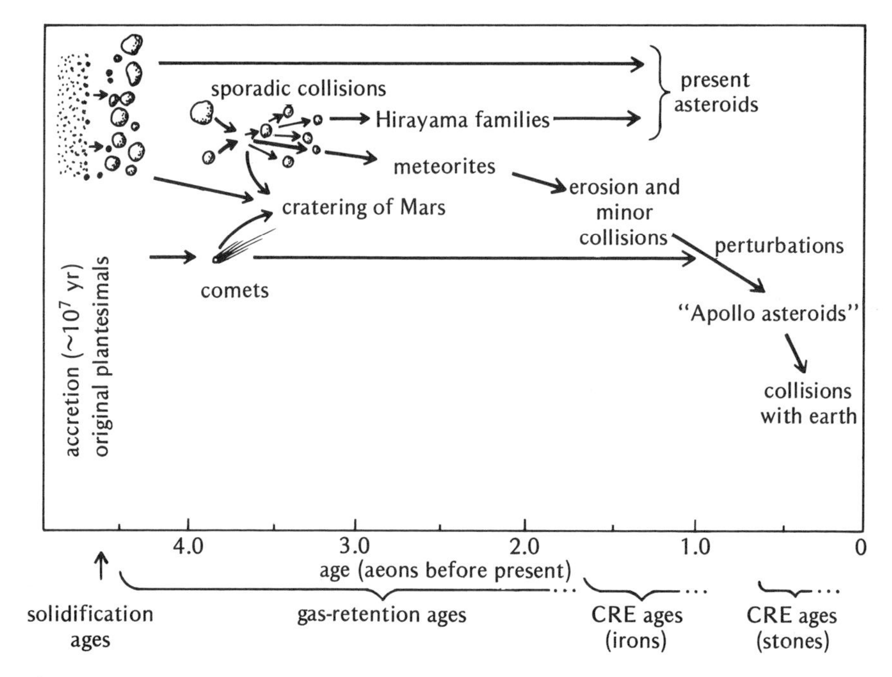

Figure 9-17

Schematic history of meteoritic material, showing interpretation of various types of measured ages.

In any case the meteoritic material is certainly a sample of the interiors of the planetesimals and earliest planetary bodies.

References

Alexander, W., C. McCracken, and J. Bohn (1965) "Zodiacal Dust: Measurements by Mariner IV," *Science, 149,* 1240.

Anders, E. (1963a) "On the Origin of Carbonaceous Chondrites," *Ann. N.Y. Acad. Sci., 108,* 514.

——— (1963b) "Meteorite Ages," in *The Moon, Meteorites, and Comets,* B. M. Middlehurst and G. P. Kuiper, eds., p. 402 (Chicago: University of Chicago Press).

——— (1964) "Origin, Age, and Composition of Meteorites," *Space Sci. Rev., 3,* 583.

——— (1965) "Fragmentation History of Asteroids," *Icarus, 4,* 399.

——— and J. R. Arnold (1965) "Age of Craters on Mars," *Science, 149,* 1944.

Baldwin, R. B. (1963) *The Measure of the Moon* (Chicago: University of Chicago Press).

Beals, C. S., and I. Halliday (1967) "Impact Craters of the Earth and Moon," *J. Roy. Astron. Soc. Can., 61,* 295.

Blander, M., and M. Abdel-Gawad (1969) "The Origin of Meteorites and the Constrained Equilibrium Condensation Theory," *Geochim. Cosmochim. Acta, 33,* 710.

Cameron, A. G. W. (1966) "The Accumulation of Chondritic Material," *Earth Planet. Sci. Lett., 1,* 93.

Ceplecha, A., J. Rajche, and L. Sehnal (1959) "New Czechoslovak Meteorite, Luhy," *Bull. Astron. Inst. Czech., 10,* 147.

Chapman, D., and H. Larson (1963) "On the Lunar Orgin of Tektites," *J. Geophys. Res., 68,* 4305.

Dietz, R. S. (1963) "Astroblemes: Ancient Meteorite-Impact Structures on the Earth," in *The Moon, Meteorites, and Comets,* B. M. Middlehurst and G. P. Kuiper, eds., p. 285 (Chicago: University of Chicago Press).

Dohnanyi, J. (1970) "On the Origin and Distribution of Meteoroids," *J. Geophys. Res., 75,* 3468.

DuFresne, E., and E. Anders (1963) "Chemical Evolution of the Carbonaceous Chondrites," in *The Moon, Meteorites, and Comets,* B. M. Middlehurst and G. P. Kuiper, eds. (Chicago: University of Chicago Press).

Eberhardt, P., and D. Hess (1960) "Helium in Stone Meteorites," *Astrophys. J., 131,* 38.

Faul, Henry (1966) "Tektites Are Terrestrial," *Science, 152,* 1341.

Fish, R., G. Goles, and E. Anders (1960) "The Record in Meteorites, III. The Development of Meteorites in Asteroidal Bodies," *Astrophys. J., 132,* 243.

Ganapathy, R., R. Keays, J. Laul, and E. Anders (1970) "Trace Elements in Apollo 11 Lunar Rocks: Implications of Meteorite Influx and Origin of Moon," *Proc. Apollo 11 Lunar Sci. Conf., 2,* 1117.

Goldberg, E., A. Uchiyama, and H. Brown (1951) "The Distribution of Nickel, Cobalt, Gallium, Palladium, and Gold in Iron Meteorites," *Geochim. Cosmochim. Acta, 2,* 1.

Goldstein, J. I., and J. M. Short (1967) "The Iron Meteorites, Their Thermal History and Parent Bodies," *Geochim. Cosmochim. Acta, 31,* 1733.

Hartmann, W. K. (1970) "Lunar Cratering Chronology," *Icarus, 13,* 299.

Heineman, R., and L. Brady (1929) "The Winona Meteorite," *Am. J. Sci.,* (5th ser.), *18,* 477.

Heymann, D. (1967) "On the Origin of the Hypersthene Chondrites: Ages and Shock Effects of Black Chondrites," *Icarus, 6,* 189.

Krinov, E. L. (1960) *Principles of Meteorites* (Elmsford, N.Y.: Pergamon Press, Inc.).

______ (1963) "The Tunguska and Sikhote-Alin Meteorites," in *The Moon, Meteorites, and Comets,* B. M. Middlehurst and G. P. Kuiper, eds., p. 208 (Chicago: University of Chicago Press).

Levin, B. Y. (1958) "Über den Ürsprung der Meteoriten," *Chem. Erde, 19,* Pt. 3, 286.

Lovering, J., W. Nichiporuk, A. Chodos, and H. Brown (1957) "The Distribution of Gallium, Germanium, Cobalt, Chromium, and Copper in Iron and Stony-Iron Meteorites in Relation to Nickel Content and Structure," *Geochim. Cosmochim. Acta, 19,* 156.

Mason, B. (1962) *Meteorites* (New York: John Wiley & Sons, Inc.).

McCrosky, R. E. (1967) "Orbits of Photographic Meteors," *Smithsonian Astrophys. Observ. Spec. Rept., 252.*

O'Keefe, J. A., ed. (1963) *Tektites* (Chicago: University of Chicago Press).

Öpik, E. J. (1966) "The Stray Bodies in the Solar System. Part 2. The Cometary Origin of Meteorites," *Adv. Astron. Astophys., 4,* 301.

Prior, G. T. (1920) "The Classification of Meteorites," *Mineral. Mag., 19,* 51.

Ringwood, A. E. (1966) "Genesis of Chondritic Meteorites," *Rev. Geophys., 4,* 113.

Shoemaker, E. M. (1963) "Impact Mechanics at Meteor Crater, Arizona," in *The Moon,*

Meteorites, and Comets, B. M. Middlehurst and G. P. Kuiper, eds., p. 301 (Chicago: University of Chicago Press).
Suess, H. E. (1949) "Chemistry of the Formation of Planets," *Z. Electrochem., 53,* 237.
Taylor, G., and D. Heymann (1969) "Shock Reheating, and the Gas Retention Ages of Chondrites," *Earth Planet. Sci. Lett., 7,* 151.
Taylor, S. R., and M. Kaye (1969) "Genetic Significance of the Chemical Composition of Tektites: A Review," *Geochim. Cosmochim. Acta, 33,* 1083.
Uhlig, H. H. (1954) "Contribution of Metallurgy to the Study of Meteorites I: Structure of Metallic Meteorites, Their Composition and the Effect of Pressure," *Geochim. Cosmochim. Acta, 6,* 282.
——— (1955) "Contribution of Metallurgy to the Origin of Meteorites II. The Significance of Neumann Bands in Meteorites," *Geochim. Cosmochim. Acta, 7,* 34.
Urey, H. C. (1967) "Parent Bodies of Meteorites and the Origin of Chondrules," *Icarus, 7,* 350.
——— and H. Craig (1953) "The Composition of the Stone Meteorites and the Origin of the Meteorites," *Geochim. Cosmochim. Acta, 4,* 36.
Vedder, J. F. (1966) "Minor Objects in the Solar System," *Space Sci. Rev.,* 6, 365.
Wahl, W. (1952) "The Brecciated Stony Meteorites and Meteorites Containing Foreign Fragments," *Geochim. Cosmochim. Acta, 2,* 91.
Wasson, J. T. (1970) "The Chemical Classification of Iron Meteorites, IV," *Icarus, 12,* 407.
Wetherill, G. W. (1967) "Collisions in the Asteroid Belt," *J. Geophys. Res., 72,* 2429.
Whipple, F. L. (1966) "Chondrules: Suggestion Concerning the Origin," *Science, 153,* 54.
Wiik, H. B. (1956) "The Chemical Composition of Some Stony Meteorites," *Geochim. Cosmochim. Acta, 9,* 279.
Wood, C. A. (1968) "Statistics of Central Peaks in Lunar Craters," *Comm. Lunar Planet. Lab., 7,* 157.
Wood, J. A. (1958) "Silicate Meteorite Structures and the Origin of the Meteorites," *Smithsonian Astrophys. Observ. Tech. Rept. 10.*
——— (1961) "Stony Meteorite Orbits," *Monthly Notices Roy. Astron. Soc., 122,* 79.
——— (1963) "Physics and Chemistry of Meteorites," in *The Moon, Meteorites, and Comets,* B. M. Middlehurst and G. P. Kuiper, eds., p. 337 (Chicago: University of Chicago Press).
——— (1964) "The Cooling Rates and Parent Planets of Sources of Iron Meteorites," *Icarus, 3,* 429.
——— (1968) *Meteorites and the Origin of Planets* (New York: McGraw-Hill, Inc.).

PLANETARY INTERIORS

AT THIS POINT in our survey of solar-system evolution, let us take stock of our position. Chapters 4 through 6 dealt with the origin of the solar system and showed how the planets must have grown from small primordial planetesimals. Chapters 7 and 8 outlined our knowledge of those small bodies that have survived. Chapter 9 discussed our only physical specimens of the interiors of the small planetary bodies.

Thus the stage is set for a direct confrontation with the problem of planetary structure, which is taken up in this chapter. The internal evolution of the planets led to various geological surface expressions, and the nature of the resulting planetary surfaces will be considered in Chapter 11. After the planets formed, interior and near-surface processes gave rise to gas emissions which altered or produced planetary atmospheres, the topic of Chapter 12. In the presence of the proper gases and liquid water, biochemical processes arose that led ultimately to the formation of living biological systems. The origin of life is the subject of the final chapter. Thus, the remainder of this book deals with the principal planets, not only in their past states of evolution but in the present states to which evolution has brought them.

The purpose of studying planetary interiors is to understand how planets are constituted. In more specific terms we want to know what the *pressure, density, temperature,* and *composition* are at any point in the interior. In addition we want to know how these parameters change as a function of *time* (i.e., how the planet as a whole evolves).

THEORY OF PLANETARY INTERIORS

Contrast with Stellar Interiors

In considering the formation of stars we discussed theoretical models of the interiors of proto-stars and stars. The mathematical theory of planetary interiors is considerably more complex than the theory of stellar interiors because stars are composed entirely of gas, which behaves in a much simpler and better-understood way than a liquid or solid. Fairly simple equations describe the behavior of all gases under the high pressures and temperatures that exist in stellar interiors, but different kinds of liquids and solids have quite different, complex behaviors. For example, the outer part of the earth may obey one equation while the inner high-density core of the earth obeys another, and the interior of Jupiter may obey still different equations.

The equation that relates any substance's *pressure, density,* and *temperature* under various conditions is the *equation of state.* Pressure, density, and temperature are sometimes called the "state variables." The preceding paragraph can be expressed by saying that liquids and solids have more complex equations of state than do gases.

Computer Models of Planets

In spite of the fact that planetary interiors are more complicated than stellar interiors, the basic theoretical treatments are the same. The hydrostatic equation,

$$dP = \rho g \, dz,$$

says that the increase in pressure dP as we descend through a layer is equal to the density ρ of the material in that layer, times the gravitational acceleration g, times the layer thickness dz (z increasing downward). A computer program can be devised that divides a planet into a thousand layers (or some other large number of layers), and then starts iterating down from the surface. At the surface, the pressure is zero (we can neglect the atmospheric pressure in most cases). We know by direct measurement the density of rocks at the surface, at least in the case of the earth and moon, and we can measure or compute the gravitational acceleration. The computer starts with these and multiplies them times the layer thickness dz to get the pressure at the bottom of the first layer. Then the computer inserts that pressure into the assumed equation of state to compute a new density for the rocks at the bottom of the first layer. Computing the new gravitational acceleration that applies at that depth, it then performs a new multiplication to get the pressure at the bottom of the second layer, and so on.

There are two obvious problems in this sort of scheme. For one thing, the equation of state involves the temperature, and we have not yet supplied the temperature to the computer. Usually a theoretical model of the temperature profile in the planet must be supplied, based on measured temperature gradients at the surface and estimates of interior heat sources. Fortunately, however, the density of a rock does not change very much as the temperature changes (in contrast to the behavior of gases). A second problem is that the equation of state must be accurate, and yet no one knows precisely the composition of any planet's interior, so the equation of state in practice has to be guessed at (see the next section).

In spite of these difficulties there are a number of observational facts that guide us in constructing our models. One fairly obvious one is that we know the *total mass* of each planet. Therefore, when the computer reaches the bottom of the last layer (i.e., the center) it must have "used up" precisely the mass of the planet. If there is mass left over, the densities computed have been systematically too low; if the allotted mass is used up in the 999th layer, the densities were too high (and the model would predict a hollow cavity at the center that would immediately be filled in a real planet by a sudden collapse). To solve this problem, an adjustment would be made to the equation of state and then the computer program could be run again. A number of other similar checks can be made—we shall discuss these later.

Figures of Planets

★ QUESTION ★

Why is a planet round?

Answer

Because it obeys the hydrostatic equation.

The hydrostatic equation describes the equilibrium-pressure distribution in a medium where the pressure and temperature are great enough to cause failure of the normal elastic properties. Under these conditions, the material deforms like a viscous, plastic fluid, and the pressure is distributed in all directions (as in the hydraulic braking system of a car), and the planet deforms as if it were plastic or liquid, approaching an equilibrium figure. The *figure* of a planet is its detailed shape.

We saw in Chapter 8 that the pressure inside the smaller asteroids is less than the strength of the rocks composing the asteroids. In such a case, we saw, the rock resists being deformed into a spherical shape, and we noted that small asteroids are indeed not round.

Planets are larger, however, and except in the outermost few kilometers the hydrostatic equation does apply. Thus self-gravitation has squeezed the earth into a round shape. It would be nearly spherical, except that rotation creates centrifugal forces radially outward in the equatorial plane, causing the familiar equatorial bulge. The flattened figure of the earth approximates the geometrical shape called an *oblate spheroid.*

Even the oblate spheroidal shapes of planets depart from equilibrium by small amounts in response to tides, interior processes, irregular mass distribution, and other phenomena. Analysis of figures of planets thus help to determine interior conditions. The earth is very slightly "pear-shaped," as was discovered by tracking artificial satellites during the International Geophysical Year, in 1957–1958.

EQUATIONS OF STATE AND DIFFERENTIATION

Rocks and Minerals

The three types of planetary samples so far obtained—from the earth, the moon, and meteorites—indicate that the terrestrial planets are composed principally of silicate and metallic compounds.* Such compounds solidify into different forms known as *minerals*. Assemblages of different minerals make up *rocks*. More detailed considerations of rock and mineral types will be given at the beginning of Chapter 11.

To deduce the conditions inside planets, we must have equations of state that describe not only relations among pressure, density, and temperature for each rock type, but also describe any *changes of state*—e.g., melting and phase changes—in the constituent minerals.

Melting Relations

Measurements in drill holes show that the temperature increases with depth inside the earth. The same probably applies to all other planets, since planet interiors contain primordial heat and radioactive materials that produce heat. If the interior temperature is high enough, rocks or metallic materials could be molten. In the case of the earth, for example, seismic evidence indicates that most of the central core is liquid, probably liquid iron. On the other hand, the high pressures at planetary centers may compress material that would otherwise be molten, forcing it back into the solid state. The innermost part of the earth's core, called the *inner core,* has been found to be solid. Any equation of state used in modeling planets must thus take into account possible sudden transitions between solid and liquid states at various points in the interior. In other words, given the planet's composition, we must know at just what pressure, density, and temperature the material will melt so that the computer can be instructed to change to the liquid state if the critical conditions are encountered while "building" the planetary model.

Phase Changes

A different and more vexing problem arises because rock-forming chemical compounds may exist in different solid forms at different pressures and temper-

* As we shall see, observations indicate that the giant planets have quite different compositions.

atures. That is to say, if the temperature and pressure change, a mineral with a certain density and crystal structure may change phase to become another mineral with the same composition but different density and crystal structure. An example is carbon's existence as graphite or diamond.

Such phase changes obviously affect the structure of a planet. They could cause a systematic layering—a structure consisting of shells of different mineralogy with discontinuities at the shell interfaces. Thus a planet could have a complicated internal structure even if it were chemically uniform.

How can we determine accurate equations of state describing the phase changes among silicate and metallic compounds? Some indications of these can come from *laboratory experiments*. If we could simulate conditions inside the earth, we could discover empirically how the silicate rocks behave. The highest *sustained* pressures that can be reached in the laboratory are in the range 3 to 5×10^{11} dynes/cm^2 (3 to 5×10^5 bars, where 1 bar $\simeq$ 1 atmosphere = the atmospheric pressure at sea level) (Owen and Martin, 1966). Such pressures correspond to depths of roughly 1200 km in the earth—less than one fifth of the way to the center.* Because temperatures are high at such depths, and because it is difficult to produce both high pressure and high temperature at the same time in the laboratory, we lack accurate simulation of silicates at such depths in the earth. The pressure at the earth's center is about 4×10^{12} dynes/cm^2, some 10 times higher than that reached in the laboratory.

To predict conditions near the center of the earth and other planets, we must rely on theoretical equations of state that can be derived from principles of physical chemistry and solid-state physics.

Differentiation

Besides melting and phase changes, a third very important process can lead to a layering of planetary interiors. Differentiation refers to any process by which homogeneous material becomes divided into masses of different chemical composition and physical properties. For example, if a rock mass melts and forms a homogeneous magma (molten rock), different minerals may crystallize at different temperatures as the mass cools. Under appropriate conditions, the final mass may be differentiated into minerals of different composition. As they form, the heavy minerals may sink and the light rise to the surface, just as happens in the smelting of metals. Complete melting is not necessary for differentiation. If even partial melting occurs, fluids can move from one region to another, altering compositions. When we ask if the moon has differentiated, we are asking in essence whether there is a low-density crustal layer overlying a higher-density interior of different composition, or at least whether local regions have compositions different from the moon's bulk composition.

* Higher pressures can be generated in instantaneous shock or explosive experiments. These give limited insight into the phases that may exist at equilibrium under the higher pressures still deeper in the earth.

Phase Changes in the Earth

Until the 1960s the topic of phase changes in the earth was clouded by controversy. Seismology had clearly shown a layered structure, but some investigators thought that the layered structure was primarily the result of phase changes, while others thought it was primarily the result of chemical differentiation into high- and low-density minerals of different composition.

The truth seems to be emerging slowly, and as often happens it seems to lie somewhere between the extremes. It has long appeared nearly certain that the earth has strongly differentiated. Large parts of the interior have evidently been molten, with the result that molten iron and other dense metallic constituents drained toward the center while the lighter, silica-rich minerals segregated toward the crust. Thus much of the earth's layered structure is due to chemical separation of compounds.

On the other hand, it appears virtually certain that some phase transitions occur within the silicate-dominated regime of the earth's outer layers. Ringwood (1956; Ringwood and Major, 1966) and others have conducted a long series of experiments demonstrating phase changes that would produce density discontinuities of about 10 percent at depths of 400 to 500 km in the earth. This corresponds closely to a discontinuity observed by seismic methods (the "20° discontinuity" discussed later in this chapter).

The famous Mohorovičić discontinuity, which separates the crust and mantle, may involve both chemical and phase discontinuities in the outer 40 km of the earth (Wyllie, 1963).*

Degeneracy in the Giant Planets

Simple calculations with the hydrostatic equation show that the pressures in the central parts of the giant planets are extremely high. Models of Jupiter give central pressures of 1.1×10^{14} dynes/cm^2, hundreds of times greater than can be sustained in the laboratory. Under such extreme pressures, the electron

* Perhaps the reader has already grasped the fact that the study of planetary interiors is laced with uncertainty and difficulties. If not, the following note by the geophysicist Birch may make the situation clearer:

Unwary readers should take warning that ordinary language undergoes modification to a high-pressure form when applied to the interior of the earth: a few samples of equivalents follow:

High-pressure form	*Ordinary meaning*
Certain	Dubious
Undoubtedly	Perhaps
Positive proof	Vague suggestion
Unanswerable argument	Trivial objection
Pure iron	Uncertain mixture of all the elements

swarms of the individual atoms are squashed together, and the atomic structure is destroyed. Many or all of the electrons may be freed to move around (as happens in a metal) while the remaining ionized atoms or nuclei attempt to locate themselves in some sort of crystal-lattice pattern. This phenomenon is called *pressure ionization,* and the electrons are said to be partially or completely *degenerate,* depending on whether the atoms are partially or completely ionized.

Since interior pressures in giant planets are beyond the reach of laboratory experimentation, we have to rely on theoretical equations of state. There is one fact in our favor, however. The very low mean densities of the giant planets (less than that of water in the case of Saturn!) imply that they are rich in the simple element hydrogen, which is fairly easy to treat theoretically.

Mean densities are thus a useful tool as quick indicators of planetary interior compositions. Figure 10-1 shows mean density vs. mass for many planetary bodies. If all planets were composed of a single material, the data points would form a smooth curve, with higher masses at higher densities due to compression of the material. Clearly this is not the case. In particular, the giant planets and some of their satellites have much lower mean densities than terrestrial rocks.

In the early years of this century, it was thought that the giant planets might be gaseous throughout. However, the English geophysicist Jeffreys (1924) showed that they surely would have cooled to a nongaseous state, and he suggested that Jupiter is a small, solid planet surrounded by a vast hydrogen-rich atmosphere. Wildt (1938) computed a more realistic Jupiter model that had a high-density inner core of mean $\rho \simeq 5.5$ g/cm^3 (the same as for the earth as a whole), an inner mantle of ice, and an outer mantle of frozen hydrogen. It was realized that Jupiter has such a strong gravity field that it could retain all its original complement of even the lightest gas, hydrogen. Since hydrogen is by far the most abundant material in the universe, attention focused on it. DeMarcus (1958) used theoretical equations of state for pressure-ionized hydrogen and helium to calculate models of Jupiter and Saturn. He found that Jupiter consists of at least 78 percent by weight of hydrogen, distributed in various pressure-modified forms, and that densities greater than 5 g/cm^3 would appear near the center of Jupiter. Observations of Jupiter's heat flow (see below) have led to refinements of these models.

Uranus and Neptune are much denser than theoretical pure hydrogen planets, of the same mass, as can be seen by studying Figure 10-1. Therefore, it is believed that Uranus and Neptune must be considerably richer in heavier elements than Jupiter or Saturn. According to models by Ramsey (1967), free hydrogen in the mantles of these planets is in the range of 10 to 15 percent by mass. Frozen methane (CH_4) and ammonia (NH_3) may be important constituents.

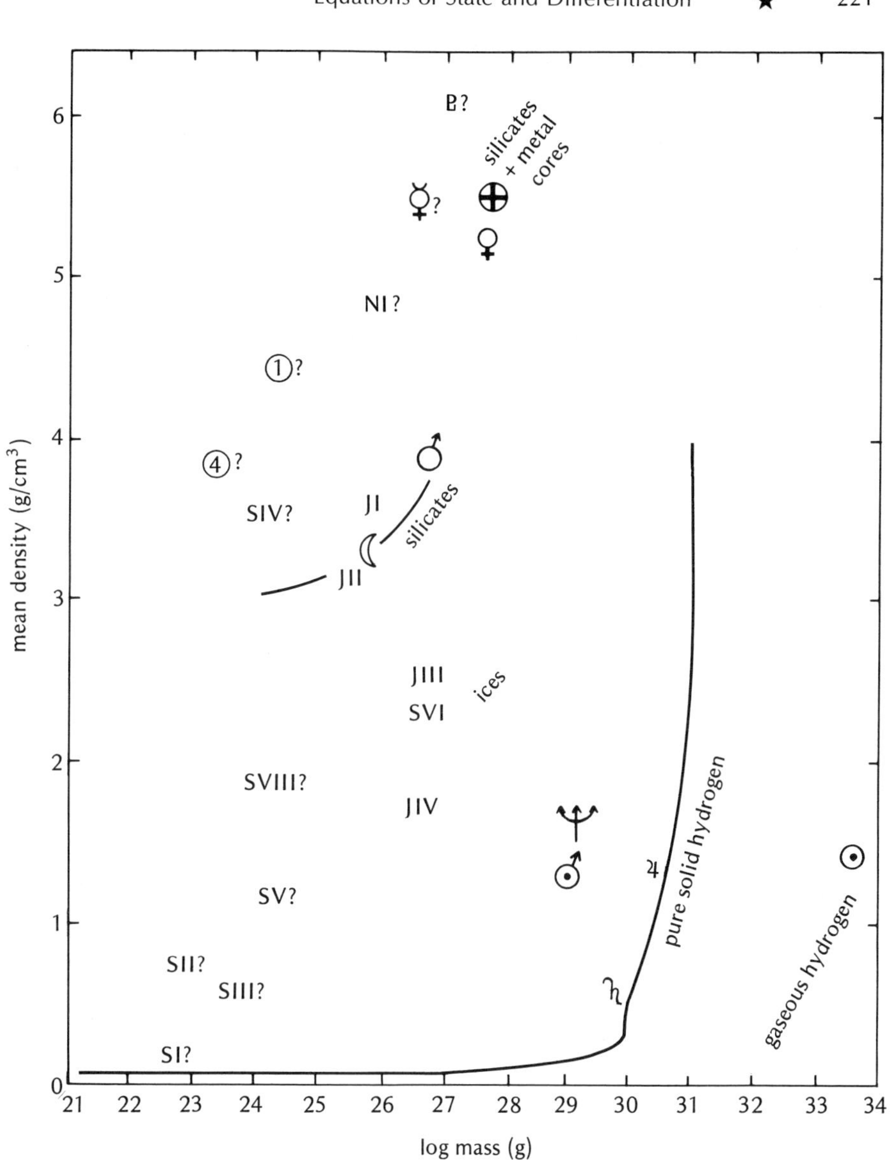

Figure 10-1

Mass versus density for various bodies in the solar system. The two curves show theoretical positions of bodies composed of pure solid hydrogen (after data given by Wildt, 1961) and a mixture of silicates. The fact that various planetary bodies do not fall on a single curve proves the diversity of compositions among the planets.

OBSERVATIONAL CHECKS ON PLANETARY MODELS

If we computed a model of a given planet according to the scheme outlined earlier, we would use as many as possible of the following 11 observational quantities as additional inputs or boundary conditions to ensure accuracy: (1) the planet's mass, (2) its radius, (3) its geometric oblateness, (4) the form of its gravitational field, (5) its moment of inertia, (6) its rotation rate, (7) the composition of its surface layers, (8) its surface temperature and heat flow, (9) the composition of neighboring planets and meteorites, (10) the form of its magnetic field, and (11) its seismic properties.

Mass

The mass of a planet is determined from its gravitational effect on other bodies. Until the 1960s, the "other bodies" were other satellites, planets, or an occasional comet or asteroid that passed close enough to the planet to suffer an observable gravitational acceleration. Because interplanetary distances are large, it was difficult to determine masses accurately, especially in the absence of satellites.

With the advent of space exploration, tremendous improvements in determination of planetary masses have been possible. Space probes near the moon, Venus, and Mars have been tracked with exceedingly great precision by means of their radio signals. Masses derived from such tracking are accurate to about 1 part in 10,000!

Pluto remains the planet whose mass is least certain. The mass of Mercury is also quite uncertain because Mercury is so close to the sun that it is difficult to observe. A space probe to Mercury is being planned for the 1970s and it will hopefully improve this datum. Present indications are that Mercury has a higher mass than expected for its size, leading to the conclusion that it has a higher concentration of iron and other heavy elements than the other planets.

Radius

There are a number of difficulties in determining radii of planets accurately. One of the most common is the *apparent* increase in radius due to an atmosphere. Venus, Jupiter, Saturn, Uranus, and Neptune all have opaque, cloudy atmospheres. In the case of Mars, photographs taken in blue light give a slightly larger radius than photographs taken in red light, since blue light is sensitive to atmospheric features, while red light reveals the Martian surface. The question then becomes: How much of the observed radius must be subtracted to allow for the thickness of the atmosphere, thus leaving the radius of the solid part of the planet?

Another problem is limb darkening. If the surface or cloud layer is

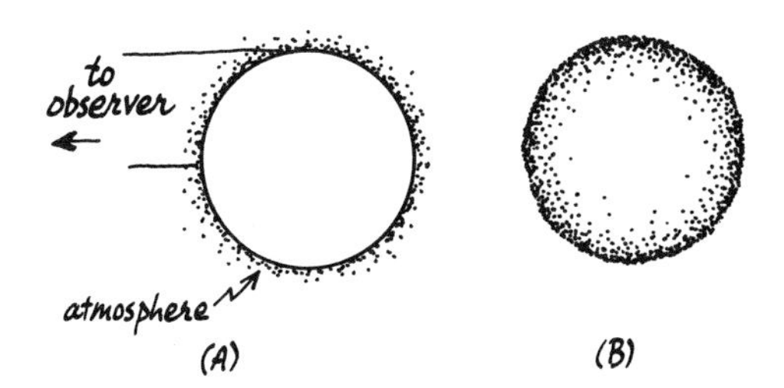

Figure 10-2
Effect of limb darkening produced by atmospheric absorption. A: Side view; B: observer's view. Limb darkening hampers measures of the planet's radius and hence mean density.

overlain by a thin layer of absorbing gas, as shown by Figure 10-2, we may be able to see straight down to the surface when we look at the center of the planet's disk, but we may find that the limb, or edge of the disk, where we look through a long tangential path, looks dark because so much light is absorbed by the gas. This limb darkening may make it difficult to tell just where the edge really is, as in the case of Jupiter.

Space probes and ground-based radar ranging have greatly added to our knowledge of planetary radii. Spacecraft have landed on the moon, Venus, and Mars, giving measures of the distance from the solid surface to the center of mass (the tracking data give the path with respect to the center of mass, which may be slightly offset from the center of the geometric figure). Closeup photography from spacecraft and *occultation measurements* when spacecraft pass behind the edge of the planet also give very accurate data on planetary radii. Flights near Mercury and passages through the satellite systems of outer planets are anticipated during the 1970s.

Geometric Oblateness

Rotating planets are oblate spheroids. Their equatorial diameters a exceed their polar diameter b. The oblateness, or flattening, is a measure of this difference:

$$\text{geometric oblateness} = \epsilon = \frac{a-b}{a}.$$

This ratio can be measured on photographs of the planets or (once we reach the planet) by the more accurate methods of geodesy. Earth-based photographs are difficult to use because of the small size of the planetary images (as in the case of Mercury) or problems with atmospheres (e.g., limb darkening as in the case of Jupiter).

The oblateness is a useful quantity because it reflects the planet's internal

Table 10-1
Value of k

Idealized cases		Observed values	
Hollow sphere	0.667		
Homogeneous sphere	0.400		
		Moon	0.40
		Mars	0.38
		Earth	0.331
		[a]Neptune	0.30
		[a]Uranus	0.30
		Jupiter	0.25
		Saturn	0.22
		Sun	0.06
Mass concentrated entirely at center	0.000		

[a] Neptune and Uranus values from Ramsey (1967).

properties, such as mass distribution or departure from hydrostatic equilibrium. It is possible to calculate the value of ϵ that will characterize a planet in equilibrium, given the planet's mass, size, rotation period, and its mass distribution. We can calculate values that would correspond to idealized cases, such as a homogeneous sphere, and then list the observed values to see how the planets compare. Rather than comparing values of ϵ, it is instructive to talk in terms of the more familiar *moment of inertia*. Theoretically, we may convert an observed ϵ into an estimate of the central concentration of mass in the planet, and hence calculate a *theoretical* moment of inertia I around the rotation axis (this is not necessarily the true I; see below). The moment of inertia of a near-spherical body of mass M and radius R may be expressed as $I = kMR^2$, where k is a numerical coefficient. For a homogeneous sphere it takes the familiar value $\frac{2}{5}$. Table 10-1 compares some observed results and shows that the smaller planets, as expected, are most nearly homogeneous, whereas larger bodies have higher concentrations of mass toward their centers.

Dynamical Ellipticity and Gravity Field

If a planet departs from spherical symmetry, either by being oblate or having an asymmetric mass distribution, or in some other way, its gravitational field will not be spherically symmetric. A nearby satellite (either natural or artificial) of such a planet will respond to the asymmetries by precession of its orbit and by departing slightly from a true Keplerian orbit. An example is the Mariner 9 spacecraft, which, after entering orbit around Mars, was found to have a variable period due to perturbations caused by the irregular shape of Mars. By measuring the satellite's motions, therefore, it is possible to estimate the departure from spherical symmetry of the mass inside the planet.

This measurement is independent of the geometric oblateness but can be usually expressed as a calculated oblateness, called *dynamical ellipticity.* If the planet is in equilibrium and axially symmetric, the geometric oblateness should be the same as the dynamical ellipticity. However, the values are often not the same, and this means that the gravitational field is revealing complexities in the planet's internal configuration. Examples of such complexities are the lunar "mascons," discussed below. Departures from spherical symmetry are often given in terms of a series of numerical coefficients known as J-values.

Moment of Inertia

The moments of inertia discussed above are calculated on the assumption that the planet's interior is axially symmetric. However, if the planet is nonsymmetric, the true moment of inertia may depart slightly from the theoretical value listed in Table 10-1. The true moment of inertia is what determines the response of a spinning body to an external torque. Such a response is always a wobble known as *precession.* For example, a spinning top wobbles (precesses) because of the torque resulting from the earth's gravity. The only significant torques that can affect a spinning planet come from celestial bodies such as the sun or a satellite. The sun and moon produce torques on the earth, causing it to describe a precessional wobble that takes 26,000 years to complete.

Fuller discussions of the application and interrelations of various geometric and dynamical constants are given by Wildt (1961) and Kaula (1968).

Period of Rotation

Rotation periods must be known in order to interpret data such as moment of inertia and ellipticity. One might expect that the rotations of the planets were determined long ago by simply looking through the telescope, but the problem is not so simple. To determine rotation visually, one must be able clearly to see surface markings. This is true only for the moon and Mars. In the cases of Jupiter and Saturn, clear markings can be seen, but the observed rotation is atmospheric and probably somewhat different than that of the underlying planet. Venus lacks visual markings because of its opaque white clouds, and Mercury is so close to the sun that it is difficult to see clearly. The other planets present such small disks that markings are very difficult to see (see Chapter 2).

Other methods of determining rotation periods of very distant planets are radar and spectroscopy. If the disk can be resolved, radar or spectroscopic data can reveal by means of Doppler shifts the recession velocity of one limb of the planet and the approach velocity of the other, thus measuring the spin rate.

Radar is the more sensitive of the methods if the planet is not too distant (see Chapter 2).

If no disk is visible (as in the case of Pluto or the asteroids), photometry can be used to determine the rotation. Assuming that there are surface markings (such as large craters, lava flows, or clouds) one hemisphere of the planet may have a different average albedo than the opposite hemisphere. In this case, rotation will be revealed by a periodic dimming and brightening (see Chapter 8).

Surface Composition

The composition of the rocks accessible at the surface is a guide to the composition of the interior. For example, on the earth we know that the crustal rocks have densities around 3 g/cm^3 and tend to be rich in silica; yet rocks that appear to come from the deepest sources (e.g., ejected from the volcanic Kimberlite diamond pipes in Africa) tend to be denser, poor in silica, and rich in heavy elements. The earth's bulk density is 5.5 g/cm^3. We thus conclude that the deep interior of the earth is richer in heavy elements, such as iron, than the surface.

The rock samples returned from the surface of the moon have a density of about 2.4 to 3.0 g/cm^3—not much less than the moon's bulk density, 3.34 g/cm^3. This is one of several indications that the moon is relatively homogeneous and lacks a large, dense core.

Jeffreys (1924) noted that under certain conditions the ratio of surface density to bulk density, ρ_s/ρ, goes as the ratio of the moment-of-inertia coefficient k to $\frac{2}{5}$. This gives another quick check on interior conditions.

Surface Heat Flow

Suppose that on a certain planetary surface we measured a certain rate of heat flow from the interior and suppose that upon measuring the radioactivity of the surface rocks we found that a layer of those rocks only 30 km thick would provide sufficient heat from radioactive decay to account for the measured heat flow. We would have to conclude that the radioactive elements were strongly concentrated near the surface. This is just the situation on the earth's continents, and similar measures on other planets should be useful in defining interior heat sources and thermal conditions. (See pp. 258–259.)

Composition of Neighboring Planets and Meteorites

Since all the planets must have formed out of a single nebula originally of cosmic composition, any adequate theory of the origin of the solar system must explain differences in composition from planet to planet. Why should the earth's neighbor Venus have no apparent water while the earth is rich in water? Why should the moon be less dense than Venus and the earth? As was shown by Figure 10-1, comparison of planetary densities reveals striking differences in composition.

The present state of knowledge is woefully inadequate to explain all these differences, but contemporary theories of planet formation are beginning to give clues. Chemical processing and fractionations in the solar nebula should be kept in mind in attempting to determine the interior compositions of the planets.

We can also be guided by meteorites, which are thought to be samples of the interior of small planetary bodies. They may include our only foreseeable physical specimens of planetary centers. It was largely because of knowledge of iron meteorites that geophysicists came to believe that the high-density core of the earth is composed of iron rather than some hypothetical high-density phase of silicate material.

Magnetic Field

The presence or absence of a magnetic field is in itself an indicator of the interior conditions in a planet, as we shall see later. Probably a magnetic field is an indication of a fluid core. Magnetic studies of planets without intrinsic magnetic fields are as valuable as studies of planets with fields from the point of view of planetary interiors. If the planet has its own strong field, as does the earth, the interaction between that field and the field associated with the solar wind occurs some distance out from the planetary surface. A "bow shock" is formed in the streaming solar wind field, and the magnetic field lines are deflected. A sort of shell thus insulates the earth's own field—the magnetosphere—from that of the passing solar wind. In the case of a planet without an intrinsic field—such as the moon—the magnetic field associated with the solar wind moves directly up to and into the planet. Deflections in the field lines reflect not an interaction with a magnetosphere but an interaction with the interior of the planet. The first magnetometers placed upon the moon in 1969 made important measures that allowed an estimate of the lunar interior conditions. Sonett et al. (1971) found from these data that the moon is solid throughout with a central temperature probably not over 1200°K. (See p. 262.)

Seismic Properties

Seismic observations give the most useful experimental data regarding the earth's interior. During the first Apollo flights in 1969, the first seismometers were deployed on another planetary body, opening the way to planetary seismology. Seismology permits us to measure positions of layers inside a planet, determine their densities, and look for special structures such as cores or mass concentrations nearer the surface. Such measures sketch the geometry of the planet's interior and put constraints on the composition. Seismology is considered in more detail in the next section.

SEISMOLOGY AND EARTHQUAKES

The Seismometer

An earthquake is a series of vibrations, swayings, or sudden jiggles of the ground. The jiggles can be so small as to be scarcely detectable; rarely, they can knock over buildings. Imagine that we wanted to record such jiggles; how could we do it? Suppose we suspended a 100-lb weight on the end of a 50-ft wire in a quiet room, and attached a scribe to the bottom of the weight so that it just touched a tabletop. Then if the earth suffered a sudden horizontal jolt, the scribe on the bottom of the pendulum would mark a line in the dust (which would have accumulated on the table by the time of the first major earthquake). This would be a record of the earthquake; the length of the line would measure the violence of the jolt. This is basically the principle of the seismometer, except that the pendulum must be designed so that its own natural swinging frequency will not be confused with the earthquake's vibrations; of course, modern seismometers are very sophisticated and much less bulky than the apparatus suggested above. Television witnesses of the moon landings will recall that the seismometers deployed by the astronauts were relatively small.

Earthquakes

What causes earthquakes? The earth is a dynamic body although its evolution is very slow. To drive an evolving system energy is required, and in the case of the earth this energy is thermal energy (i.e., heat). The earth continually tries to readjust its temperature distribution; radioactivity produces new heat while the interior as a whole tries to cool.

The earth might be compared to a house cooling in the night; occasionally it creaks or pops. Because of contraction or expansion or slow movements of materials, stresses build up in the solid outer part of the earth. The rock may stretch elastically but if the stress is too great, the rock may fail, like a twig bent too far. The sudden splitting and slippage of the rock to a new position of minimum stress is an earthquake. This process is called faulting, and any rock fracture along which there has been an offset motion is called a *fault*. Many of California's earthquakes are caused by slippage along the famous San Andreas fault. An example is the famous San Francisco earthquake of 1906, whose aftermath is shown in Figure 10-3.

Figure 10-3

Destructive effects of the San Francisco earthquake of 1906. A major active fault in the earth's crust passes near the downtown urban center. (S. Larson.)

A less important source of earthquakes is volcanic activity at the earth's surface. The spewing forth of billions of tons of lava during a volcanic outburst may be locally accompanied by vibrations and shocks that can literally shake the ground to pieces.

A typical large earthquake may release 10^{24} ergs of energy. The annual release of earthquake energy is estimated to be about 5×10^{24} ergs, most coming from a very few large quakes.

The point inside the earth where the earthquake occurs is called the *focus*. The point on the earth's surface immediately above the focus is called the *epicenter*. The greatest destruction of life or property in a major earthquake occurs if a populated area lies near the epicenter.

From an ecological point of view it is depressing to note that by dropping one of the earliest atom bombs on a city in 1945, we produced casualties comparable to the largest natural disasters. Today's nuclear weapons are hundreds of times as energetic as the Hiroshima bomb. Man is just beginning to be able to tamper with energies big enough to affect a substantial portion of the planet. Yet the earth can occasionally still dwarf us. Warfare in Vietnam killed some 45,000 Americans and several hundred thousand Vietnamese in the course of 10 years; yet 70,000 persons died in a single day in the May 31, 1970, Peruvian earthquake, and roughly 150,000 Japanese died in the 1923 Tokyo earthquake. Table 10-2 lists some of history's most energetic earthquakes and volcanic eruptions.

Not all earthquakes are disastrous. The lower the energy, the more common the earthquake. "Garden variety" earthquakes cause a slight vibration of the ground or a rattling of windows and walls. Vibrations detected by seismographs range in energy all the way down to tremors produced locally by

Table 10-2
Selected destructive earthquakes and volcanic eruptions

Date	Location	Event[a]	Log energy (ergs)	Deaths (approx.)[b]
Aug 24 79	Pompeii	V		15,000
Feb 2 1556	Shensi, China	E		800,000?
Dec 16 1631	Resina, Italy	V		18,000
Nov 1 1755	Lisbon, Portugal	E	~24.5	60,000
Feb 5 1783	Calabria, Italy	E		40,000
1815	Subawa Island, E. Indies	V		75,000
Oct 18 1828	Sanzyo, Japan	E	21.8	1,443
Mar 26 1872	Owens Valley, California	E	⩾23.8	~36
Aug 26 1883	Krakatoa	V		37,000
May 8 1902	St. Pierre, Martinique	V		29,000
Apr 18 1906	San Francisco	E	23.7	600
Jan 31 1906	Colombia	E	24.8	?
Dec 28 1908	Messina, Italy	E	22.7	80,000
Jan 3 1911	Tien Shan	E	24.0	?
Dec 16 1920	Kansu, China	E	24.1	~150,000
Sep 1 1923	Kwantō, Japan	E	23.8	~150,000
May 18 1940	El Centro, California	E	22.6	7
Feb 29 1960	Agadir, Morocco	E	20.0	~12,000
Jul 26 1963	Skopje, Yugoslavia	E	20.4	1,200
Mar 27 1964	Alaska	E	24.1	114
May 31 1970	Peru	E	23.0	~70,000
Feb 9 1971	San Fernando, California	E	~21.0	62
Aug 6 1945	Hiroshima	Atom bomb	20.9	78,000
1969–1970	Total annual lunar seismic activity		~14	—

[a] V, volcanic eruption; E, earthquake.
[b] Various sources; see Tazieff (1964). Figures include all related destruction, fire, etc.

wind, traffic, etc. Such tremors are not detectable by humans. The smallest earthquakes are called *microseisms.*

Studies that utilize the seismic signals from naturally occurring earthquakes are called *passive seismology.* In order to better analyze the probing seismic signals, it is necessary to know the exact time of initial shock and the exact location of the focus; for these purposes, man-made explosions are used as sources. Such studies are called *active seismology.*

Passive seismology has shown that earthquakes are concentrated in belts of activity along geologically young areas of mountain-building activity, island arcs, and often in conjunction with volcanic activity. The geographical distribution of earthquakes (Figure 10-4) reveals the activity pattern of the earth's crust. We will discuss the significance of this geographical distribution later in this chapter. The depth distribution of earthquakes, shown in Figure 10-5, reveals provocative information about the interior structure of the earth. Earthquakes are not distributed at random depths but are concentrated in certain depth zones with peak activities near 100, 200, 375, and 600 km (Ritsema, 1954). Seismological tradition divides earthquakes into shallow-focus (down to 70 km), intermediate-focus (70 to 300 km), and deep-focus (300 to 700 km) groups, but the physical significance of these groups is uncertain. Stresses or weaknesses are apparently concentrated at certain depths, and the absence of earthquakes below 700 km implies that stresses are relieved by faulting only in the outer part of the earth.

"Moonquakes"

In contrast to the earth, the moon is disturbed only rarely by natural earthquakes (sometimes called "moonquakes"). Microseismic background activity is absent, allowing very sensitive seismographs to be used on the moon. During the first year of lunar observation, major lunar quakes rated only 1 to 2 on the well-known Richter scale, compared to ratings of 5 to 8 for occasional major terrestrial quakes.

A totally unexpected finding of the first lunar seismic experiments was that the tremors caused by impacts (including man-made impacts) did not have the sequence of wave arrivals familiar from terrestrial impact experiments, but showed a mixed series of vibrations that persisted as long as an hour. This may reflect heterogeneous structure of the near-surface layers of the moon, causing multiple scattering of the seismic waves from buried rock fragments and debris, with additional waves generated by fallback of ejecta from the impact (Latham, 1971).

Another source of moonquakes is tidal strain in the moon. This was discovered when the first lunar seismographs registered tremors grouped near the time of the moon's passage through perigee, a time when tidal forces induce the maximum flexing of the moon's body. Lunar quake swarms have the fur-

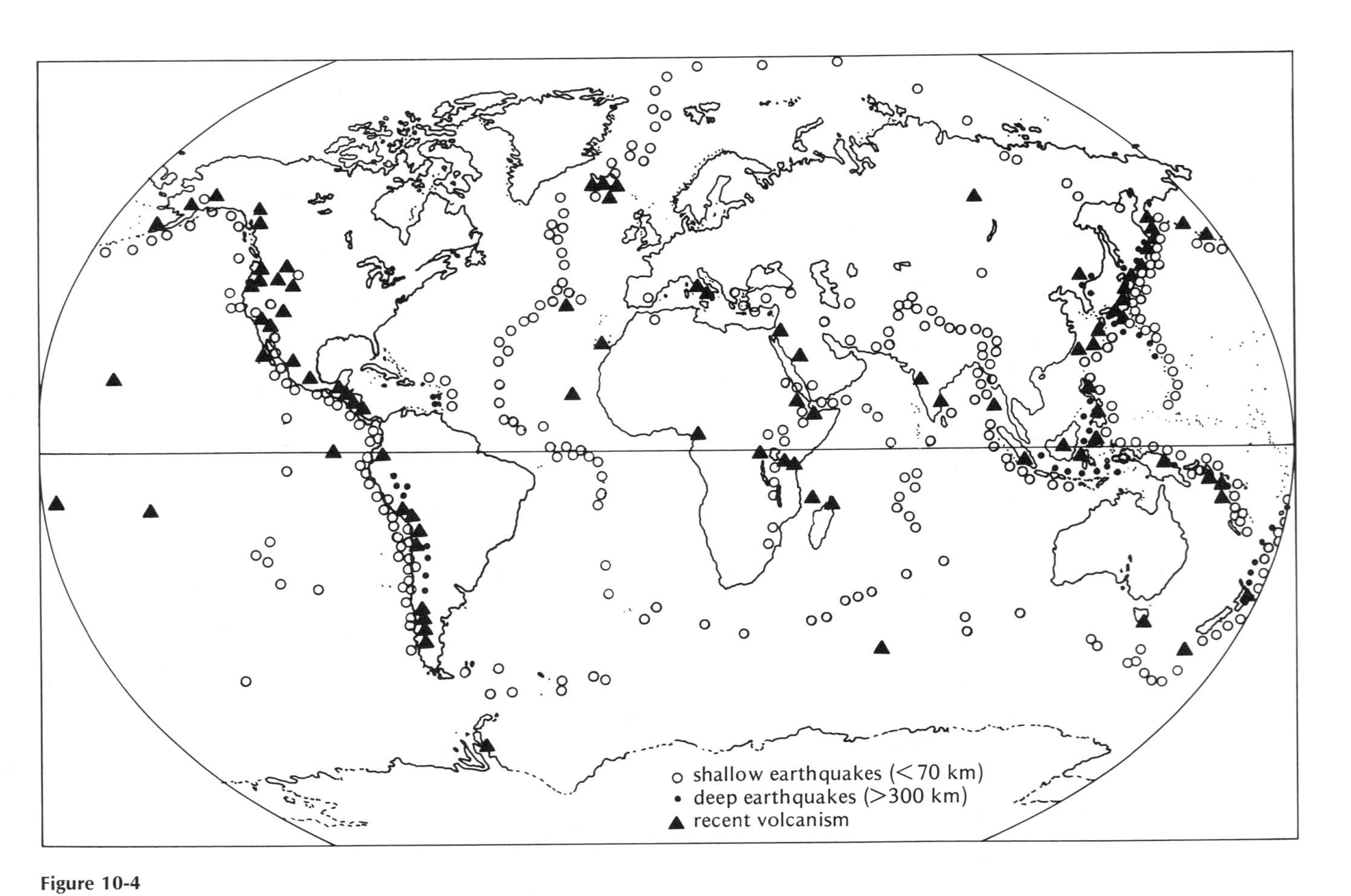

Figure 10-4

Distribution of earthquakes and volcanism on the earth. Note that deep earthquakes usually occur on the inward side of island arcs, that shallow earthquakes outline the oceanic ridge system, and that recent volcanism is usually associated with earthquakes.

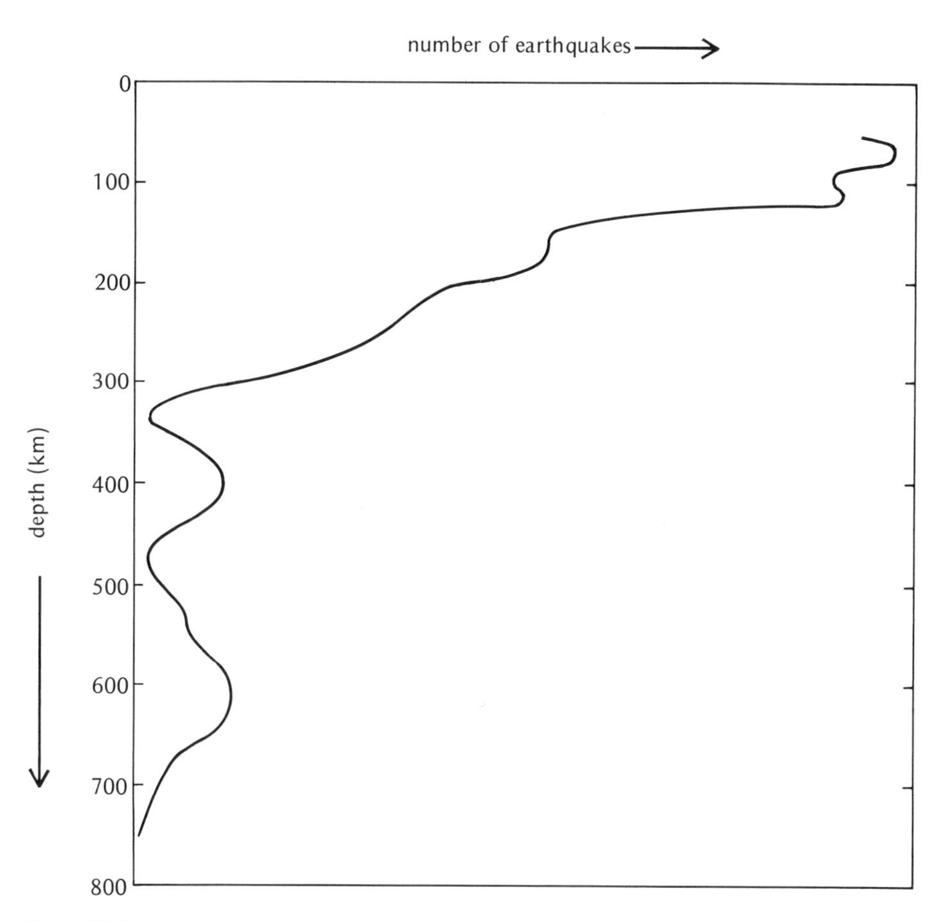

Figure 10-5
Depth distribution of earthquakes, showing clustering into shallow, intermediate, and deep groups.

ther interesting property of running as deep as 800 km, a greater depth than terrestrial earthquakes at their deepest.

The moon's total annual seismic energy release is estimated to be only about 10^{14} ergs (Latham, 1971), compared to a terrestrial annual seismic energy release of about 5×10^{24} ergs. This is a measure of the extent to which the moon has become a "dead world," by virtue of the faster cooling of its interior than has occurred in the earth. Since major quakes are rare on the moon, active seismology will be an important tool for studying the moon's interior.

Wave Types and Early Seismic Observations

The vibrations felt during an earthquake are waves propagating along the earth's surface. The nineteenth-century physicists Poisson (1829) and Rayleigh (1887) predicted theoretically the main types of seismic waves that could be propagated from earthquakes. The most important of these are the P and S waves predicted by Poisson (Figure 10-6). P waves are *P*ressure waves, such

Figure 10-6
Longitudinal (P) and shear (S) seismic waves.

as sound waves, where the motion of individual particles is along the direction of the wave's motion (mnemonic: think of P waves as *P*ush–*P*ull action). S waves are *S*hear waves, where the motion of individual particles is at right angles to the wave's motion; an example of such motion occurs in a wave along a rope or garden hose.

Because P and S waves have different velocities they reach the seismic observer at different times. If the observer is located close to the focus, the interval between the P and S *arrival times* would be short, but if he is farther away it would be longer. The time interval between the P and S waves is thus a measure of the distance to the focus.

★ QUESTION ★

If an observer saw *only* the P or S wave, could he tell how far away the focus was?

Answer

He would not know how long the wave had traveled, and so he could not tell how far it had come.

A minimum of three observing stations in a triangular array is needed to locate an earthquake. Each observes the P-S *arrival-time interval* and thus computes the radial *distance* to the focus. On a map a circle of this radius can be drawn around the station to indicate possible locations. If three stations provide three circles, there will usually be only one mutual intersection. This marks the earthquake's epicenter. Thus an array of three simultaneously operating seismometers on the moon would be far more valuable than a single seismometer.

Oldham (1900) was first to distinguish the P and S waves in seismic records, and he first constructed useful tables of travel times for the two waves.

Seismologists quickly discovered that they could observe even earthquakes that occurred on the other side of the globe. In 1906 Oldham found that waves arrived at the anticenter (180° from the earthquake*) later than would be predicted by his travel-time curves. This meant that the center of the earth contained a low-velocity region, as shown in Figure 10-7. This was the first evidence of a distinct *core*.

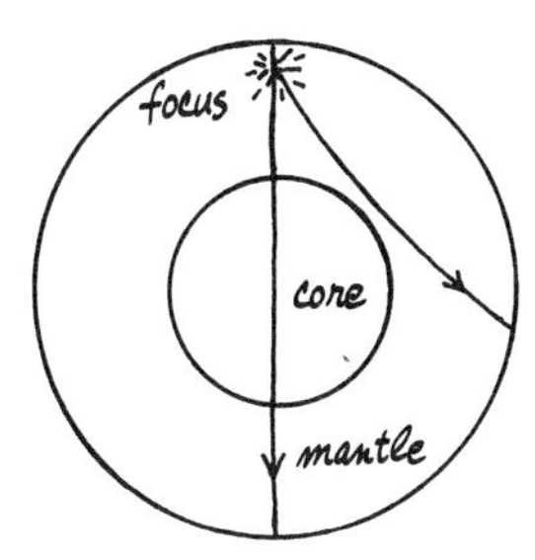

Figure 10-7
Paths of seismic waves through the earth. Waves that arrive on the far side of the earth from the earthquake focus have been intercepted by the core.

Another famous discovery was soon made by the Yugoslavian seismologist Mohorovičić (1909). In studying shallow earthquakes, he found that there were two sets of P waves and two sets of S waves. One set of P and S waves traveled at a high velocity and was traceable to large distances; the second was a low-velocity set that was best observed near the epicenter. Apparently the shallow rocks transmitted waves slower, as shown by Figure 10-8.

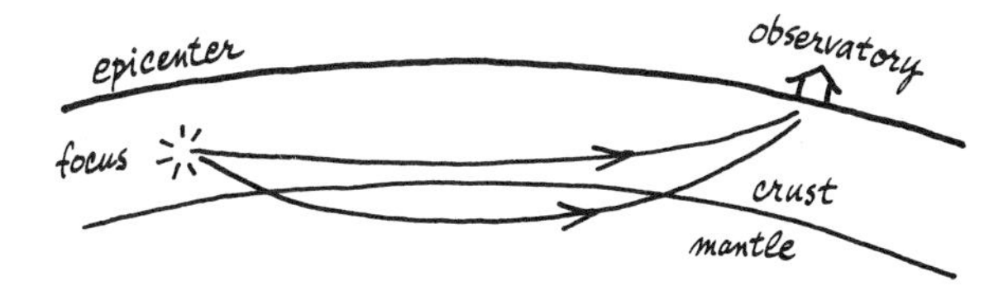

Figure 10-8
Discovery of the Mohorovičić discontinuity by observation of two sets of seismic waves with different velocities, coming from a single shallow earthquake.

* Angular distance along the earth's surface, measured at the earth's center (as with latitude).

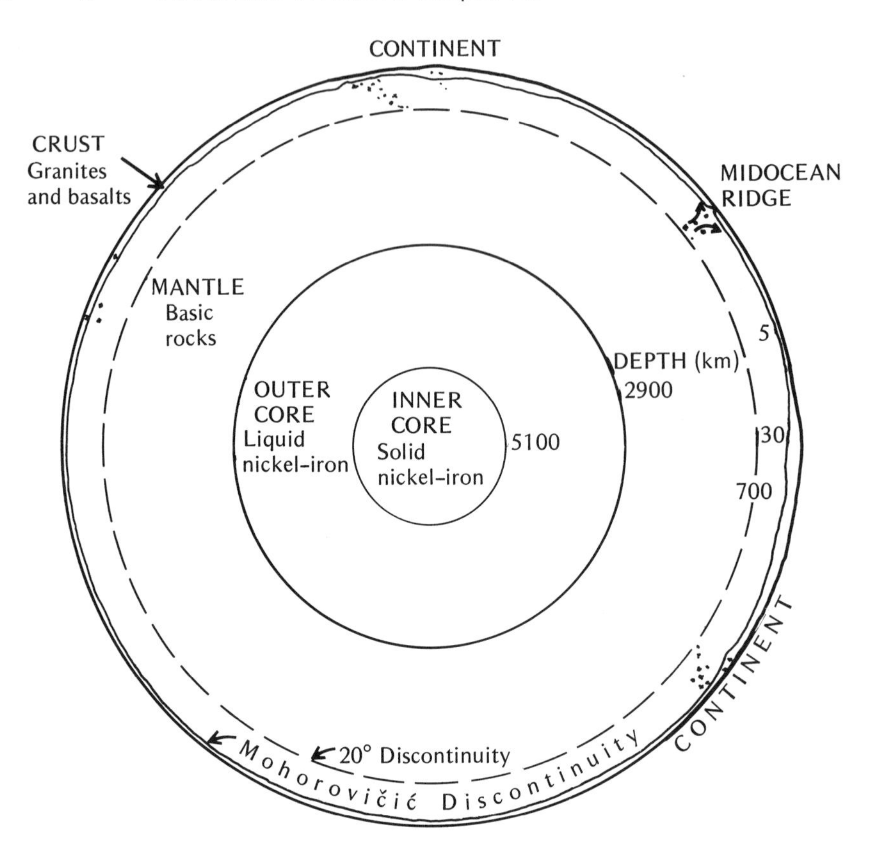

Figure 10-9
Structure of the earth's interior as revealed by seismology, with compositions inferred from additional evidence. Earthquake positions are indicated by dots.

The interpretation was that a discontinuity in wave velocity—hence rock type—occurred at a depth of some 30 km. This is called the *Mohorovičić discontinuity* (sometimes called the "Moho" for short), and its discovery was the first proof of a distinct *crust* overlying the interior of the earth.

About 1930 another discontinuity in wave arrival times was noted at a distance of 20°* from the epicenter of earthquakes. This corresponds to a discontinuity in rock properties at a depth of several hundred kilometers and is called the *20° discontinuity*.

In 1937 a Danish seismologist, Miss I. Lehmann, detected structure inside the core. P waves traveled faster in the interior part of the core than in the outer part. Studies of the outer part of the core revealed that it did not transmit

* Angular distance along the earth's surface, measured at the earth's center (as with latitude).

any S waves, indicating that it has zero rigidity. Therefore, the *outer core* must be liquid. The *inner core,* however, was found to be solid.

It is important to note that the structure of a planet's interior is revealed through seismology only as a pattern in wave velocities, such as discovered by Oldham, Mohorovičić, and Lehmann. Any other properties, such as chemical composition, must be inferred.

Figure 10-9 shows the interior of the earth as revealed by seismology. Of course, much more subtle structure has been detected than is shown, but the main divisions are clear. The *crust* is the region down to the Mohorovičić discontinuity; the *mantle* is the region between the crust and the *core.*

HETEROGENEITY IN PLANETS

Homogeneous, "Plum-Pudding," and Shell Models

The earth as revealed by seismology is built in layers or shells; an inner iron core is surrounded by a silicate mantle, surrounded by a crust of lighter silicate rocks. Why should a planet be so layered?

Shell structure may be an indication of melting in a planet. It is usually assumed that immediately after formation planets are essentially homogeneous, although localized irregularities may reflect accretion of different planetesimals with different densities. If the planet heats and partially melts, iron may separate chemically from silicate minerals in the geochemical process known as differentiation. In such a planet pools of iron may have formed without completely draining toward the center. Such planets would then have a heterogeneous, "lumpy" structure represented by the so-called plum-pudding model (Figure 10-10). After more complete melting, iron could drain toward the planet's center and a layered structure could result.

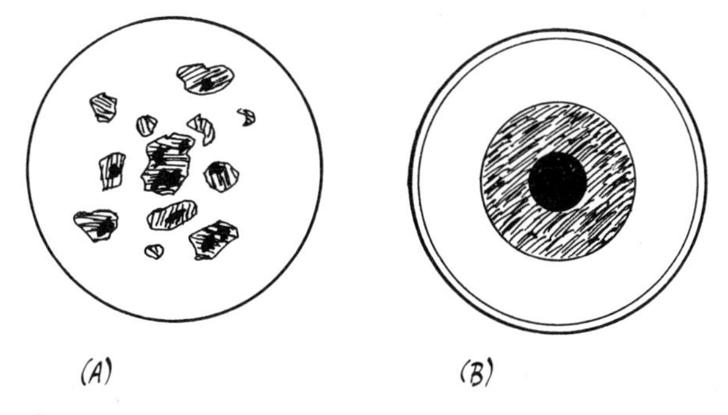

Figure 10-10

Cross sections of hypothetical planetary bodies, showing "plum-pudding" structure (A) and core-mantle structure (B).

Which would be most likely to melt and differentiate into symmetric shells: a large planet or a small planet? We might intuitively answer "large" because the larger planet would have a more pronounced gravity field and because small planets could cool faster, thus inhibiting processes requiring a fluid state. This intuitive judgement is correct and is borne out in our earlier Table 10-1, since we find that the smaller planets appear to be more homogeneous. Minor departures from homogeneity are very important, however, because they suggest limits on the degree of fluid behavior in the past. For example, we have seen some evidence that the parent bodies of the meteorites had plum-pudding structures (Chapter 9), and we shall discuss below the lunar mass concentrations ("mascons") that lie near the lunar surface and indicate that at least the surface layers of the moon are rather heterogeneous.

Inhomogeneities in the Earth

Everyone realizes that the earth's crust is highly inhomogeneous. Thick continental blocks appear in some places, here broken by large underground intrusions of granite, here marked by volcanism, and there covered by layers of sediments, while on the ocean floor the crust thins into a 5-km basaltic layer.

For many years, however, it was tacitly assumed that the *mantle* was laterally homogeneous (i.e., containing variations in the radial direction but lacking in any structure in circumferential directions). This assumption was made partly because of lack of observational evidence to the contrary, but also because it seemed reasonable that the mantle should be thoroughly mixed if the earth had been strongly differentiated. However, MacDonald (1964) showed that more heat was being transported out of the subocean mantle than out of the subcontinental mantle, indicating that the mantle under the continents is different from the mantle under the oceans. Seismic evidence supports this conclusion, and further research is being devoted to the problem (Moberly and Khan, 1969). Such studies were among the first clues leading to the revolutionary recent discovery of plate tectonics and continental drift (see below).

If recent work on the earth is an indicator, we shall have to be prepared to find that planetary interiors, even deep interiors, may be more complex than is apparent.

Gravimetric Analysis

One way of detecting inhomogeneities is by gravimetry. In discussing dynamic ellipticity of planets, we noted that measurements of the gravity field could be converted into a model of the planetary structure. The same can be done on a much more local scale. Figure 10-11 shows how a *gravimetric traverse* could be made across the region of a buried intrusion of, let us say,

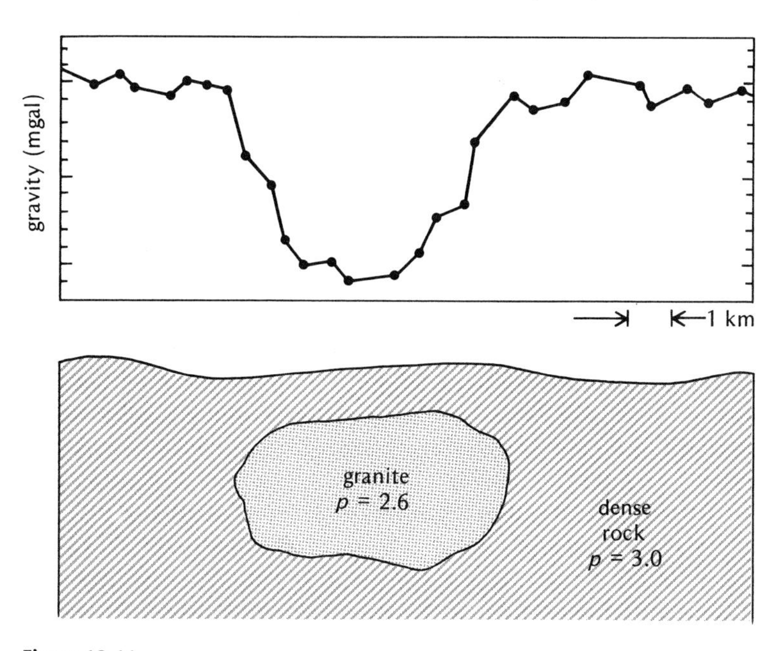

Figure 10-11
Gravimetric traverse (top) used to reveal subsurface structure corresponding to buried masses of different density (bottom cross section).

granitic composition. The granite in the figure has a lower density than the surrounding rock, and so there is less mass between the gravimeter and the center of the earth when the gravimeter is over the intrusions; hence the local gravity field is slightly less than the ambient field at that point. This method is valuable for probing the internal structure in the outer few kilometers of a planet. It has yielded, for example, information on the subsurface structure of large craters, helping to distinguish impact craters from volcanic features. Gravimetry from low satellites could reveal larger-scale, deeper inhomogeneities.

Lunar Mascons

The Orbiters were satellites placed in orbit around the moon. In 1968 Muller and Sjogren published an analysis of the motions of Orbiter V. They found that the satellite underwent unexpected accelerations when it passed over each of the five large, circular, lava-filled basins on the moon's near side. The analysis indicated that each of these was the site of a mass concentration. For these, Muller and Sjogren coined the term *mascon*.

Urey (1952) had long before suggested that the huge circular basins might be the impact site of large, dense asteroids that might still lie buried beneath the

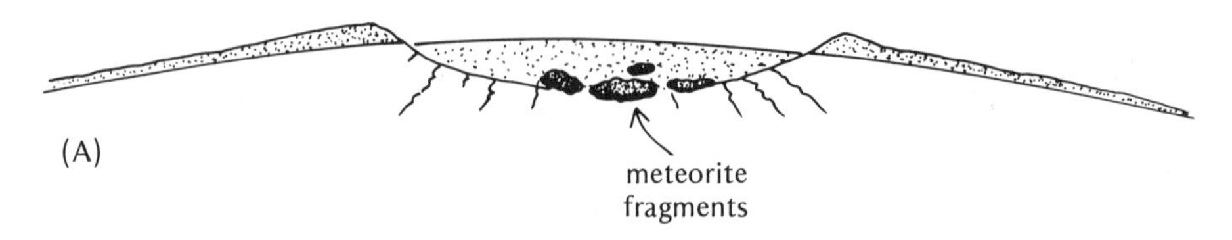

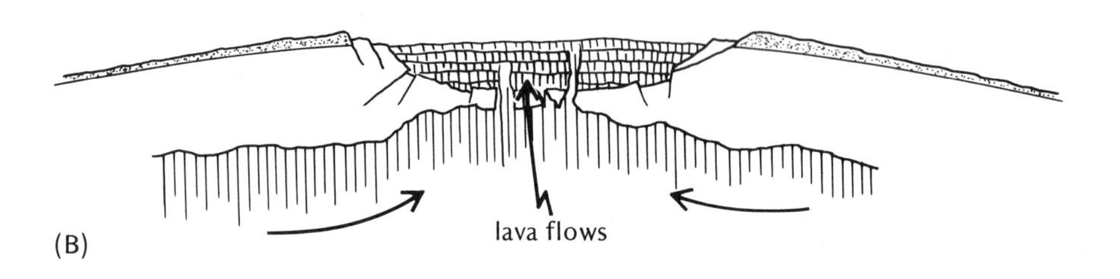

Figure 10-12
Hypothetical cross sections of a major lunar basin, showing an early proposed explanation for lunar mascons (A: high-density meteorite fragments buried beneath the basin), and a more widely accepted recent explanation (B: postimpact intrusion and extrusion of high-density magmas).

lava-covered plains. At first, the Muller–Sjogren discovery was taken as confirmation of this hypothesis, but more recently other, more promising, hypotheses have been suggested. The principal problem is to suggest a means by which excess mass can be deposited near the moon's surface in the column between the surface and the moon's center. Kaula (1969) has discussed some mechanisms and implications. The lava material itself may be responsible, perhaps having degassed because of its initial high temperature and therefore being more dense than the surrounding lunar rock, as shown in Figure 10-12.

Apart from its intrinsic interest, the concentration of masses near the moon's surface implies that the moon's outer part has had a certain long-term rigidity, for if the moon had been highly plastic after the mascons formed, the mascons should have sunk toward the center, displacing material until no extra mass was in the column from the surface to the center. This state is called *isostatic adjustment;* if it had been achieved the mascons would no longer be gravitationally detectable.

THE EARTH'S INTERIOR

Let us now put together all the various sources of information in order to describe the interior of the earth.

★ QUESTION ★

List the sources that we have discussed.

Answer

There are three main areas of information. The *theoretical models,* with sufficiently good equations of state, allow us to estimate the pressure and density and predict that the earth may be layered by composition changes and by changes of state. A number of *observational checks* help identify the proper equations of state, provide boundary conditions for models, and indicate the degree of central compression of mass. *Seismology* identifies the positions of discontinuities.

The Earth's Core

Not much is known about the earth's core because the most direct observational evidence is only seismic data—not samples. Not even the deepest volcanic activity can bring up samples of core material, and iron meteorites are at best only analogs of the core material. The composition is thought to be nickel-iron with 0 to 20 percent of some lower-density material, probably silicon or sulfur. The core has one extremely important observable characteristic, however: It is evidently the source of the earth's magnetic field. The process by which the magnetism is produced will be discussed at the end of this chapter. One aspect of the process is a slow circulation of material in the liquid outer core. The motions of the liquid iron are sluggish, and estimates of the flow velocity are in the neighborhood of 0.02 cm/sec (less than $\frac{1}{1000}$ mph).

Why is the inner core solid if the outer core is liquid? An answer must take into account the equation of state of the liquid iron (more accurately, we should call it liquid iron-nickel-etc.). Under the high pressures in the center, the melting point is probably higher than it is partway out in the core. Thus, even though the temperature is highest in the center, it is not high enough to exceed the melting point, leaving the inner part of the core solid.

The Earth's Mantle

Although the mantle lies only some 30 km beneath the continents and only about 5 km below the ocean floor, it is almost as remote scientifically as the core. No hole has been drilled deep enough to penetrate the mantle. A national effort, called the Mohole Project, was launched in the 1960s to drill through the ocean crust, but the project was eventually canceled in a political–scientific controversy. As a result, the only *direct* evidence we have about the mantle comes from certain kinds of volcanic vents, such as the Kimberlite diamond pipes in Africa, which have brought to the surface certain kinds of rocks thought to be representative of the uppermost mantle materials. Representative of these rock types are *dunite, peridotite,* and *eclogite* (see discussion of rock types in Chapter 11).

The mantle is composed of rocky material denser and more rich in ferrous metals than most of the rocks familiar to us on the surface. The mantle rocks may be similar to achondrite stony meteorites, which are probably fragments of the interior of planetary bodies smaller than the earth (see Chapter 8). In this context it is interesting that the lunar lavas show certain chemical and isotopic similarities to the assumed composition of mantle rocks.

The mantle is solid, but a zone from about 70 km to 200 km in depth has seismic-wave velocities markedly lower than in the region above and below it, and many seismologists believe this is a region of near or partial melting. The region is more pronounced in areas of volcanic and mountain-building activity. We can speculate that this depth zone may be the region from which much volcanic lava originates. Support for this comes from the lavas of the Hawaiian volcano Kilauea, which originate in an earthquake-active zone roughly 60 km deep (Eaton, 1962).

Earthquakes occur at depths as great as 700 km but not beyond that. This fact indicates that rocks are stressed and brittle enough to fracture in the upper part of the mantle. How are these stresses re-created time and time again, even after earthquakes relieve them? This question brings us to the dynamics and evolution of the earth's mantle. It is probable that the upper mantle is in a state of slow, plastic deformation and is trying to flow in circulation patterns driven by the escape of heat energy from the earth's interior, just as boiling water is driven by the heat flow from a stove.

We have said that the mantle is solid and brittle enough to fracture and produce earthquakes. Yet we have just suggested that it is flowing. Does this seem paradoxical? There are numerous examples of this phenomenon in everyday experience. Pitch or tar, for example, can be shattered with a hammer, yet if a cannonball were left for a day resting on a barrel of the same pitch it would sink into the pitch. Given enough time, pitch can deform like a fluid. Another example is a glacier. If you walk up to it you can chip off icy splinters, yet the whole thing, on a much slower time scale, is flowing downhill. The earth's mantle has analogous properties. Stresses can be relieved on a short time scale by fracturing; yet on a long time scale the material deforms like a fluid. Earlier in the chapter we asked why the earth is round. The answer was that the interior can behave like a hydrostatic fluid because the pressure on the rocky material exceeded the crushing strength of the material. Here we are pointing out that the resulting fluid behavior requires long time periods to be manifest.

Rheology is the branch of science that studies materials behaving in this way—as solids over the short term but as fluids over a sufficiently long term.

Such materials are known as *rheids*. Rheology is finding increasing application to geology and geophysics (Carey, 1962). Only if the proper, time-dependent, rheological equation of state is known can deformations of a planetary interior be fully understood.

The Earth's Crust

The crust is the outer shell of the earth—the stage for man's activities. Is it a permanent, stable stage? No. If the outer few hundred kilometers of the mantle is in motion, then the crust, which is only a 5- to 30-km-thick skin on the mantle, ought to show the disruptive effects of this motion. Indeed it does, except that the time scale of the crustal disruptions is much longer than the lifetime of a man, so that as individuals we are not aware of the disturbance. But if we could set up a time-lapse movie camera in space and take one frame every few thousand years we could see a movie of the crust of the earth warping, rifting, splitting apart in some regions and squeezing together into series of folds in other regions. Our film would show chaotic churning, with great mountain belts rising up, eroding, and gradually disappearing in some regions while other regions remain stable for long times.

Figure 10-13

Deformation of the earth's crust by folding: the Appalachian Mountains in central Pennsylvania. Erosion has produced ripple-like parallel ridges marking the underlying folded strata.

Even the seemingly solid, permanent rock layers beneath our feet can deform like fluids if given enough time. We are all familiar with highway roadcuts that reveal cross sections of immense folds that as well could have been made out of putty (Figure 10-13). Even more dramatic is the behavior of salt domes such as those found around the Gulf of Mexico (Figure 10-14). Low-density masses of salt—and even masses of rock, such as granite—can rise up through denser rock masses much as submerged blobs of oil will rise through water. Crustal mobility is furthur illustrated by the immense movements involved in continental drift, discussed in the next section.

The crust is made up of two major parts. The lower part is a global layer of basalt (a type of rock common in lava flows). This basalt layer is about 5 km thick. In the continents, sediments, granites, and many other kinds of rock* overlie the basalt layer, building the total crustal thickness to some 30 km or more, depending on whether we are in a thick mountainous section or in a thin region near the edge of a continent. In the ocean, the continental rocks are lacking, and the 5-km basalt layer makes up the entire thickness of the crust except for scattered thin layers of sediments. Another 5 km or so of water obscures the ocean floor, which must be studied by seismology, soundings, etc.

* These surface-rock types will be discussed in more detail in Chapter 11.

Figure 10-14

Deformation of the earth's crust by isostatic adjustment: cross section of salt domes along the Gulf coast. Masses of low-density salt ascend through and disrupt overlying rock strata.

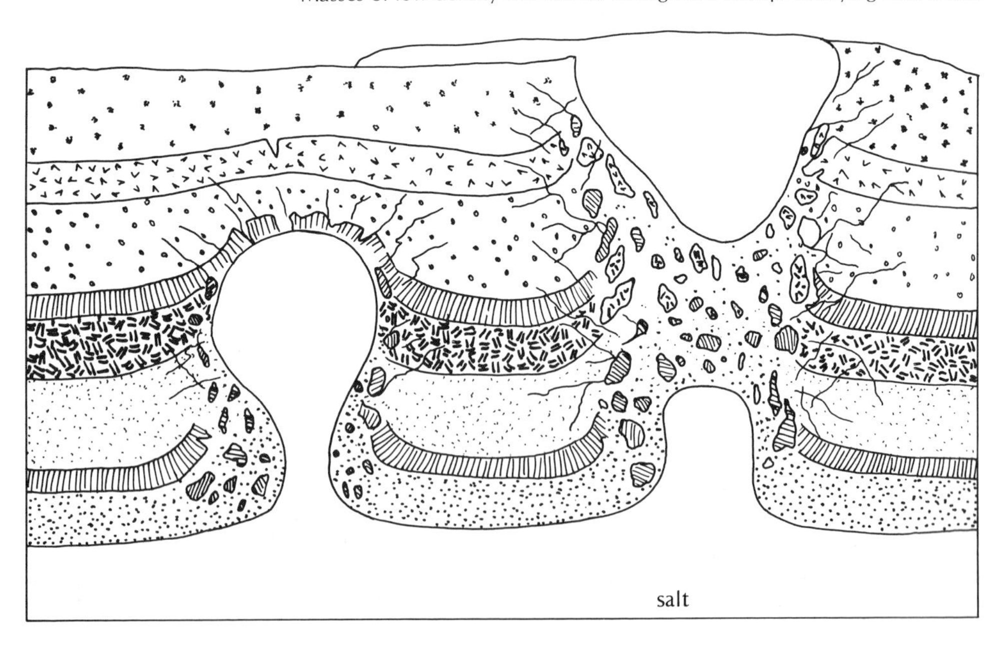

PROCESSES AND FEATURES OF THE EARTH'S CRUST AND UPPER MANTLE

The Continents

For a long time men have pondered features of the continental crust. How do immense mountain ranges form? Why do certain regions sink and become invaded by shallow seas? Why are only certain areas shaken by earthquakes while other areas are quiet? Careful studies of mountains, coastlines, seismic zones, and other regions, instead of giving clearcut answers, have revealed only a seemingly endless tangle of complexities with very few diagnostic patterns.

One fact early recognized was that the continental blocks have lower density than the heavier mantle rocks, and that therefore the continents can be loosely said to be "floating" on the upper mantle. This is an example of Archimedes' principle. One observational support is seismic evidence, which shows that the crust is thickest under high mountains, which must have "roots" of low-density material to support the weight of the part that sticks up above the mean continental surface. An analogy is an iceberg, which must have a large volume of ice below the ocean to support the small volume that protrudes.

If the continents were perfectly supported by low-density roots, they would be in *isostatic adjustment* (i.e., floating in equilibrium). In actual fact, some areas are not in isostatic adjustment. One example is the Scandinavian region, which was depressed by the weight of glaciers during the Ice Ages and has been rising ever since at the rate of about 1 cm/year. Gravimetric analysis helps to identify the locations and causes of regions of disequilibrium and geological activity, as was shown in Figure 10-11.

The oldest preserved regions of continents are the *shields,* so named because they are flat lying and often roughly circular. These are areas that have not been disturbed recently by mountain-building activity. They yield the oldest known terrestrial rocks, which solidified about 3.3 billion years ago. Shield areas are often found to be areas where erosion has worn down former mountains and masses of lava. They are often the central "cores" of continents. An example is the Canadian Shield, which surrounds Hudson's Bay and has been mostly quiescent for at least 1 billion years.

At the opposite extreme are the major mountain ranges, which are the youngest, most distinctive, and impressive features of the earth's continents (Figure 10-15). The moon possesses no features such as the Himalayas, Alps, Andes, or Rockies. Instead of being churned by mountain-building activity, known as *orogeny,* its static surface perserves ancient craters. The "mountain ranges" of the moon are really rings of piled-up ejecta and uplifted rocks deformed by ancient catastrophic impacts. Lunar "mountains" are rims of immense craters, fundamentally different from terrestrial mountain ranges. Mars may be intermediate between earth and moon. Whereas parts of its surface

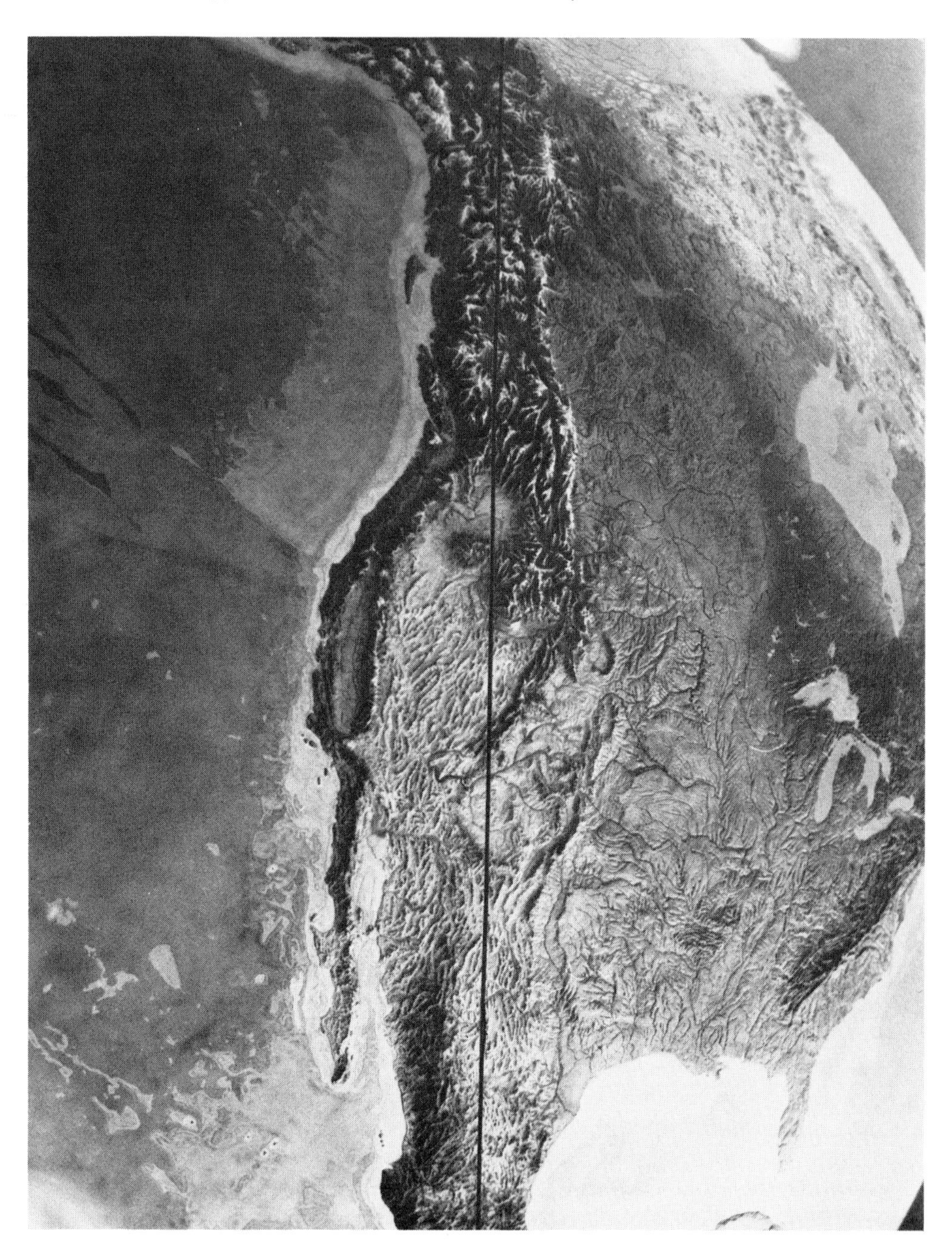

Figure 10-15
The Rocky Mountains, a major linear mountain belt on the earth. (Copyright © by Rand McNally & Company, R.L.)

are relatively flat and cratered, reminiscent of the moon, other regions are broken by complex valley systems and high regions. Mars shows surface relief of at least 13 km, compared with the earth's 19 km (counting ocean depths).

Whereas the earth's mountain belts are usually arranged in linear or arcuate belts, illustrated by Figure 10-15, present topographic data on Mars reveal no such patterns. A broad, elevated region of Mars, about 2000 km across, contains several mountains (one measured to be about 3 km in height) capped by caldera-like craters, as shown in Figure 10-16. These formations suggest volcanic activity in an area of incipient orogeny on a planet where no stream and glacial erosion could transform an uplifted area into jagged peaks such as those produced on earth.

Major mountain belts of the earth's continents usually consist of enormous thicknesses of intensely contorted sediments (Figure 10-17) with injected igneous rocks. For example, the Appalachians contain a sedimentary rock section six to eight times as thick as the sedimentary section of equal age in the neighboring Mississippi Valley. When we contemplate the folded layers occasionally exposed along highway roadcuts and recall that sediments are initially laid down in flat beds, we realize the tremendous forces at work to deform the continental crust. Because many mountain ranges are made from thick layers of sediments, it is believed that the deformation preceding the growth of a mountain range must be a downwarping of the earth's crust to form shallow seas, where thick sediments could accumulate. Such a downwarping of the crust is called a *geosyncline*. (In any folded rock stratum, the U-shaped part of an individual fold is called a *syncline;* the hill-shaped part of the fold is an *anticline*.)

Modern studies of geosynclinal systems, using radioisotopic dating of rocks, have revealed that orogeny is not a continuous process or a randomly intermittent process but occurs in spurts of activity. For example, the Appalachians were built during a series of orogenic events punctuating the Paleozoic era (roughly 5 to 2×10^8 years ago). The Rocky Mountains from Mexico through Canada were very active during the "Laramide" revolution at the end of the Mesozoic era about 7×10^7 (70,000,000) years ago. There is some evidence that we are in the midst of a worldwide orogenic phase at the present time, and there is increasing evidence that these periods of crustal unrest alter climates, encourage new species, and thus have a major effect on the evolution of life.

The geosynclinal description of orogeny, such as given by the Soviet geologist Beloussov (1962), demonstrates the various stages of development of many mountain systems, but it does not give the fundamental key to the question of what causes mountains.

Indeed, the realization that mountains formed by evolutionary contortions of the earth's crust came long ago. The English scientist Robert Hooke (1635–1703) noted (*Posthumous Works,* 1705) that ancient marine sediments

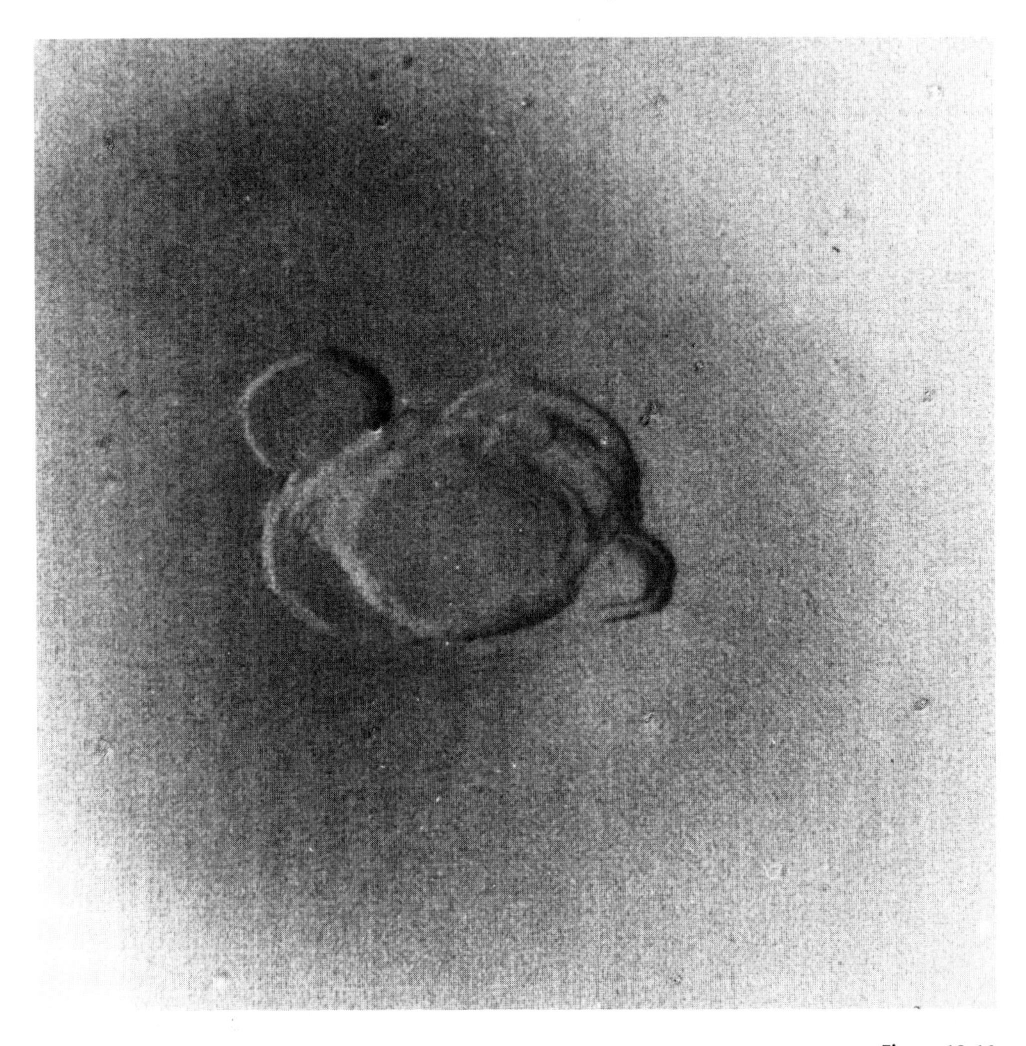

Figure 10-16

Complex Martian crater, resembling a terrestrial volcanic caldera. Photographed by the Mariner 9 spacecraft, this crater is seen only dimly through the great Martian dust storm of 1971. Bright areas on all sides are totally obscured by airborne dust. The crater is believed to lie on a mountain whose summit protrudes through the dust clouds. (NASA.)

were now high and dry in the mountains. He thought the upwarpings and downwarpings must be due to "eruptions . . . or earthquakes."

> It seems not improbable that the tops of the highest and most considerable Mountains in the World have been under Water, and they . . . seem to have been the Effects of some very great Earthquake. . . .

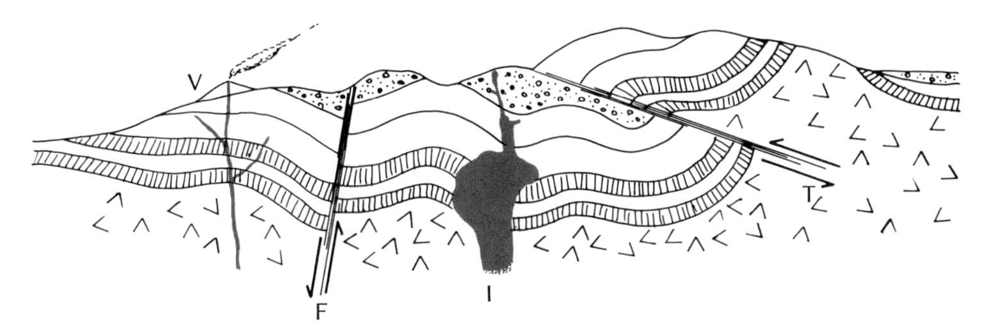

Figure 10-17
Schematic cross section of a terrestrial mountain belt, showing folds and tectonic structures. V, volcano; F, normal fault; I, intrusive body; T, thrust fault.

Although we may regard Hooke's use of the term "earthquake" as naive, we must question whether the views described above are really more profound than Hooke's 267-year-old statement.

What causes mountain ranges? What is the significance of the thick folds and elongated patterns of relief? Are the earth's mountain ranges unique among the terrestrial planets, and if so, why? The lack of mountain ranges on the moon suggests one clue to the causes of orogeny. Smaller planets may have already lost the heat required to drive orogenic activity in their mantles. Yet the earth's mantle, as we have seen, contains flowing currents and suddenly released stresses.

We might expect to learn best about orogenic, crustal, and upper-mantle processes by studying the very active continental regions, where the effects of the activity have been most dramatic. On the contrary, the continental crust is so exceedingly complex, contorted, and active that the patterns are almost lost in the chaos; any message that is there is so cryptic that we can hardly read it. But what of the other 71 percent of the earth's surface area that lies under water?

The Ocean Floor

Curiously enough, it is the ocean floor that holds the keys to the evolution of the earth's crust. Only in the decade of the 1960s were these keys revealed, causing a revolution in our ideas about the nature of crust and its continents.

What happened, briefly, was this. Prior to the late 1940s it was more or less tacitly assumed by most geologists that the ocean floor was an inundated and somewhat modified version of the continental crust, a fixed and stable divider of the thicker continental blocks. The continents were supposed to be fixed, evolving regions of crustal activity where mountains and plateaus protruded above the ocean.

In 1947 an expedition from the Woods Hole Oceanographic Institution recognized that the ocean floor was not simply a submerged version of the continental landscape but was marked by different features, such as abyssal plains. *Abyssal plains* were defined as essentially flat areas of the ocean floor. It was found that they were covered by sediments rarely older than 1×10^8 years. This raised a mysterious puzzle: How could it be that the supposedly ancient ocean basins had no sediments older than 100 million years (i.e., only 2 percent of the age of the earth)?

In 1948 another discovery was made. Winding through the various ocean basins were features first described as midocean canyons. They were about 5 to 8 km wide and often hundreds of meters deep. Then it was realized that these canyons marked the crests of *oceanic ridges*. Ewing and Heezen (1956) mapped ridge interconnections from ocean to ocean and pointed out that the ridges were centers of shallow seismic activity, as illustrated by Figure 10-18. At about the same time it was shown that the amount of heat flowing out of the earth in the ridges was several times the heat flow measured in either the continents or the abyssal plains.

By 1959 the oceanic ridges were described as possibly representing a third province of earth's geography, neither continent nor ocean basin. This was an indication of the importance they were to play in understanding the earth's upper mantle and crustal structure. However, the origin of the oceanic ridges was still in doubt.

The next advance came with *paleomagnetic* studies of rocks from the oce-

Figure 10-18

Schematic cross section of a terrestrial oceanic ridge. Central rift, high heat flow, and increasing age of rocks outward from ridge are accounted for by ascending, outward moving magma currents. Note exaggeration of vertical scale. Compare Figure 10-14.

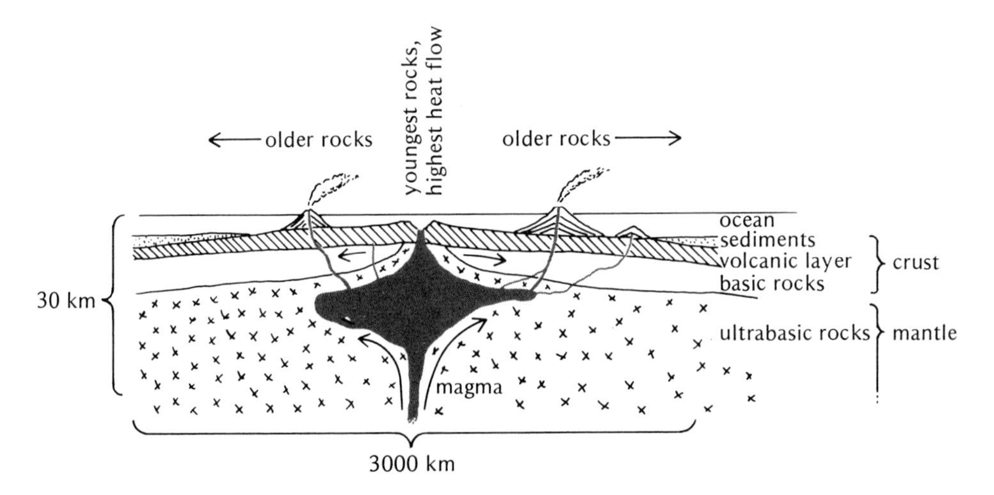

anic ridges. This type of work relies on the fact that when a rock crystallizes, the magnetic elements in the rock are aligned according to the earth's magnetic field direction at the time and place where the rock crystallized. Thus it was possible by dating the rocks and measuring their paleomagnetism to determine the direction of the earth's field when they crystallized. The astonishing result of these complex measures was that (1) the earth's magnetic field has been reversing its polarity rather sporadically every 500,000 years or so, and (2) the ocean-floor rocks are originating in the central canyons and moving outward in both directions away from the canyon, as determined by mapping the dates and magnetic polarities of oceanic ridge rocks.

The first suggestion of this concept, which is called *ocean floor spreading,* came from Vine and Matthews in 1963. At the time it was highly unorthodox, but by the late 1960s the facts had been abundantly confirmed and a revolution in geology was underway.

The idea that the ocean floor was an active source of new rock and that the rocks were spreading outward toward the continents was only a part of the geological revolution.

Continental Drift

With the discovery of sea floor spreading, a hoary and controversial geological idea was reexamined. In 1620 Francis Bacon had noted that the east and west coastlines of the Atlantic could fit in the manner of a jigsaw puzzle. He made the remarkable suggestion that the Americas may once have been in contact with Europe and Africa. By the late 1800s paleontologists had noted the similarity of Carboniferous fossil plants (about 3×10^8 years old), and the famous Austrian geologist Eduard Suess had noted jigsaw-puzzle fits of geological formations across the Southern Hemisphere. Suess therefore hypothesized a primeval southern continent which he named Gondwanaland (from a key geological province in India, Gondwana). Gondwanaland was supposed to be the original parent of the present southern continents, which formed when Gondwanaland broke into pieces.

The theory that the modern continents were drifting fragments of a primeval continent or continents came to be called the theory of *continental drift.* In 1908 the American geologist F. B. Taylor considered possible mechanisms to keep the continents moving. This problem led to a major objection to the theory, since obviously enormous friction must be overcome if we are to force rock to drift through rock.

Continental drift came to be championed by the German meteorologist Alfred L. Wegener (1922) in the decades following 1910. Wegener's thesis, that the continents broke up and began drifting only a few hundred million years ago, became the center of an international storm of controversy. Why, sceptical geologists asked, did such a radical change in the pattern of the earth's evolution suddenly start in only the last 6 percent of the planet's history? What drove the continents on their courses? Wegener stuck to his guns, not with answers to all the critics' questions, but with empirical evidence such as the cross-ocean fits of geological structure and fossil flora and fauna.

New evidence came in the 1920s and 1930s when South African geologists mapped distinctive glacial deposits in South America, Africa, Australia, India, and Madagascar, indicating that these areas were once closer together and grouped near the south pole of the earth's rotation, where the glacial ice sheets originated. In the 1930s the geophysicist F. A. Vening Meinesz carried out extensive gravity measurements which led him to suggest that convection currents (slow flowing of material due to dissipation of heat) in the upper mantle might be sufficient to drag the continents along.

Opinions on continental drift became polarized between Southern Hemisphere field geologists (pro) and Northern Hemisphere theoreticians and geophysicists (con). In retrospect we can see one reason why this happened: The southerners had much of the best evidence in their hemisphere.

Perhaps the most consistent alternative to the theory of continental drift was based in part on the work of the Canadian geologist J. Tuzo Wilson (1954). According to this alternative theory, the continents were permanently fixed throughout time and grew by an orderly process. Wilson identified the ancient continental shield areas as the old nuclei of the continents, and he proposed growth of continents along the active *island arcs* at the continental margins where volcanic outpourings add to the continental land area. This picture was more in accord with the traditional views of Northern Hemisphere scientists. As recently as 1962 the famous English geophysicist Sir Harold Jeffreys echoed their views:

> [Some say] that if evidence from paleontology and meteorology proves that continental drift has taken place, evidence from geophysics that it is impossible is beside the point . . . it is remarkable that the advocates of continental drift have not produced in thirty years an explanation that will bear inspection . . . if evidence is conflicting, the scientific attitude is to look for a new idea that may reconcile it.

The new ideas and new evidence were on their way when Jeffreys wrote those words. Mapping of the Atlantic floor led to proof that new material was coming up along the mid-Atlantic oceanic ridge, pushing to the east and west, creating a very young ocean floor only a few hundred million years old, and apparently pushing the continents away from each other.

The year 1966 was the turning point in acceptance of the continental drift theory. In a meeting of the Geological Society of America in San Francisco a number of papers gave convergent but independent evidence which put the opponents of continental drift on the defensive for the first time [as remarked by

Hurley (1968) in his article "The Confirmation of Continental Drift"]. Among scientific results announced were proof of sea-floor spreading, the fact that the continental shelf boundaries could be fitted together in jigsaw-puzzle fashion with mean errors of less than 1 degree over most of their length, and the detailed fitting of geological provinces across the Atlantic by means of radioisotopic dating of rocks. Figure 10-19 shows the fit of South America and Africa according to the rock ages. These data do not disprove certain older ideas. It is undeniably true, for instance, that old continental shields are surrounded by younger material—especially in arc-shaped alignments of islands—so aspects of the continental growth theory may have to be synthesized into the continental drift theory.

Present evidence indicates that 300 million years ago there were two major continents, called *Laurasia* (in the north) and *Gondwanaland* (in the south). Rifting of these continents may have started 300 million years ago but

Figure 10-19

Evidence for continental drift: the fit of two different age provinces on the African and South American coasts when the continents are fitted together. (After Hurley, 1968.)

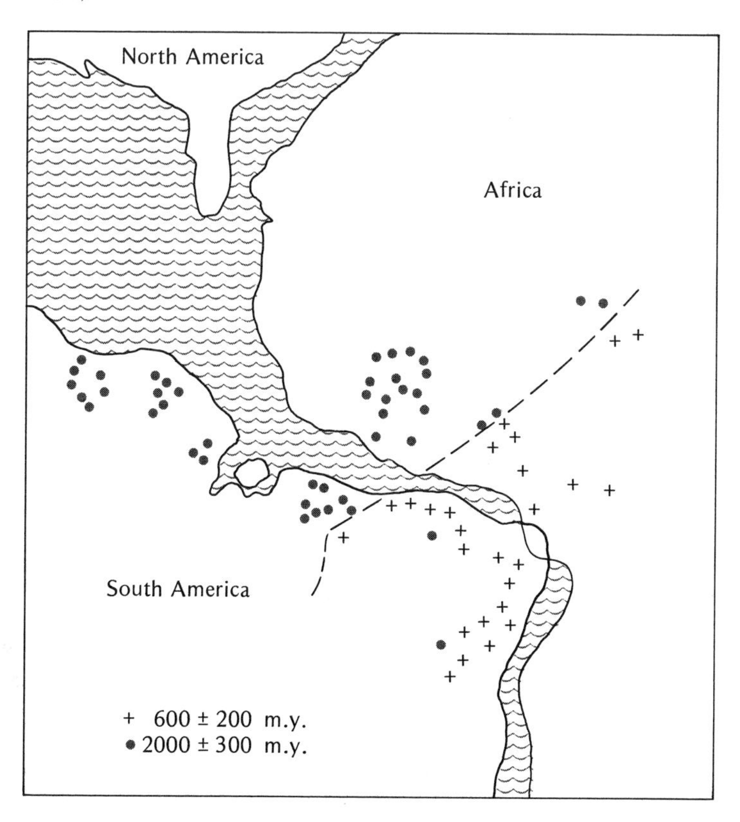

Figure 10-20
Island arcs off the Asian coast. (Copyright © by Rand McNally & Company, R.L.)

significant separations date from about 100 million years ago. A still earlier single protocontinent, called *Pangea,* is sometimes hypothesized as a predecessor of Laurasia and Gondwanaland.

Island Arcs

Features of terrestrial geography that have long fascinated geologists are the island arcs, such as the Aleutians, the Japanese Islands, and the Sumatra–Indonesian arc, shown in Figure 10-20. These have consistent characteristics which indicate that the arcs are a significant, repeated phenomenon in the evolution of the earth's crust. Figure 10-21 shows features of the island arcs, starting from the ocean side and working toward the continent, including the following. Offshore from the islands the ocean floor is depressed in an *oceanic trench,* typically 100 km wide, several kilometers deeper than the adjacent sea floor, and often 1000 km or more long. Next comes the arc of islands, usually highly volcanic, with a type of lava not as rich in iron as the midoceanic lavas that may be derived from the upper mantle, but more rich in silicates, as if they were derived at least partly from melted sediments. Enclosed by the island arc is a shallow sea, such as the Sea of Japan. It meets the continental shore, under which the crust thickens to its usual continental dimensions. Sloping down at roughly a 45° angle from the trench beneath the island arc and the shallow sea is a zone along which earthquakes are common down to a depth of 400 to 700 km.

The concepts of continental drift and sea-floor spreading have helped explain these observations. It is now thought that the island arcs mark places where a drifting continent is overriding the ocean floor layer, which may be spreading toward the continent from a nearby oceanic ridge. The ocean block buckles downward, creating a trench, and then sinks out of sight beneath the continent, creating a slanting fault surface where each new fracture is accom-

Figure 10-21

Cross section through a typical island arc. Vertical scale exaggerated (see the text).

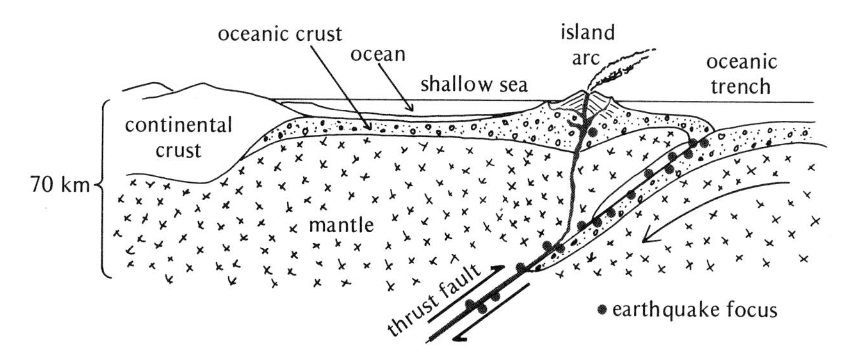

panied by an earthquake. Sediments are scraped off the ocean block and pile up to create the volcanic mountains of the island arc.

The New Global Tectonics

Tectonics is the field of geology that deals with movements in the planetary crust and interior. These may range from the smallest local faults to large-scale crustal deformations. A contemporary theory of global tectonic activity, based on observations during the last decade, is often called the *new global tectonics*. The principal feature of this new theory is known as *plate tectonics,* since it involves the novel concept that the crust is divided into a number of "plates" that move independently. This view is more sophisticated than the older idea of continental drift, which simply pictured the continents as blocks drifting on a more-or-less homogeneous mantle. In terms of plate tectonics, a drifting continent and associated ocean margins, or an oceanic province such as the Pacific basin, is a crustal "plate." Plates move independently and may collide. Boundaries of plates may be continental margins, oceanic trenches, or oceanic ridges. Depending on whether the plates collide at high speed (say 10 cm/year) or low speed (say 4 cm/year), various forms of island arcs or mountain ridges may be created at the point of contact between plates (Menard, 1969; Dewey and Bird, 1970).

The cause of plate–drifting is associated with sea–floor spreading and the ascent of convection currents that bring up hot mantle material under the oceanic ridges and spread it laterally, pushing the continents apart. The evidence for ascending convection currents includes raised oceanic ridges with tension cracks at their crests, earthquakes beneath them, young volcanic rocks along them, and very high heat flow at their crests.

THERMAL HISTORIES OF PLANETS

Definition of the Problem

The above details of terrestrial geology, as well as general theoretical considerations, show that planets are evolving systems. Whereas the first part of this chapter stressed *statics* (i.e., the description of planetary interiors in their present state), we must now come to grips with the fact of planetary evolution, *dynamics*. The heating or cooling of the planet is one of the driving factors in its evolution. The problem in the study of thermal histories is to discover the temperature and related properties (motions, liquid or solid state, etc.) of planetary interiors as a function of time.

Theory of Heat Transport

Heat is one form of energy. If a body is hot, its atoms and molecules have a higher amount of energy than if it is cold. The atoms and molecules of a hot body move around and vibrate faster, and hit each other harder, than in a cold body. *Temperature* is a measure of this thermal energy.

Why are the interiors of planets hot? Some of the heat is original; the material from which the planet formed had a certain temperature, and more heat was generated as the planetesimals crashed and transferred their kinetic energy into the growing planet. Other heat is added to the planet over the course of time by radioactivity. Radioactivity is the decay process of a very few unstable kinds of atoms, whereby the atom emits a small particle and thereby changes its mass. The small emitted particles speed out into the surrounding atoms, hitting them and thereby adding to the total thermal energy.

Heat can be transported only from hot regions to cold regions—never from cold to hot. *Heat transport* is the process of transferring the heat and equalizing temperatures. It can occur in three ways. First, *conduction* is the process by which atoms or molecules in the substance strike each other, preferentially transferring energy from the fast-moving atoms in the hot region to the slow-moving atoms in the cold region. Second, *radiative transfer* is the process by which high-energy atoms emit radiation which travels through the surrounding material and is absorbed preferentially by low-energy atoms in a cooler region. *Convection* is the process by which an entire hot region—because it has expanded to a lower density than a cold region and is therefore lighter—rises upward as a unit through the colder material while nearby colder material sinks. An example of this is the hot smoke from a burnt match rising through the air, and a more impressive example from a geological point of view can be seen by heating a shallow pan of cooking oil, which will set up a cellular pattern of convection currents reminiscent of the ascending and descending currents postulated in the theory of plate tectonics.

To apply these concepts to the problem of planetary thermal histories, we must have a great deal of quantitative information. First we must have an adequate mathematical theory describing each of the three modes of heat transport. While conduction and radiative transfer have long been well understood, an adequate theory describing convection is still being developed. Next, we must have observational measurements of each of the parameters required in the theory. All the theories require a numerical estimate of the *temperature gradient*—the rate of increase of temperature with depth inside the planet, in °K/km—because heat transport, regardless of the mechanism, depends on temperature differences. The temperature gradient in a planet's interior is, of course, difficult to estimate.

Next we must determine which mode of transport is most important. If heat transport is by conduction, we need to know the rock conductivity at different depths; if radiative transport, we need the transparency of the rock to the kind of radiation involved; if convection, we need to know the mobility of the material.

Another important quantity to be estimated is how much heat is being added at each moment by radioactivity. The chemistry of the main radioactive elements—uranium, potassium, and thorium—is such that they tend to be concentrated in the silica-rich minerals of the planetary crusts, not in the iron-rich core material, which sinks to the centers of planets. We have measures of the radioactive content of surficial earth rocks and moon rocks, but it is difficult to estimate the amount of radioactivity deep inside a planet even if we have samples of surface rocks.

There is a final additional complication. Meteorites indicate that early short-lived heat sources were present during or shortly after the formation of the solar system, and that the early planetary bodies melted within a few 10^8 years of the formation of the solar system (see Chapter 9). Uranium, potassium, and thorium have *half-lives* of several billion (10^9) years and would produce heat too slowly to melt a planet in the first few 10^8 years. Either short-lived radioactive isotopes or some other heating process must be assumed to account for very early planetary melting.

Heat-Flow Measurements

An important boundary condition that can be used to test theoretical calculations of thermal history is the rate of heat flow. The heat flow can be measured in the following way on the planet's surface. First, a hole is drilled and then a thermal probe is lowered into the hole. The hole must be deep enough to get away from *diurnal* (day–night) surface changes of temperature. This probe is several meters (yards) long and has a thermometer to measure the temperature at each end. Next, we must wait till the temperatures in the hole reach equilibrium values, because the drilling of the hole created heat by friction. After some hours the frictional heat will have been dissipated and we can measure the environmental temperatures in the hole. The difference between the temperature at the top and bottom of the probe, divided by the length of the probe, gives the temperature gradient. For example, the difference in temperature might be 0.006°K and the probe 2 m long, giving a temperature gradient of 0.003°K/m, or 3°K/km.

The heat flow [i.e., the amount of energy coming through a square centimeter of the surface in 1 second (ergs/cm^2-sec)] is directly proportional to the temperature gradient and to the conductivity of the rock. If we take a sample of rock from the drill hole and measure its thermal conductivity in the laboratory, we can multiply the conductivity times the temperature gradient and get the heat flow coming out in the region of the hole. Such experiments have been done all over the earth. The first extraterrestrial heat-flow measurements were made on the moon in August 1971 using equipment emplaced by Apollo 15 astronauts. The heat flow at that site was found to be about 33 ergs/cm^2-sec.

Heat flow on the terrestrial continents and the flat ocean floors averages about 40 to 80 ergs/cm^2-sec but over the oceanic ridges ranges about 80 to 320 ergs/cm^2-sec. Heat flow from oceanic ridges is higher probably because ascending currents in the mantle carry hot material toward the surface along the ridges.

Results of Thermal History Calculations

Once a theoretical model for a planet has been assumed, a computer program can be set up to start from some chosen initial conditions, iterate over a large number of small time intervals, and thus compute the thermal history. As a check, this calculation should predict the correct value of the heat flow.

Figure 10-22 shows some results of computer-calculated models for the earth and moon. It is interesting to note that this particular lunar model predicts a melting of the outer parts of the moon roughly 1.5 billion years after the formation of the moon; lunar rock samples from the great lava plains on the moon indicate that the lava flows in these plains occurred about 1 billion years after the moon formed, in fair agreement. However, recent measurement of lunar heat flow and central temperature will allow still better models.

Effects of Convection

Since the 1930s there have been quantitative theories asserting that convection is the dominant mode of heat transport in the earth. The theories appeal to two

Figure 10-22

Theoretical study of the thermal history of the moon. Curves give temperatures at various points inside the moon as a function of time. Heat is produced by radioactivity in a moon of chondritic composition with heat transport by conduction and radiation. Differentiation follows after melting occurs (hatching). (After Levin.)

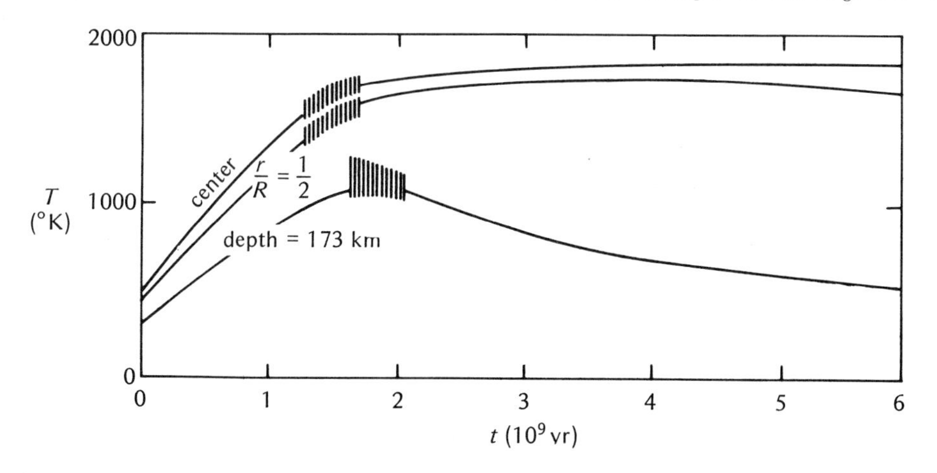

lines of evidence: structural features such as the oceanic ridges (which are supposed to be the sites of ascending currents), and theoretical arguments which indicate that neither conduction nor radiative transfer would transport enough heat to give the observed heat flow.

The theoretical arguments are based on the fact that a high temperature gradient favors convection. Material of negligible strength will start convecting if the temperature gradient exceeds a critical value called the *adiabatic temperature gradient*. In the earth the adiabatic temperature gradient is about 0.2°K/km, while the *observed temperature gradient* in the crust is about 3 to 30°K/km. While rocks seem to have a great deal of strength, we have already seen that if they are stressed for long enough periods of time, they can deform by flowing. Thus it appears theoretically possible for slow convection currents to exist in the upper mantle. Sea-floor spreading rates suggest the currents have velocities of a few centimeters per year.

Convection has markedly different ramifications than either of the other two modes of heat transport, because in convection huge masses in outer parts of the planet are in motion, whereas there is no large-scale motion in the case of conduction or radiative transfer. Long-term motion in the outer part of a planet ought to produce observable surface features, and, as we have already remarked, the "new tectonics" indeed indicate that the upper mantle is in motion.

What is happening in the other planets is not so clear. The absence of continental "plates," island arcs, trenches, and so on, on the moon and Mars suggests that convection is not important there, although there have been suggestions that the moon's triaxial shape might be related to a pattern of convection currents in the moon's interior. Probably the moon, Mars, and Mercury, because they are small, have cooled to such a degree that convection has little effect on the surface features. On the other hand, recent observations of the heat balance of Jupiter suggest that convection is proceeding in its interior.

Heat Balance of Planets

If a planet were in equilibrium with its surroundings, it would radiate just as much heat as it receives from the sun. If it radiated less than it received, it would be heating up; if more, cooling down. We have just pointed out, however, that in addition to sunlight, a planet contributes a certain amount of heat of its own, from radioactivity and possibly from early initial heat.

★ QUESTION ★

Which is more important in heating the earth's surface: sunlight or heat from radioactivity in the interior?

Answer

Everyday experience tells us that it gets warm in the daytime when the sun is up and cold at night when the sun goes down. Therefore, we might intuitively answer that sunlight dominates the earth's heat budget. But let us make a quantitative comparison. We have just quoted the observed value of heat flow coming out of the interior: about 60 ergs/cm^2-sec. In Chapter 2 we noted that the rate of arrival of sunlight energy on the earth is called the "solar constant," and it has a value about 1.4×10^6 ergs/cm^2-sec. Thus the heating of the earth's *surface* by solar energy is about 2×10^4 (20,000) times greater than that from internal heat. In the earth's deep *interior,* of course, internally generated, trapped heat is far more important than the negligible amount of heat conducted in from the surface, and the central temperature reaches about 3900°K. Fortunately, the surface is well insulated from this internal heat!

In the case of the other planets (excepting the moon), we cannot yet measure heat flow by drilling holes. However, there is another method of measuring the heat balance of planets. All bodies with a nonzero temperature must radiate. Cool bodies radiate *infrared* light that our eyes cannot detect.* Astronomical detectors can measure the infrared radiation and thus allow us to calculate the energy coming out of the distant planets.

Surprising results have been found in the case of Jupiter and Saturn. Kuiper (1952) concluded that the Jupiter atmosphere was heated primarily by internal sources because the spectacular cloud activity did not appear to be driven by solar energy. Öpik (1962) and Taylor (1965) found that the radiation from Jupiter and Saturn considerably exceeds the radiation they receive from the sun by 20 to 60 percent, indicating again that the internal heat sources are not at all negligible. A recent study concludes that the excess internal heat may be two to three times as much as is supplied by the sun (Trafton and Wildey, 1970)! This heat is possibly a remnant of that created during the formation of Jupiter by the accumulation of its large mass (Hubbard, 1970), or more probably heat released due to a slow contraction of the planet (Bishop and DeMarcus, 1970). The probable amount of radioactive material in Jupiter, judged by the high percentage of hydrogen (60 percent by mass) necessary to explain Jupiter's low density, seems insufficient to explain the excess heat.

* The sun has a surface temperature of about 5800°K and radiates "white" light (a mixture of all colors). Our eyes are most sensitive to this kind of radiation. No doubt natural selection gave us eyes most sensitive to the light radiated by the star around which our planet revolves. If we were in orbit around a cooler, more red star, our eyes would probably be sensitive to redder light.

Hubbard's work indicates that convection is necessary in Jupiter to account for the observed heat flow.

MAGNETISM OF PLANETS

Earth's Field and Its Reversals

★ QUESTION ★

How does a compass work?

Answer

A compass is a magnetized needle, and if a magnetic field is present, the needle aligns itself in what we call the direction of the field. Arbitrarily we define a "north-seeking" end of the needle, and then say that that end points toward the north direction of the magnetic field. The stronger the field, the stronger the tendency of the needle to align itself. Magnetic fields, then, have both strength and direction.

To make matters conceptually easier, physicists often speak of imaginary "lines of force," which are lines paralleling the direction of the field (the direction in which the compass needle comes to rest). This makes it possible to represent a magnetic field in a drawing such as Figure 10-23, showing a series of lines representing the lines of force.

Any simple magnetic system has two "poles," a north magnetic pole and a

Figure 10-23

Analogy between dipole magnetic field of a bar magnet (A) and the earth's dipole magnetic field (B). Drawing C shows the distortion of the earth's dipole field by the onrushing solar wind.

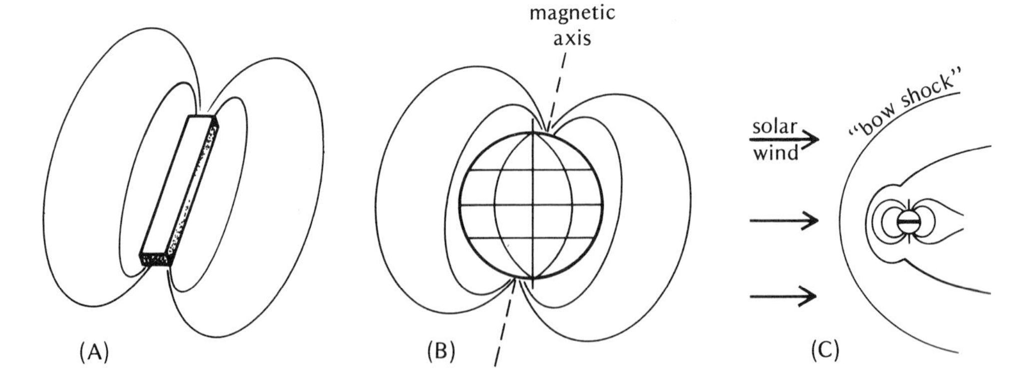

south magnetic pole. At any place, the compass will point toward the north and away from the south. Each magnetic system is thus like a small bar magnet with one end marked "north" and the other "south." One could create a very complex field by making a random pile of bar magnets with varying strengths and orientations.

The earth's field is similar to that of a single bar magnet. Such a field is called a *dipole field* because it is dominated by two poles. However, there are local complex, nondipole components of the field, about one tenth as strong as the main field.

The magnetic poles do not coincide with the poles of rotation but depart at an angle of $11^\circ\!.5$. This is why surveyors, navigators, and hikers must apply a "magnetic correction" to directions determined with a compass.

One might suppose that a characteristic as major as a planetary magnetic field would be relatively permanent. On the contrary, measurements show the field is continually changing strength, drifting so as to change its pole positions, and even reversing direction! About 2.5×10^8 (250,000,000) years ago the north magnetic pole was located in the present north Pacific. Measurements show that the present field is fluctuating at a rate of about 0.1 percent/year, so that we would expect major changes in strength on a time scale of 1000 years. According to a summary of data by Kaula (1968), the earth's field in 500 B.C. was 50 percent stronger than it is now.

Most mystifying is the fact that the direction of the field has reversed itself at sporadic intervals. No one yet knows why this happens, although evidence for reversed magnetism has been known since 1906. In the last few million years the reversals have occurred every few hundred thousand years, but there have been intervals as long as 50 million years when the direction remained constant or nearly constant. The process of reversal takes less than 10,000 years, and some believe that the field may reach nearly zero strength for a short time during this interval. There has been speculation that without a strong field to deflect cosmic ray particles, the incidence of cosmic rays at the surface may have gone up during the times of reversal, increasing biological mutation rates and affecting the course of biological evolution.

Paleomagnetism

The determination of past magnetic fields is the subject of a scientific field called *paleomagnetics*. Whenever a rock crystallizes from molten lava in the presence of a magnetic field, the magnetic elements in the rock, like compass needles, are frozen into position aligned with the field. The rock therefore retains a record of that original field for as long as the rock remains solid and undisturbed (e.g., barring melting, movement, or mechanical shocks).

If a geologist can locate an ancient undisturbed outcrop of rock and very carefully remove a sample, noting its original orientation, he can take it back to his laboratory and measure the very slight magnetism of the rock itself, which is called *remanent magnetism.* This gives a measure of the direction and strength of the ancient magnetic field.

Origin of Planetary Magnetic Fields

Could the earth's magnetic field be due to a magnetization of the solid inner core, making the inner core a huge analog of a bar magnet? In a series of papers from 1945 to 1956, Elsasser and others showed that neither this hypothesis nor several others would adequately explain the behavior of the earth's field (see review by Parker, 1970). Two arguments against the magnetized core are that (1) the core is too hot to retain any magnetization, and (2) the dramatic changes of the field indicate that it results not from a static situation but from a dynamical process. Further studies indicate that the process has to be able to regenerate the field, since the characteristic decay time for fields in a body of the earth's dimensions can be calculated to be of the order 3×10^4 (30,000) years, close to the observed time scale for major changes in the field.

The upshot of modern theoretical investigations is that the earth's field is produced through the combination of planetary rotation and convection currents in the core. The liquid part of the core is thought to be in convection, with a number of temporary individual cells operating at a given time. This number may be about 10 to 20, as suggested by patterns detected in the earth's field. Rotation of the earth produces *coriolis forces*—forces that affect the motions of any fluid in a rotating system. The effect of the coriolis forces is to cause a shearing of convective cells in the highly conductive core.

The core is thus supposed to work like a dynamo. The loss of heat from the core by convection provides a continuous energy input to keep the circulation going. As long as the material keeps moving within the magnetic field, currents are set up which perpetuate the field. This theory thus predicts that slowly rotating planets or planets without active convection would not have strong magnetic fields (see the next section).

Magnetic Fields of the Sun and Other Planets

Table 10-3 summarizes available data on magnetic fields of other bodies in the solar system. The sun has a magnetic field in some ways similar to the earth's except with a shorter time scale. The sun's magnetic field reverses every 11 years, in step with a cyclic behavior in the number of sunspots. Every 22 years the sun goes through a complete magnetic cycle and two cycles in sunspot numbers.

Spacecraft measures near Venus, the moon, and Mars indicate that none of them have significant magnetic fields comparable to the earth's. This sup-

Table 10-3
Planetary magnetic fields[a]

	Planetary magnetic dipole moment (emu)	Inclination of dipole to rotation axis	Local measured field (gauss)	Source of data
Venus	$< 10^{23}$			Mariner 5
Earth	8×10^{25}	$\sim 11°$	~ 0.5 (surface)	
Moon	$< 10^{21}$		4×10^{-4} (surface)	Apollo 12
Mars	$< 2 \times 10^{22}$			Mariner 4
Jupiter	$\sim 8 \times 10^{30}$	$\sim 11° \pm 5°$	0.1–1.0 ($3R_{♃}$)	Radio
Saturn	$\ll$ Jupiter			Radio
Sun		$\sim 6°$	~ 2 (surface)[b]	Spectra

[a] The ambient interplanetary magnetic field among the terrestrial planets averages a few gammas (1 gamma = 10^{-5} gauss) but is time variable by an order of magnitude.
[b] Highly variable, up to 10^3 gauss in sunspots.

ports the theoretical ideas explained above, since Venus rotates very slowly and the moon and Mars are so small that they have probably cooled to a point where convection currents are no longer operative in transporting heat in their interiors.

Jupiter has a very strong magnetic field and rotates $2\frac{1}{2}$ times faster than the earth. The evidence for the strong field comes from radio signals emitted as a result of interactions between high-energy atomic particles and the magnetic field. Like the earth, Jupiter apparently has a swarm of such particles trapped and moving in belts around the planet. In the case of the earth, these belts are called the *Van Allen belts,* after physicist James Van Allen, who discovered them in one of the first experiments with artificial satellites (see the review by Van Allen, 1961).

Effect of Io on Jupiter's Radio Emission

Jupiter has strong radio emission at various wavelengths. For the various kinds of strong, erratic radio signals there are various explanations, although the signals are not well understood. Radiation at wavelengths of about 10 cm is *synchrotron radiation,* which is produced by charged particles caught in Jupiter's magnetic field (Field, 1959).

At longer wavelengths of tens of meters [frequency about 20 megacycles (20 MHz according to recent convention)], Jupiter has a different sort of radiation. This *decameter radiation* is strong, erratic, and comes in bursts. It comes predominantly from a specific source area on Jupiter, which has been measured as less than 10,000 km in dimension (Dulk, Rayhrer, and Lawrence, 1967). A surprising discovery about the decameter radiation was made by

Bigg (1964): The bursts of radio noise are strongly correlated with specific positions of the satellite Io in its orbit around Jupiter.

Io, which is rather similar in size to our own moon, is the innermost of the four large Galilean satellites of Jupiter. Why should such a satellite control Jupiter's decameter radio noise? Neither the cause of the radiation nor the effect of Io is well understood, but it is widely thought that Io may disturb the magnetic field of Jupiter through a magnetic link along the "lines of force" in such a way as to produce beams of radiation that are intermittently pointed toward the earth as Jupiter rotates (Duncan, 1966; Dulk, Rayhrer, and Lawrence, 1967).

Two Pioneer spacecraft that will be well instrumented to study the magnetic field are being readied for a flight to Jupiter in the mid-1970s. Perhaps by 1975 we shall have better data to discuss Jupiter's magnetic field, satellites, and internal and atmospheric processes.

CONCLUSION: MODELS OF PLANETARY INTERIORS

Mercury, Venus, and Earth

Of Mercury and Venus there are few observations that illuminate the observational desiderata discussed earlier. Consequently, little work has been done in constructing theoretical models of the interiors of these planets. It is probable that spacecraft observations made within a decade will greatly improve the observational data. It will be of special interest to compare the interior of Venus with that of the earth, since the two planets have nearly equal masses and radii but very different surface chemistries and atmospheres.

The interior of the earth is relatively well known, as discussed earlier and shown in Figure 10-9.

The Moon

As shown in Table 10-1, the moon has very nearly the moment of inertia of a homogeneous body. Its mean density, 3.34 g/cm^3, is so similar to that of ordinary rocks that a massive nickel-iron core like the earth's can be ruled out. The magnetometer deployed on the Apollo 12 mission enabled Sonett et al. (1971) to make the best model yet available for the lunar interior, shown in Figure 10-24. According to this model, the lunar core (out to 82 percent of the radius) is nonmolten primitive material, at a temperature less than 1200°K. The lunar mantle, about 340 km thick, has undergone partial melting that produced a thin lunar crust composed mostly of light-colored "terra material" with scattered darker lava flows, or "maria." Analysis of lunar samples in-

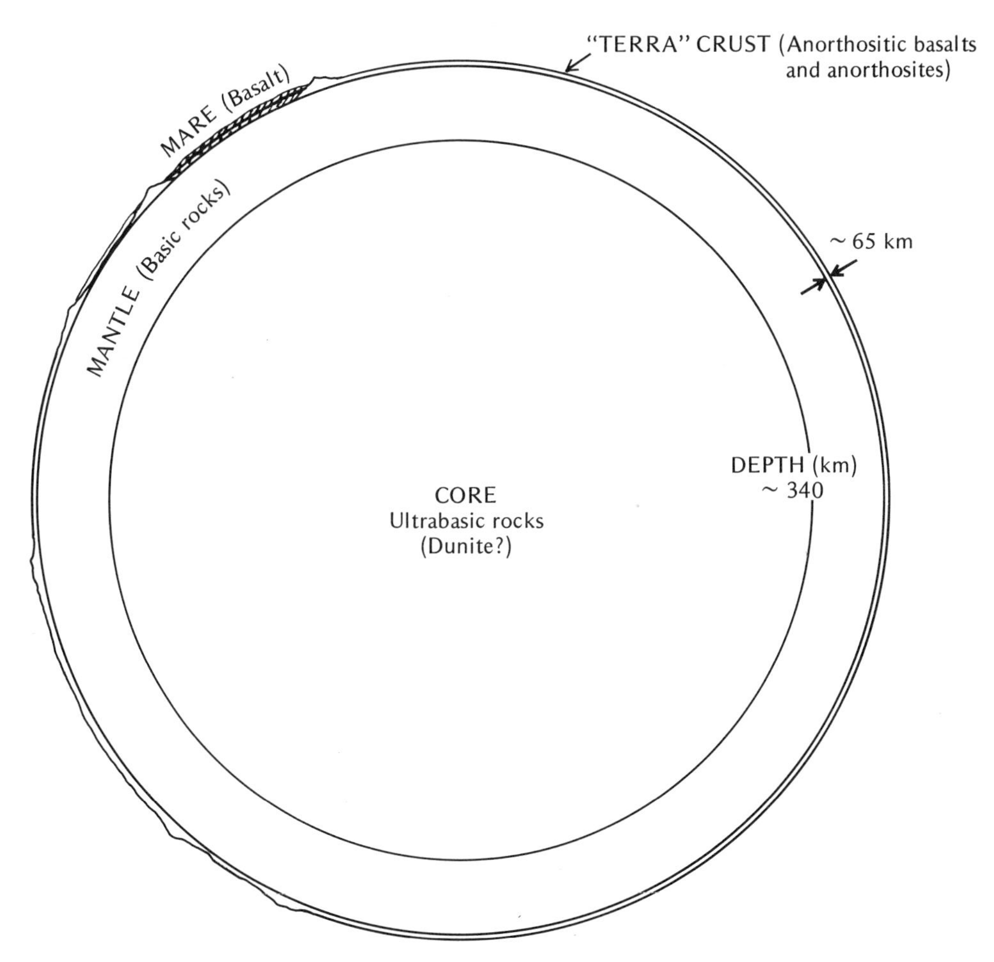

Figure 10-24

A model of the moon's interior based on rock samples and magnetometer measurements obtained during early Apollo flights. (After Sonett et al., 1971, and other data.)

dicates that the terra material may be rich in *anorthosite,* a rock type consisting of plagioclase feldspar minerals (see Chapter 11) and common in deep-seated massive terrestrial inclusions.

Sonett's original interpretation of the lunar data was that the core had never been melted, and that heating sources had been external, melting the moon from the crust and mantle inward. An alternative interpretation is that the lunar core was once melted and somehow cooled to its present state. This is more consistent with the strong evidence that bodies both smaller than the moon (meteorite parent bodies) and larger (the earth) melted internally. Final interpretation of data on the lunar interior is still being debated.

Mars

Mars presents a still more difficult case. Table 10-1 shows that it has nearly the moment of inertia of a homogeneous body; and its mean density, about 4.0, is much less than the earth's, 5.52. Early theorists, such as the English geophysicist, Jeffreys, discussed two models, one with a small iron core and a second with nearly homogeneous composition, slightly higher in density near the center. The difference is crucial, as it carries an implication as to whether Mars ever melted and differentiated. Recent models do not rule out an iron or nickel-iron core. A model by Binder (1969) concludes that Mars has a mean composition rather like that of the earth's mantle (and similar to the moon), with an iron core of 23 to 28 percent of the planet's radius. This model suggests that Mars went through an early differentiation process and once had a hot, molten core and possibly an associated magnetic field, which disappeared as the core cooled.

On the other hand, some investigators have concluded that Mars has *not* yet gone through a partial melting or differentiation process. Sharp (1971) points to so-called "chaotic terrain" on Mars, found in the 1969 Mariner spacecraft photos, as a possible example of incipient orogeny. On the basis of certain theoretical models, he suggests that Mars may still be heating up and just reaching a state of thermal maturity in which orogenic processes are now beginning.

Giant Planets

Models of the giant planets have been calculated by means of the techniques described earlier. Following the realization that the interiors of Jupiter and Saturn must be predominantly frozen hydrogen and helium—not containing large silicate cores as had earlier been proposed—DeMarcus (1958) computed theoretical models composed of solid hydrogen and helium. Öpik's (1962) review of data on Jupiter found these models best satisfied available information, but subsequent observations of Jupiter's heat balance disclosed an enormous heat flow from the interior and necessitated new models. Hubbard (1970) concluded that the entire interior of Jupiter was in convection, implying no chemically distinct central core, but a chemically mixed interior with a hydrogen content about 60 percent by mass and a central density around 4 g/cm^3. Bishop and DeMarcus (1970), on the other hand, proposed a solid conductive core similar to that in DeMarcus's 1958 models, with a convecting outer mantle overlying the core. Figure 10-25 shows the salient features of these recent models.

Current models of Saturn are less certain, the heat balance being less well determined than Jupiter's. Hubbard applied his assumption of complete convection to Saturn and computed models with central densities of 2.3 to 2.4 g/cm^3, but found that these did not satisfy observed dynamical considerations.

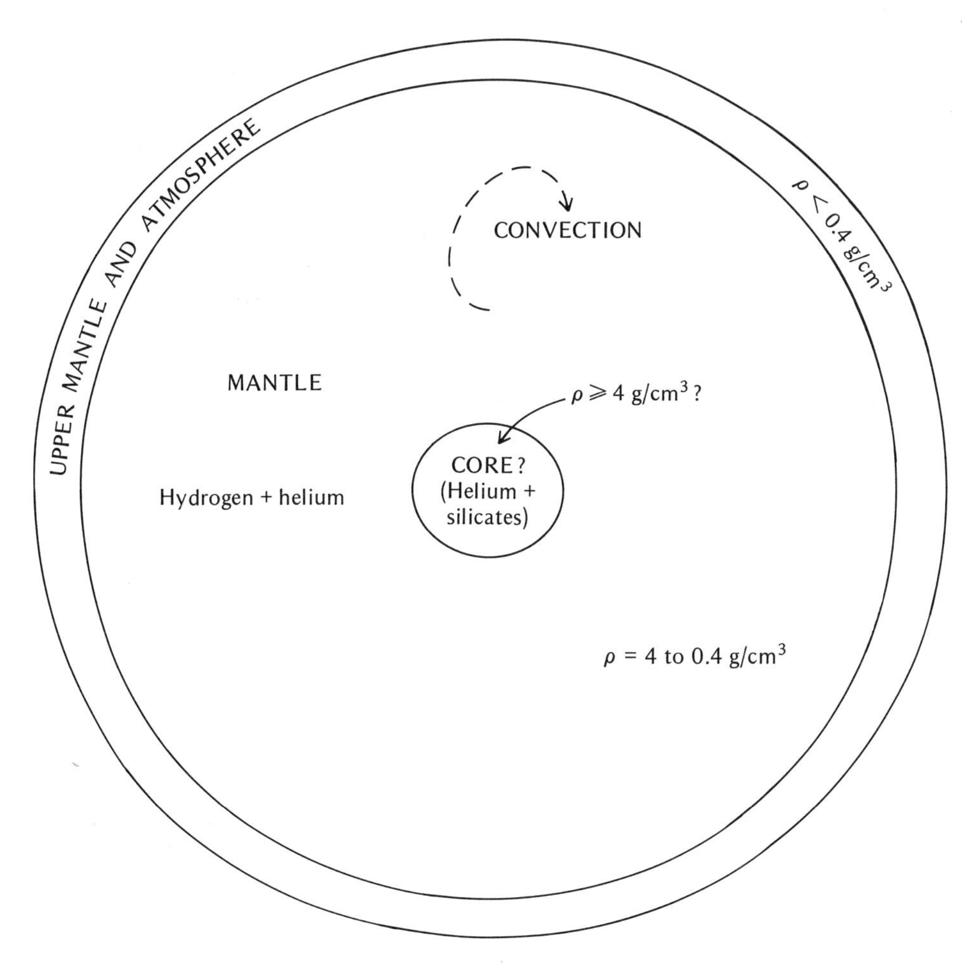

Figure 10-25

A model of the interior of Jupiter, based on calculations by DeMarcus (1958), Öpik (1962), Hubbard (1970), and Bishop and DeMarcus (1970).

As is clear from their mean densities (Figure 10-1), Uranus and Neptune are composed of more dense materials than Jupiter and Saturn. Ramsey (1967) proposed models in which central cores of earth-like composition are surrounded by uniform mantles of hydrogen and mixed, frozen, hydrogen-saturated compounds of C, N, and O, such as methane, metallic ammonia, and water ice, with an additional admixture of neon. Ramsey shows that the hydrogen content lies in the range 10 to 15 percent by mass, contrasted to the higher values found for Jupiter and Saturn. Neptune has a substantially larger core than Uranus. Central densities for Uranus and Neptune are found by Ramsey to be about 2.8 and 3.6 g/cm^3, respectively.

References

Bacon, Francis (1620) *Novum Organum* (London).

Beloussov, V. V. (1962) *Basic Problems in Geotectonics* (New York: McGraw-Hill, Inc.).

Bigg, E. K. (1964) "Influence of the Satellite Io on Jupiter's Decametric Emission," *Nature, 203,* 1008.

Binder, A. B. (1969) "Internal Structure of Mars," *J. Geophys. Res., 74,* 3110.

Bishop, E., and W. DeMarcus (1970) "Thermal Histories of Jupiter Models," *Icarus, 12,* 317.

Carey, W. (1962) "The Scale of Geotectonic Phenomena," *J. Geol. Soc. India, 3,* 97.

DeMarcus, W. C. (1958) "The Constitution of Jupiter and Saturn," *Astron. J., 63,* 2.

Dewey, J. F., and J. Bird (1970) "Mountain Belts and the New Global Tectonics," *J. Geophys. Res., 75,* 2625.

Dulk, G., B. Rayhrer, and R. Lawrence (1967) "The Size of Jupiter's Decameter Radio Source," *Astrophys. J. Lett., 150,* L117.

Duncan, R. A. (1966) "Comments on the Modulation of Jovian Decametric Emission by Jupiter's Satellites," *Planet. Space Sci., 14,* 173.

Eaton, J. P. (1962) "Crustal Structure and Volcanism in Hawaii," in *Crust of the Pacific Basin, Geophys. Monograph, 6,* p. 13 (Washington, D.C.: American Geophysical Union).

Ewing, M., and B. C. Heezen (1956) *Some Problems of Antarctic Submarine Geology, Am. Geophys. Union, Monograph, 1,* p. 75 (Washington, D.C.: American Geophysical Union).

Field, G. B. (1959) "The Source of Radiation from Jupiter at Decimeter Wavelengths," *J. Geophys. Res., 64,* 1169.

Hooke, R. (1705) "Lectures and Discourses on Earthquakes," *Posthumous Works* (London).

Hubbard, W. B. (1970) "Structure of Jupiter: Chemical Composition, Contraction and Rotation," *Astrophys. J., 162,* 687.

Hurley, P. M. (1968) "The Confirmation of Continental Drift," *Sci. Am., 218* (April), 52.

Jeffreys, H. (1924) "On the Internal Constitution of the Earth," *Monthly Notices Roy. Astron. Soc., 84,* 534.

______ (1962) *The Earth* (New York: Cambridge University Press).

Kaula, W. M. (1968) *An Introduction to Planetary Physics* (New York: John Wiley & Sons, Inc.).

______ (1969) "Interpretation of Lunar Mascons," *Phys. Earth Planet. Interiors, 2,* 123.

Kuiper, G. P., ed. (1952) "Planetary Atmospheres and Their Origin," in *The Atmospheres of the Earth and Planets* (Chicago: University of Chicago Press).

Latham, G. V. (1971) "Lunar Seismology," *Trans. Am. Geophys. Union, 52,* 162.

MacDonald, G. J. F. (1964) "The Deep Structure of Continents," *Science, 143,* 921.

Menard, H. W. (1969) "The Deep-Ocean Floor," *Sci. Am., 221* (Sept.), 127.

Moberly, R., and M. A. Khan (1969) "Interpretation of the Sources of the Satellite-Determined Gravity Field," *Nature, 233,* 263.

Mohorovičić, A. (1909) *Jb. Met. Observ. Agram., 9,* 1 (Zagreb, Croatia).

Muller, P. M., and W. L. Sjogren (1968) "Mascons: Lunar Mass Concentrations," *Science, 161,* 680.

Oldham, R. D. (1900) *Phil. Trans. Roy. Soc. London, A194,* 135.

______ (1906) "The Constitution of the Earth," *Quart. J. Geol. Soc., 62,* 456.

Öpik, E. J. (1962) "Jupiter: Chemical Composition, Structure, and Origin of a Giant Planet," *Icarus, 1,* 200.

Owen, N. B., and J. E. Martin (1966) "The Selection, Orientation, and Mounting of Diamonds for Use as Bridgman Anvils," *J. Sci. Instr., 43,* 197.

Parker, E. N. (1970) "The Origin of Magnetic Fields," *Astrophys. J., 160,* 383.

Poisson, S. D. (1829) *Mem. Acad. Sci. Paris, 8,* 623.

Ramsey, W. H. (1967) "On the Constitutions of Uranus and Neptune," *Planet. Space Sci., 15,* 1609.

Rayleigh, Lord (J. W. Strutt) (1887) "On Waves Propagated along the Plane Surface of an Elastic Solid," *Proc. London Math. Soc.* (1), *17,* 4.

Ringwood, A. E. (1956) "The Olivine-Spinel Transition in the Earth's Mantle," *Nature, 178,* 1303.

——— and A. Major (1966) "High-Pressure Transformations in Pyroxenes," *Earth Planet. Sci. Lett., 1,* 351.

Ritsema, A. R. (1954) "A Statistical Study of the Seismicity of the Earth," *Meteorol. Geophys. Serv., Indonesia, Verhandl., 46.*

Sharp, R. P. (1971) "The Surface of Mars: 2. Uncratered terrains," *J. Geophys. Res., 76,* 331.

Sonett, C., B. Smith, D. Colburn, G. Schubert, and K. Schwartz (1971) "Preliminary Assessment of the Lunar Lithospheric Thermal Gradient, Heat Flux, Deep Temperature, and Compositional Gradation," presented at 1971 Lunar Science Conference, Houston.

Taylor, D. J. (1965) "Spectrophotometry of Jupiter's 3400–10,000 Å Spectrum and a Bolometric Albedo for Jupiter," *Icarus, 4,* 362.

Tazieff, H. (1964) *When the Earth Trembles* (London: Hart-Davis).

Trafton, L. M., and R. L. Wildey (1970) "Jupiter: His Limb Darkening and the Magnitude of His Internal Energy Source," *Science, 168,* 1274.

Urey, H. C. (1952) *The Planets: Their Origin and Development* (New Haven, Conn.: Yale University Press).

Van Allen, J. A. (1961) "The Geomagnetically Trapped Corpuscular Radiation," in *Science in Space,* L. V. Berkner and H. Odishaw, eds., p. 275 (New York: McGraw-Hill, Inc.).

Vine, F. J., and D. H. Matthews (1963) "Magnetic Anomalies over Ocean Ridges," *Nature, 199,* 947.

Wegener, A. L. (1922) *Die Entstehung der Kontinente und Ozeane,* 3rd ed. (Braunschweig: Vieweg und Sohn).

Wildt, R. (1938) "On the State of Matter in the Interior of the Planets," *Astrophys. J., 85,* 508.

——— (1961) "Planetary Interiors," in *Planets and Satellites,* G. P. Kuiper and B. M. Middlehurst, eds. (Chicago: University of Chicago Press).

Wilson, J. T. (1954) "The Development and Structure of the Crust," in *The Earth as a Planet,* G. P. Kuiper, ed. (Chicago: University of Chicago Press).

Wyllie, P. J. (1963) "The Nature of the Mohorovičić Discontinuity, A Compromise," *J. Geophys. Res., 68,* 4611.

PLANETARY SURFACES

THE TOPIC OF PLANETARY SURFACES is perhaps the most intriguing of topics in planetary science because it addresses the questions we ask even as children: What is it like to stand on another planet? The formal goals of studying planetary surfaces are to describe the general conditions (rock types, temperature, surface structure, etc.) and understand the processes acting (volcanism, impacts, erosion, deposition, etc.) on the surfaces of the planets.

The surfaces of the nearer planets are composed of rocky materials, and hence we shall begin with a necessarily sketchy survey of rocks and minerals, stressing definitions of common terms. Only armed with a knowledge of rocks and rock types can we begin discussions of complications such as interactions with atmospheres. As the Rev. John Fleming noted in 1813 (Geikie, 1905),

> He who has the boldness to build a theory of the earth without a knowledge of the natural history of rocks will daily meet with facts to puzzle and mortify him.

PETROLOGY

Petrology is the study of rocks, their description, classification, evolution, and origin. This vast subject is reviewed in many geology textbooks. The present introduction is intended only as a guide for students with no geological background. An excellent discussion of the physics and chemistry of terrestrial rocks is given by Mason (1966), and a related work on the implications of lunar rocks is by Mason and Melson (1970).

Minerals

Minerals are inorganic substances that compose planetary crusts. They are characterized by forms assumed by their constituent elements or compounds and they are defined by composition and structure. Rocks are composed of minerals, usually in the form of crystals arranged in various mixtures, but sometimes in glassy or amorphous forms which may occur under proper conditions, such as rapid cooling.* Many minerals are defined by simple, fixed atomic composition (e.g., SiO_2, which is called silica or quartz), but they do not all necessarily have such simple formulas. For example, olivine has the compositional formula $(Mg,Fe)_2SiO_4$, which means that ions of either magnesium (Mg) or iron (Fe) may *substitute* in the crystal lattice. The governing factor in substitution is often the size of the atom or ion; the situation is analogous to a structure made of packed Ping-Pong balls in whose interstices BB's, but not golf balls, could fit. The number of substituting ions is, of course, constrained by the requirement of electrical neutrality. Many minerals are thus defined in terms of ranges of composition.

Although they are defined chemically, minerals are visually identified in the field by a series of semiquantitative tests of such properties as hardness (on a 0-to-10 scale called "Moh's" scale), streak (color of powdered form produced when sample is rubbed onto a small, hard tile or "streak plate"), shape, luster, and density.

In typical rock specimens, the mineral crystals range from easily visible to microscopic. In the latter case the rock seems homogeneous but is really a mass of fine mineral crystals. Microscopic examination is often required to identify the various minerals present.

Minerals and rocks are sometimes loosely classified in terms of composition, ranging from *acidic* or *siliceous* (high silica content) through *basic* or *mafic* (high content of heavy elements such as iron and elements with chemical affinity for iron) to *ultrabasic* or *ultramafic* (very rich in heavy elements and low in silica). A *very rough* rule of thumb is that the darker or denser the rock type, the more basic it is. In a planet that has been at least partially molten, crustal minerals and rocks tend to be siliceous and deep-seated minerals and rocks tend to be ultrabasic. Volcanism brings up basic minerals and rocks from depth.

The vast majority of rocks are composed mostly of a few important minerals. Examples of important minerals include the following:

Quartz (silica): SiO_2. By far the most common simple rock-forming chemical compound at the earth's surface, constituting roughly 60 percent of igneous and sedimentary rocks.

Feldspars: $(K, Na, Ca)AlSi_3O_8$. *Orthoclase* feldspars are rich in potassium; *plagioclase* feldspars are rich in sodium and calcium. Accurate identification of the feldspar crystals is necessary in classifying igneous rocks—see below.

Olivine: $(Mg, Fe)_2SiO_4$. Greenish-colored, ultrabasic mineral; often found as inclusions in basalt lava.

* Glasses are noncrystalline; that is, the atoms and molecules have not been able to arrange themselves in the characteristic, orderly, geometric arrangement of a crystal. The difference between a crystal and a glass, as George Gamow once remarked, is like the difference between a carefully built brick wall and frozen caviar.

Pyroxenes: A group of Mg, Fe, Ca, and Al basic silicates, including specific minerals such as *augite* (probably the most common), *enstatite, hypersthene, pigeonite,* and *diopside.* Found in all basic rocks. These minerals are common in meteorites and basic terrestrial rocks.

Amphiboles: A group of Mg, Fe, Ca basic silicates—of different crystal structure from the pyroxenes—including *hornblende.*

Micas: K, Al, and Mg silicates, including *biotite,* $K(Mg, Fe)_3AlSi_3O_{10}(OH)_2$.

Examples of minerals important in extraterrestrial samples are

Troilite: FeS, accessory mineral in most meteroites, typically constituting 5 to 6 percent by weight in chondrites.

Limonite: Hydrous amorphous ferric iron oxide of variable composition and orangish color. Widely discussed as a possible constituent of the Martian surface, it contains a mixture of other minerals, such as hematite and goethite.

Rocks and Rock Types

Rocks are divided into three groups. *Igneous rocks* are rocks formed directly from the cooling of molten magma. *Sedimentary rocks* are formed by deposition and cementing of small particles, either of biogenic or nonbiogenic origin. *Metamorphic rocks* are rocks originally formed in either of the above modes but changed to a new rock form by high pressure, high temperature, or addition of new chemicals.

Compositions in all these categories may range from siliceous to basic. An example of a siliceous igneous rock is *granite.* An example of a basic igneous rock is *basalt,* dark-brown or black craggy rock common in lava flows.

Magma. The parent stuff of rocks is *magma,* molten rock material with variable content of volatiles that may escape as the magma crystallizes. *Lava* is exposed magma (or, loosely, the rock crystallized from exposed magma). As they cool, magmas crystallize in a *sequence* of events, from the appearance of the first solid phases to the disappearance of the last watery solution. Certain parts may separate out, or the sequence may be interrupted at any time. In Mason's words (1966, p. 196), magma is characterized "(a) by composition, in that it is predominately silicate, (b) by temperature in that it is hot (. . . from 500 to 1200°K), and (c) by mobility, in that it will flow."

Where does terrestrial magma come from? At one time magma was assumed to issue from molten regions of the earth, but modern seismic evidence indicates that the crust and mantle, where the magma must originate, are predominantly solid. The answer to this question will be discussed later in this chapter. Is there a single, parent, "*primary magma*" from which all rocks and other magmas are created? Decades ago geologists spoke of two distinct types of primary magmas (which were supposed to produce grantitic and basaltic rocks). Evidence cited for the two types of magmas includes: (a) two distinct frequencies of silica content in igneous rocks, corresponding with (b) two principal classifications of rocks observed in the field, granitic and basaltic. Such a bimodal distribution of magma composition might be related to inhomogeneities in the mantle (see Chapter 10). Contemporary geologists are inclined to view all magma as ultimately derived from mantle material, but with the more siliceous magma having been recycled several times in the erosion and mountain-building cycles of the continents. Basic as the issue may be, it cannot be regarded as entirely settled. Rocks from different regions of the moon and Mars may make the situation clearer by showing what processes those planets have experienced.

Igneous rocks. Igneous rocks are rocks that are the direct result of crystallization of magma (either remelted surface material or "primary magma"). The classification scheme is shown in Figure 11-1. If a rock is slow to cool, as in a deep-seated *intrusive* body, the crystals can grow to visible size; then the rock is coarse grained, or *phaneritic.* In certain circumstances, when the mobility of the matrix is great, often in the presence of fluids (as around the margins of intrusives that have cooled and mostly crystallized) certain crystals may grow very large (up to 15 meters); then the rock is unusually coarse grained and called a *pegmatite.* When crystallization is more rapid (e.g., in a surface, or *extrusive,* flow), the crystals are microscopic and the rock is fine grained, or *aphanitic.* Rocks with fine-grained matrices may incorporate prominent large crystals, called phenocrysts; such rocks are called *porphyries.*

Igneous rocks and their associated geologic structures were defined in a classic book by Daly (1933), but the nomenclature has been modified and multiplied by succeeding authors. Probably the best current reference for such nomenclature is the *AGI Glossary of Geology* (Howell, 1960).

In the context of planetary science, it should be remembered that the magmas and rocks of the earth's surface are products of a highly complex differentiation process (Chapter 10). Terrestrial rock compositions reflect more highly selective chemical processing than occurred on some other planets. Figure 9-12 showed that terrestrial rocks exhibit extreme departures from cosmic abundances. Although some two thirds of the mass of the universe is hydrogen, the planets are composed mainly of heavier elements, volatiles having been lost in the early planetesimal stages. Vestiges of the "cosmic" abundance curve can be seen, but terrestrial surface rocks show erratic fluctuations, produced by the chemical effects of melting and recrystallizing of the earth's rocks. (See Chapter 9 for discussion of similar effects on meteorites.)

The kind of rock produced by a given melt depends on what happens during the cooling. As the temperature drops, minerals react with each other and with the melt. Therefore, the resulting rock types depend on the degree of *fractionation* during the crystallizing process (i.e., the degree to which solid, early-crystallizing minerals are separated out of the melt). Bowen (1928) expressed this concept in his famous *reaction series* (also described by Mason, 1966, p. 124), which shows the sequence of minerals that will crystallize as the temperature drops if the melt is fractionated. If the more basic minerals, which crystallize early, are removed, the last stages of crystallization will be the

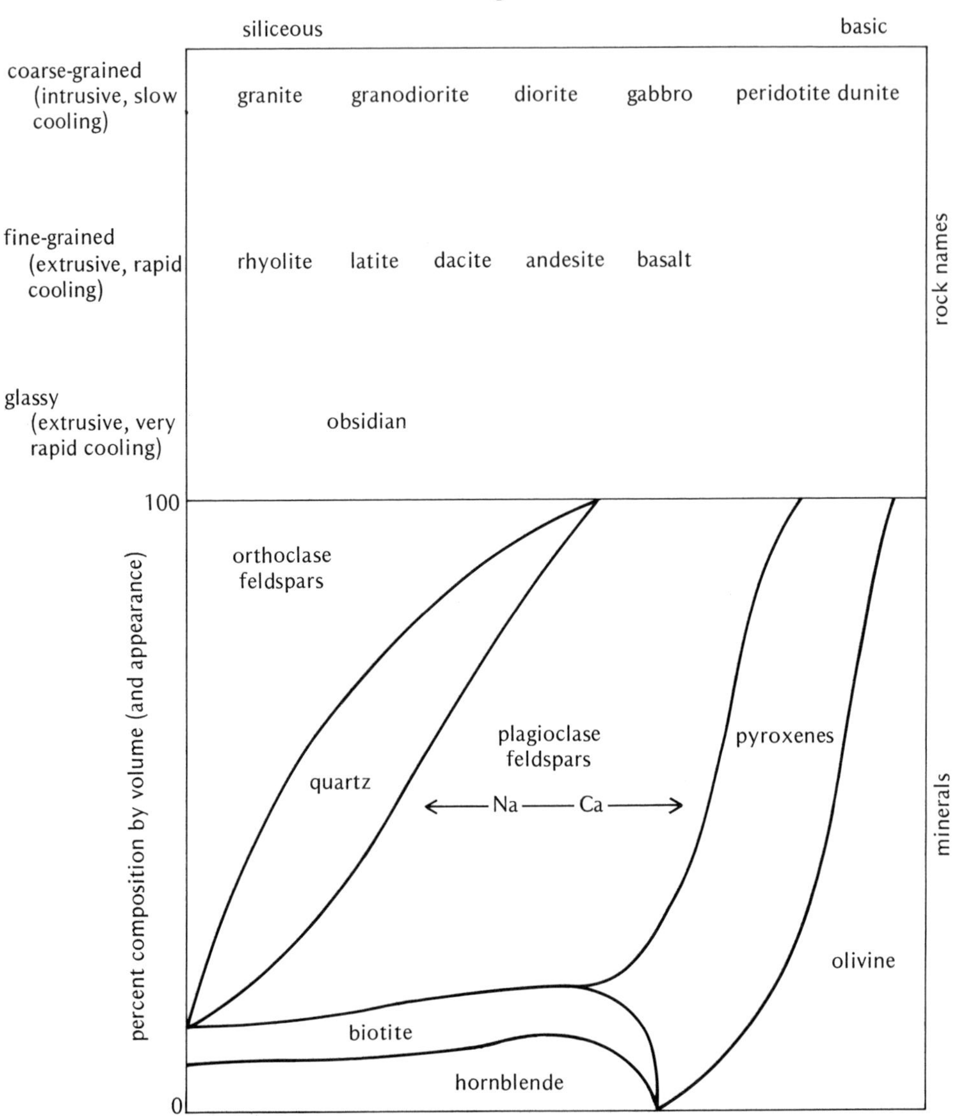

Figure 11-1

Classification scheme for igneous rocks. The bottom diagram shows composition by mineral content, and the upper diagram shows the corresponding rock names, depending on grain size of the mineral grains of which the rock is composed.

production of quartz, in watery solution of hydrothermal minerals. Conversely, if the final igneous rock is again heated, the first stages of melting will produce a hydrothermal mineral solution, then fusion of quartz, and lastly certain basic minerals.

The geochemistry of igneous rocks is very important in planetology. The meteorites and the moon, for example, show evidences of igneous processes. Fundamental questions about the nature of magma and the importance of planet-wide melting and differentiation were reopened as a result of study of the igneous rock samples, particularly gabbros and anorthosites, brought back in Apollo missions (see The Moon Issue of *Science,* January 30, 1970). *Anorthosite,* a rock type lighter in color than the gabbroic lavas and rich in plagioclase feldspars, is thought by many to be a major constituent of the light-colored lunar uplands.

Igneous rocks, such as dunite (olivine-rich), peridotite (olivine-pyroxene), and eclogite (pyroxene-garnet), are thought to represent the earth's mantle.

Sedimentary rocks. Sedimentation is essentially a planetary skin effect, "the interaction of the atmosphere and hydrosphere on the crust of the planet" (Mason, 1966). Chemical and mechanical *weathering* constantly attack rocks exposed at the surface of a planet with an atmosphere. Chemical weathering involves chemical reactions that break down the rocks; mechanical weathering involves nonchemical breakdown of rocks, for example by abrasion in blowing sand or in stream beds. The rock particles are transported away in suspension and in solution. New rocks of different compositions and texture are created in other places from their remains. Sedimentary rocks are the most common rocks outcropping on the earth's land *surface,* although by volume they constitute only a few percent of the total crust.

As indicated, sedimentary rocks can be produced either by chemical processes (e.g., salt deposits) or as fragments of other rocks (e.g., sandstones, shales). The first rocks are called *chemical sediments* and the second, *clastic sediments.*

In the weathering and transport processes there is chemical separation in addition to that discussed under igneous rocks. Certain minerals, especially quartz, are hard—resistant to solution during the low-temperature weathering processes—and are dropped as fragments; sandstones may result. Other minerals are broken down; the aluminosilicates give rise to clays, muds, and shale; ferrous iron is oxidized to the ferric state, which may produce hydroxide precipitates abundant enough to form ore deposits; calcium minerals produce calcium carbonate solutions, which may yield limestones. Certain elements that remain in solution are carried preferentially from the land to ocean basins; sodium, the best example, is constantly being added to oceans, and the ocean's saltiness is increasing.

Sedimentary rocks are probably absent on the smallest planets because of the absence of adequate transport and compaction processes. We can speculate that Mars is the smallest planet with true sedimentary rocks because yellowish dust clouds are observed on Mars, indicating that transport and deposition of wind-blown dust must be effective there.

Metamorphic rocks. Metamorphism is the net effect of all processes acting to alter and recrystallize solid rock material beneath a planet's surface. These processes are a result of variable pressure, temperature and chemical environments. A given environment *and a given composition* imply equilibrium only for a certain assemblage of rock minerals, and these particular minerals may or may not be present. If they are not present, adjustments in the rock will tend to produce them.

The presence of chemically active volatiles is perhaps the most important factor in metamorphism. They may carry off or introduce new material to the system, a process called *metasomatism.* Examples of other metamorphic processes are creation of high-density minerals by increases in pressure and physical rearrangement of minerals (e.g., along parallel bands by vertical pressure or shearing).

Some common metamorphic rocks are listed in Table 11-1. Pressure plays an important role in producing the banding and cleavage characteristic of the first three entries.

If temperature rises too high during metamorphism, rocks will of course undergo complete melting, called *anatexis,* and a new magma is produced.

Controversies over granite. So widespread is granite on the surface of the earth that the eighteenth-century geologist Werner (1749–1817), one of the early petrologists, speculated that granites originated by crystallization out of a world-wide ocean. This idea was overturned when James Hutton (1726–1797), the Scottish naturalist who is regarded as one of the founders of modern geology, first concluded in the late 1700s that many rocks, including granite, had introduced into their present positions and crystallized directly from magma. Many of Hutton's views, derived from hours of tromping around in the field, have stood the test of time, but his view that all granites stem from magmas has been again questioned in recent decades because of field observations of, for example, sedimentary rocks that grade continuously into granites. In some granitic zones, vestiges of the sedimentary structure (e.g., stratification)

Table 11-1

Metamorphic rock	Common examples of parent rock	Description
Gneiss	Granite	Marked parallel band, coarse grained
Schist, phyllite	Shale, granite	Banded, fine grained
Slate	Shale	Fine grained, splits into thin sheets
Quartzite	Sandstone	Massive quartz rock, breaks through quartz grains
Marble	Limestone	Familiar in architecture

can still be traced. How can a sediment grade continuously into an igneous intrusion? What happened to tremendous volumes of sediments which, according to some interpretations, were replaced by intruded granitic magmas? These questions led to a gradual realization that granites can be produced both by direct crystallization of a granitic magma and by metasomatism, through which existing rocks can be altered in place by invasion of other fluids through them (see Mason). The second process, also called *granitization,* has been highly controversial but is becoming accepted. There is still argument over relative importance of the different modes of origin.

The origin of granite is related to the origin of the continents. Can the continents be viewed as a granitic "scum" floating on a basic interior? Weathering and transport, operating on igneous and metamorphic rocks, leave residual accumulations of silica-rich sediments from which surface processes produce granite. Jacobs, Russell, and Wilson (1959, p. 310) have thus argued that the continents (i.e., accumulations of siliceous rocks) originally grew by this sort of process.

EROSION, DEPOSITION, DEBRIS, AND DUST

Once rocks are exposed on planetary surfaces, what processes act on them? If they are exposed to any atmospheric or oceanic agents, there will be processes tending to wear them away. Even if there is no atmosphere, certain agents will alter the rocks, as we will see in the next section. Such processes can be divided into two broad categories, erosion and deposition.

Erosion

Erosion includes all processes breaking down, or *weathering,* and *transporting* rock. On the earth, running water is the most important erosive agent, both in streams and wave action along coastlines. In arid regions, wind is an important agent in wearing down rocks by driving sand and in transporting the resulting particles. Further examples of weathering include chemical weathering and the action of repeated freezing and thawing when moisture is present. The latter process fractures rock because of water's unusual property of expanding when it freezes. Water drains into tiny fractures in rocks and upon freezing expands, acting like a wedge to split the surrounding material.

Deposition

Deposition includes all processes depositing fragmented rock products on the surface. The forms of resulting deposits and rock types vary widely and depend on the planetary surface conditions. Most deposition on earth is associated with settling of fine material after transport by water. Compaction and cementing of such deposits produces sedimentary rocks, often in distinct layers reflecting different deposition episodes. Deposition of windblown fine material can produce shifting dune deposits. Less familiar on the earth but common on the moon are massive sheets of material deposited as fallout after ejection from impact craters.

Mass Motions

Once transported and deposited, the products of rock erosion are not necessarily static. In addition to possible further transport by wind or water, an interesting process is movement of loosely consolidated ground cover in response to gravity, even under conditions of seemingly insignificant stress. Various kinds of mass motions of this type are described in standard textbooks on physical geology. Of special interest are motions such as *slumping, earthflow,* and *creep,* whereby material on gentle slopes (considerably less than the angle of repose) migrate downhill. The motion can be discontinuous, as in *landslides,* or a series of small slumping movements, or virtually continuous creep. On earth, this process is aided by water, not acting as a lubricant underneath the flow, but as a destroyer of the surface tension created by the normally small amounts of moisture between grains of the soil.

Photographs from spacecraft, such as the examples in Figure 11-2, have shown that lunar hillsides often have profiles and soil patterns characteristic of downslope mass wasting (Milton, 1967). No water is available on the moon to aid the process, but vibrations from sporadic meteorite impacts may stimulate such activity.

The Lunar Regolith—Debris on the Moon

For decades men have wondered what the surface of the moon would be like. Inferences in the 1950s were that it would be like volcanic slag or a frothy rock with a consistency like crunchy snow. Gold (1955) proposed that vast amounts of mobile dust migrated and collected in low places on the moon to make the flat lunar plains. Although this hypothesis was not widely accepted, it was widely publicized because it led to concern that spacecraft might sink out of sight in a sea of dust. This concern was in part responsible for an intensive effort to analyze the lunar surface prior to Apollo landings.

After the first high-resolution pictures of the surface were transmitted by television from the Ranger VII probe in 1965, Shoemaker (1965) concluded that the entire lunar surface was blanketed by a layer of pulverized ejecta caused by repeated impacts. This ejecta would range from boulder-sized rocks to finely ground dust. After the remaining Ranger shots, VII and VIII, the Ranger investigators were agreed that a debris layer existed, although interpretations varied from virtually bare rock with only a few centimeters of debris (Kuiper, Strom, and Le Poole, 1966) through estimates of a layer from a few to tens of meters deep (Shoemaker, 1965).

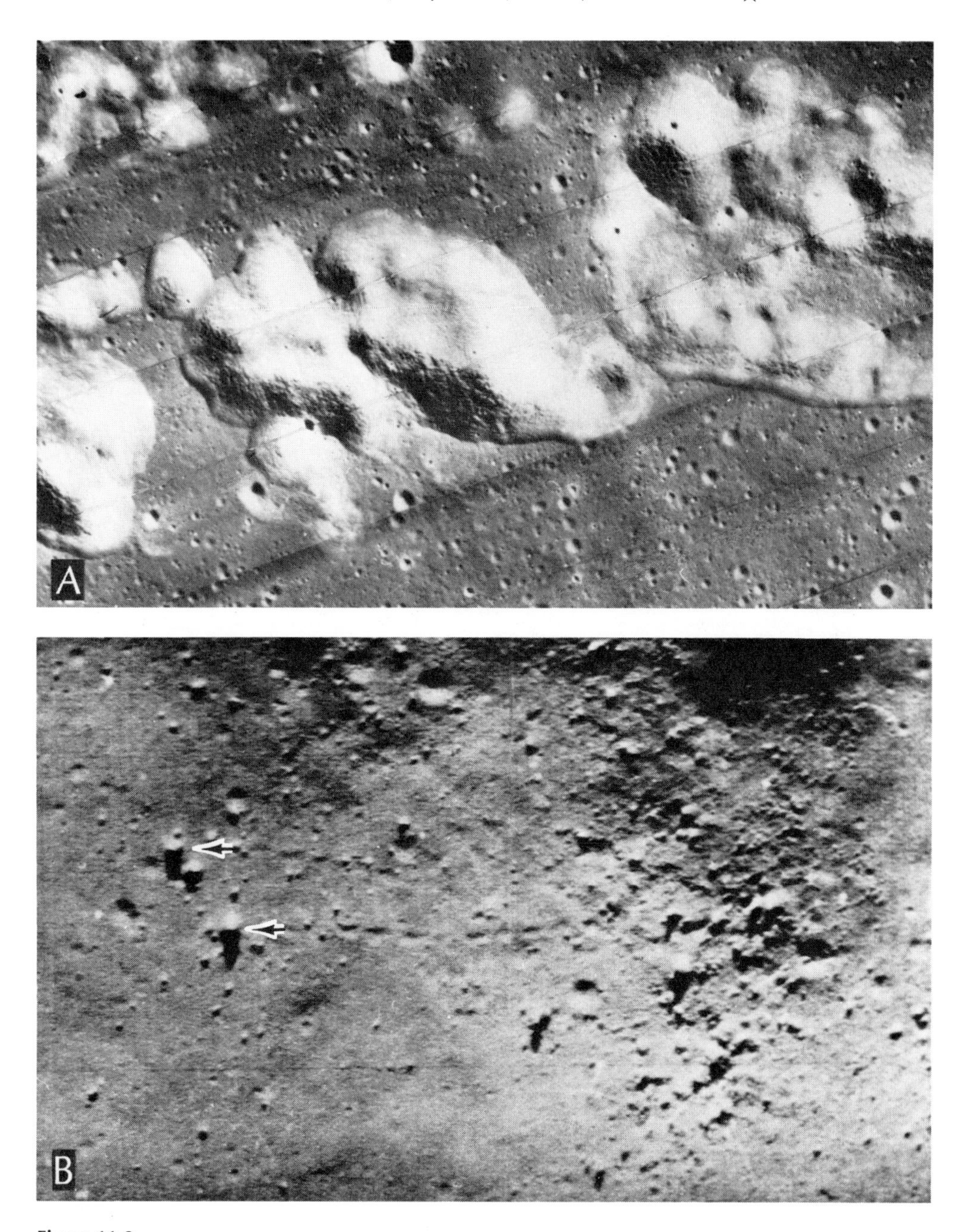

Figure 11-2

Examples of downslope motion on the moon. A: Lunar hills with profiles suggesting downslope mass motions, either by flow or mass wasting. A "toe" lies at the base of each slope. These are part of a light-toned ring of hills known as the Flamsteed ring, which lies in the darker lunar maria. B: Lunar boulders (arrows) which have rolled downslope leaving tracks down the wall of the Hadley rille, not far from the Apollo 15 landing site. (NASA.)

Figure 11-3
The surface of the moon, photographed during the Apollo 12 mission. (NASA.)

This blanketing layer is called the *regolith*. Recent estimates, based on Orbiter and Surveyor photographs, indicate an apparently universal regolith averaging several meters deep in the lunar plains and possibly much deeper in the older lunar uplands. Shoemaker et al. (1970) estimated that it was 3 to 6 m (3 to 6 yards) deep at the Tranquillity Base, site of the first moon landing. Beneath the regolith lies a "rocky substratum," either solid or fractured crystalline rock. This substratum presumably is the uppermost lava flow (or flows) from which originated the lava rock samples brought back by the Apollo 11 astronauts. Once developed, the regolith tends to shield lower layers from all but the biggest meteorite impacts.

The origin of the regolith can be understood from meteorite impact measurements. As long ago as 1964 McCracken and Dubin estimated the total cumulative amount of meteoritic matter on the moon to be equivalent to a layer about 3 cm (1 inch) deep, a figure approximately borne out by recent measures. Each meteorite dislodges many hundred times its own mass (Gault, Heitowit, and Moore, 1964). Thus we can account for several meters of debris produced by meteoritic impacts.

It is interesting to note that in spite of the historical controversies surrounding this subject, a quite accurate concept of the lunar surface had emerged (out of considerable argument) even before the Ranger photography. Salisbury and Smalley (1964) summarizing a 1963 conference on the lunar surface, wrote

> It is concluded that the lunar surface is covered with a layer of rubble of highly variable thickness and block size. The rubble in turn is mantled with a layer of highly porous dust which is thin over topographic highs, but thick in depressions. The dust has a complex surface and a significant, but not strong, coherence.

As Figures 11-3 and 11-4 may suggest, a more accurate verbal and qualitative description could hardly have been written. Optimistically viewed, this

Figure 11-4
Imprint of the footpad of a Surveyor spacecraft in the lunar "soil." The mechanical properties evidenced here are characteristic of fine powders and can be simulated with dry portland cement. (NASA.)

suggests that scientific consensus may often be more nearly correct than hypotheses that lead to irreconcilable controversy.

Martian Dust

The large Martian craters photographed by Mariners 4, 6, and 7 appear to have been softened in relief, compared to lunar craters of the same size, as shown in Figure 11-5. The Martian wind and observed "dust storms" may carry the necessary material to fill in craters (Öpik, 1965; Hartmann, 1966). Figure 11-6 shows the result of deposition in a volcanic crater in a terrestrial arid region—a scene that may be in some ways not unlike scenes on Mars. The physics of the transport process has been studied in detail (e.g., Bagnold, 1941). Sagan and Pollack (1969) applied Bagnold's data to show that in the Martian environment, with observed windspeeds of about 10 m/sec (22 mph), dust particles smaller than a few hundred microns (0.3 mm) in diameter can be airborne for large distances, while larger dust grains can hop erratically over the surface in a well-known process called *saltation,* familiar to anyone who has lain on a windy, sandy beach.

The above discussions of the moon and Mars only hint at the variety of phenomena under the heading of erosion and mass transport facing students of the surface morphology of planets.

Figure 11-5
Comparison of the lunar (A) and Martian (B) surfaces at similar solar lighting angles. The Martian landscape is less crowded with craters and deficient in small craters, relative to most regions of the moon.

Figure 11-6
A terrestrial volcanic caldera partially filled with sand, silt, and ash deposits. The desert environment and filling of this crater may be analogous to conditions affecting Martian craters. (MacDougal crater, diameter approx. 1 mile; Sonora, Mexico.)

EFFECTS ON AIRLESS SURFACES

The surfaces of some planets (e.g., the moon, Mercury, and the asteroids) are exposed to space. They all exhibit a certain degree of uniformity, apparently brought on by external influences (Gehrels, Coffeen, and Owings, 1964). As school children are taught, the earth's atmosphere protects us from many such external influences which would be indeed hazardous to life. The effects of these influences on unshielded surfaces are discussed in this section.

Micrometeorites and Meteorites

All planets are being peppered by a rain of solid objects from space. Atmospheres intercept and break up the smaller objects but atmosphereless planets are struck by objects of all sizes. In this case it is useful to distinguish semantically several effects.

Sandblasting is the integrated smoothing effect of micrometeorites of size comparable to the particle size of the planetary surface. *Gardening* is the tilling or ploughing effect of intermediate-sized bodies (e.g., centimeter scale) that dislodge, fracture, and mix the surface rocks. This creates the lunar *regolith*. *Cratering* is the formation of discrete pits by individual impacts, regardless of meteorite size.

Since each meteorite dislodges 100 to 1000 times more local material than its own mass, there will be an admixture of only a small fraction by mass of meteoritic material in planetary regolith, as predicted by Gault, Heitowit, and Moore (1964). Study of the lunar soil confirms this, with the finding that meteoritic material comprises only about 2 percent of the total soil and has a total influx rate of only 4 g/cm^2/aeon on the moon (Anders et al., 1971) in the last few aeons.

The spectacular cratering effects of larger meteorites have, of course, long been known, and the majority of large craters on the earth, moon, and Mars are widely held to be impact craters (see Chapter 9 and below). A milestone in the acceptance of this idea was the work of Baldwin (1949), who analyzed the lunar surface in terms of repeated impacts.

Radiation Damage and Colorimetry

The surfaces of the moon, Mercury, and the asteroids are stony material exposed directly to space. All these surfaces are characterized by low albedo and "reddening" (increase of albedo toward long wavelength; Figure 11-7). They are, for the most part, darker than granitic rocks common to the earth and resemble dark basalts. The colors on the moon are very monotonous compared to those of the earth, as shown by the colorimetric measures of Sytinskaya (1957) and confirmed in color photos by lunar astronauts. Hapke (1965) proposed a widely discussed "solar darkening" effect, whereby irradiation by solar protons darkened and reddened rocks after about 10^5 years. However, this theory was brought into grave doubt by the discovery that the uppermost surface layers are *lighter* than the layers a few centimeters down (Figure 11-4). Investigators next searched for some "bleaching" process that would slightly lighten the surface! The Apollo astronauts discovered abundant glassy coatings and beads, produced and splattered about by meteorite impact melting of rock fragments. This lunar glass apparently helps to brighten the surface layer, affects lunar photometry, and is possibly the cause of the "rays"—bright deposits that radiate from young lunar craters.

Sputtering

Atmosphereless surfaces of planets lacking strong magnetic fields are struck continually by particles of the solar wind (e.g., protons and electons). Effects on planets, proposed as early as 1959, by Whipple have been investigated experimentally (Wehner, 1964). *Sputtering* is the sandblasting caused by low-

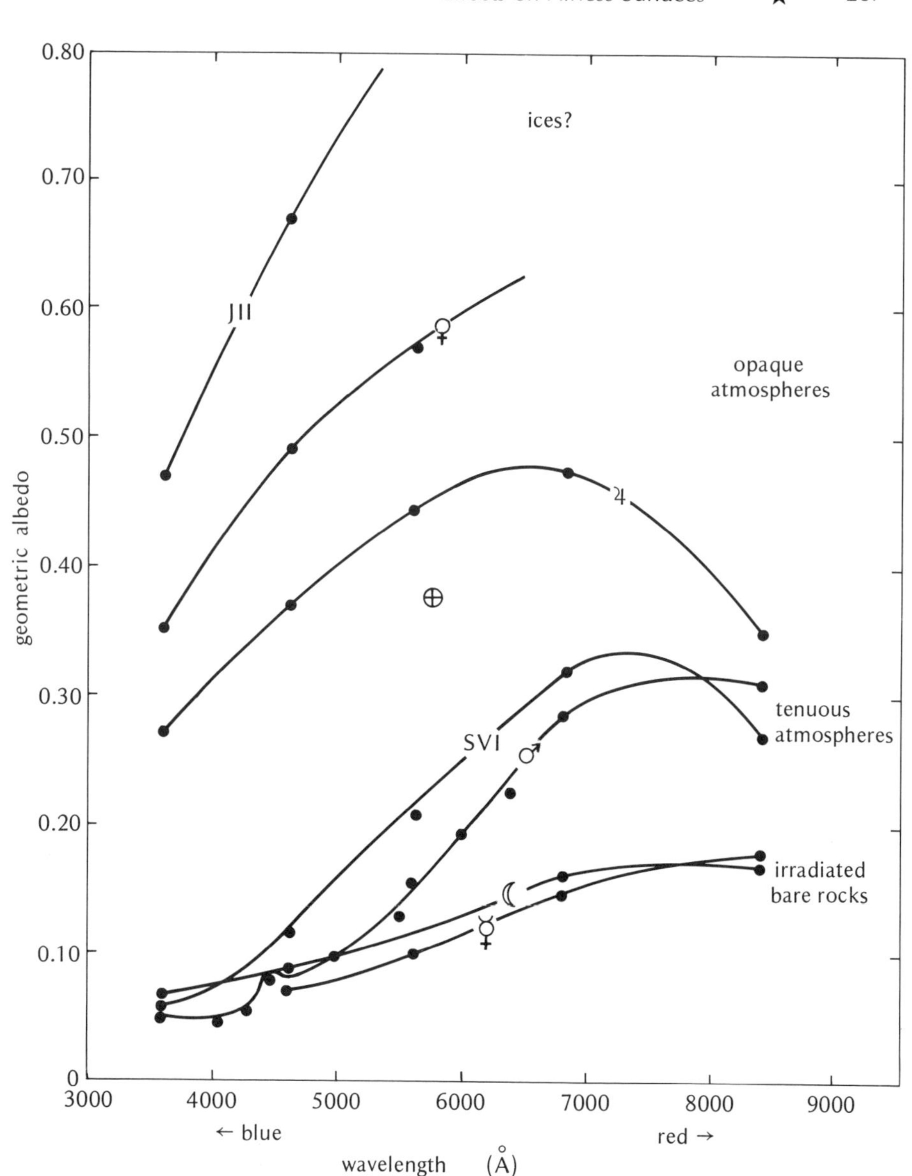

Figure 11-7

Geometric albedo versus wavelength for various planetary bodies (see the text).

energy ions striking solid surface. (High-energy particles penetrate farther and produce other effects.) Principal agents in the planetary environment are solar protons and α particles in the energy range 1 to 20 keV. Sputtering tends to smooth macroscopic irregularities by dislodging atoms on projecting surfaces and removing them to less accessible spots. On the other hand, polished smooth surfaces are microscopically roughened by differential crystalline response to the sputtering process. A cementing of granular material also results.

The moon and other small bodies lose mass to the sputtering process because some rock particles are blasted off into space. KenKnight, Rosenberg, and Wehner (1967) estimated that exposure of the lunar surface for 4.5×10^9 years has resulted in loss of a layer a few centimeters thick.

Isotopic Effects

Not only chemical and mechanical effects result from exposure to space environment; nuclear effects occur, too. Here we describe only two examples.

^{3}He production in meteorites. Bauer (1947) pointed out that some of the helium in meteorites is produced by cosmic-ray bombarding atoms near the surface of meteoroids while they are in space. Because there is strong absorption of the cosmic rays, this is a skin effect in approximately the outer 1 m of a meteoritic body. Measurements on meteorites (e.g., Fireman, 1959) confirm the decreasing amount of ^{3}He toward the (shielded) interiors of large meteorites.

^{7}Li production in planetesimals. Fowler, Greenstein, and Hoyle (1962) presented an argument for surface effects that occurred during the earliest history of the solar system. They proposed a theory in which slow "thermal" neutrons, produced by spallation reactions with high-energy solar particles struck other atoms and created an excess of the lithium isotope, ^{7}Li, in the outer layers of icy, metric-sized planetesimals. Although details of the theory are now disputed, the concept shows how surface chemistries might have been altered in the early solar system.

PHOTOMETRY, POLARIMETRY, AND SPECTRAL STUDIES OF PLANETARY SURFACE ROCKS

This section discusses the results of some diverse methods that have been used to determine from the earth the composition and rock types of surfaces of other planets.

Photometry, Polarimetry, and the Behavior of Rock Powders

Prior to the sampling of the lunar surface, much effort was expended to determine lunar-surface composition and structure from photometric and polarimetric observations. The methods, although not definitive, were indicative and are still of interest when applied to Mercury, the asteroids, and Mars.

These methods make use of various photometric, or light-reflecting, properties. Because of the universality of processes acting to pulverize the surface rocks of planets, planetary regoliths are probably common, and, as expected, many of the important photometric properties observed on planets are found to be characteristic of rock powders.

The *opposition effect* is an intense *backscattering* of light from the illuminated surface back toward the light source. The moon is a strong backscatterer, which is why the full moon looks so bright.* Another phenomenon, characteristic of the moon's photometry is uniform brightness across the face of the full moon (i.e., independence of this brightness from the angles of incident and reflected light). It has been known that these photometric properties imply extreme roughness on some scale. Minnaert (1961) traces theoretical and empirical studies of these properties back to 1921, with all conclusions indicating a very open, porous lunar surface structure. The good insulating properties of the lunar surface were recognized as compatible with this hypothesis. Telescopic studies indicated that the megastructure (1 kilometer scale) was not rough enough, and radar studies of the moon indicated no great roughness at scales about 1 m, so that by the late 1950s the roughness was known to be small in scale (see the review by Dyce, 1970).† Gold's (1955) hypothesis of deep, finely divided, and highly mobile dust led to experiments with rock powders which showed that the microscopic structure of such powders could produce the necessary characteristics and even some of the later-observed mechanical characteristics (Hapke, 1967). One such property is *fairy-castle structure,* a loosely packed, porous microscopic latticework assumed by the powders as they are deposited. Fairy-castle microstructure in rock powders would account for a number of photometric properties of the moon and asteroids. Observations on the surface of the moon confirm that the opposition effect, or heiligenschein, is pronounced even locally (as can be seen in Figure 11-8) and indicates that the porous, or fairy-castle structure is smaller than 1 mm in scale. The particle sizes in the lunar dust range down to microscopic. Most of the mass of the lunar dust is concentrated in particles of about 0.01 mm (Duke et al., 1970).

The similarities of the photometry of the moon, at least some asteroids, Mercury, some satellites, and to some extent Mars (Harris, 1961; Gehrels, Coffeen, and Owings, 1964) suggest that similar surfaces have been created on those bodies, presumably by erosion and fragmentation due to external agents, especially repeated small-scale impacts.

* At full moon an observer on the earth is nearly on a line between the sun and the moon. This produces a zero *phase angle* and the terrestrial observer sees the intense backscattering.

† Thus artists who during this era depicted the moon's surface as exceedingly rough, with great stone spires and pinnacles, should have known better!

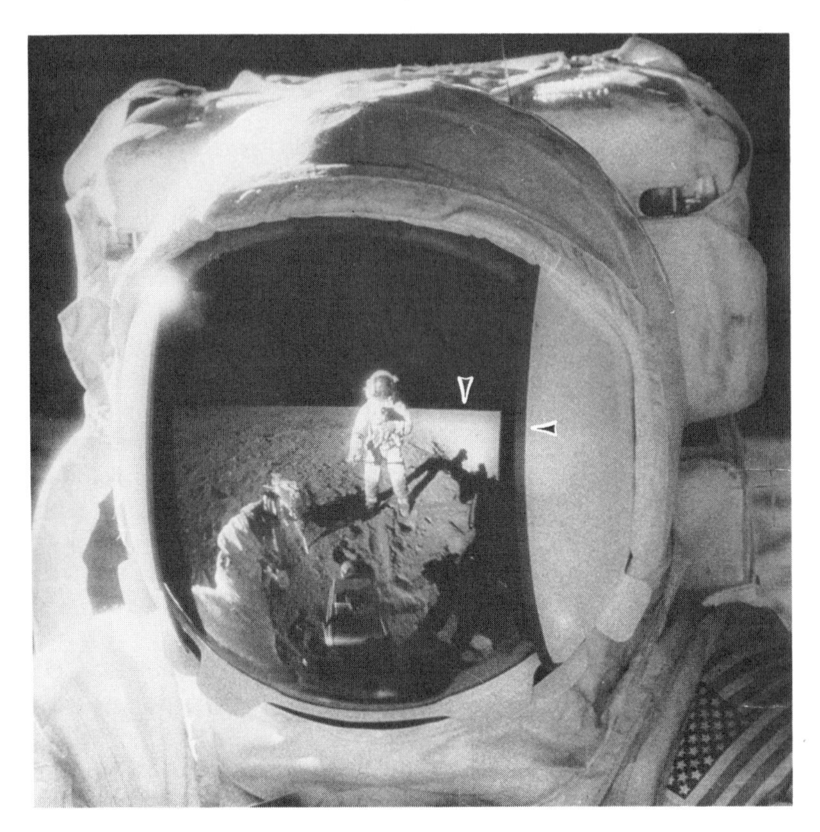

Figure 11-8

The lunar heiligenschein (arrows) as seen in the reflected wide-angle view of the lunar landscape in an Apollo astronaut's faceplate. The heiligenschein is a pronounced brightening of a surface at a zero-phase lighting angle (antisolar direction; near the shadow of the astronaut's head). (NASA.)

Polarization is another property of light, familiar to wearers of polarized sunglasses. Different surfaces polarize light in different amounts, depending on the surface properties and lighting conditions. Polarization measurements of planetary bodies thus give information on surface properties. The measurements accord with photometric results and also show similarities among the surfaces mentioned. For example, both the moon and Mercury show greater polarization in their dark maria than in their brighter regions (Dollfus, 1961, p. 372). The polarimetric observations of Martian yellow clouds support their interpretation as dust "storms" (Dollfus, 1961, p. 386). Coffeen (1969) has used polarization measures to show that the scattering particles in the Venus atmosphere are not pure water droplets.

Colorimetry, the study of the dependence of reflectivity on color (i.e., wavelength of light), is another method of estimating the properties of surface rocks. As mentioned earlier, the atmosphereless planets show some colorimetric similarities which may be due to radiation damage.

Unfortunately, the methods of photometry, polarimetry, and colorimetry are not as powerful as we might wish. Although they may indicate some structural properties of the surfaces and may rule out some constitutents, they do not uniquely specify the type of rock reflecting the light.*

Spectral Studies of Planetary Rocks

A more critical test of rock composition comes from spectrometric studies. In principle certain minerals may absorb certain colors of light, and the sunlight reflected off rocks containing these minerals would be missing these colors—i.e., the spectrum would reveal *absorption bands*. In practice it is difficult to observe such absorption bands because they are very subtle and often at wavelengths difficult to observe (e.g., the infrared).

The spectral method for determining planetary rock types has thus only recently become practical. An important survey of planetary surfaces by this method is being carried out by McCord and his associates (McCord, Adams, and Johnson, 1970; Lebofsky, Johnson, and McCord, 1970). Among the results to date are the discovery of a ferrous iron absorption band probably resulting from pyroxene minerals in the surface of the asteroid Vesta. The implication is that Vesta has a surface composition similar to basaltic achondrite meteorites (see Chapters 8 and 9). Ten other studied asteroids lack this band. The rings of Saturn also lack the spectral characteristics of these minerals. The rings may be composed of ice particles or frost-covered silicate particles and are probably composed largely of H_2O ices, but other materials must also be present (see Chapter 6).

CHEMICAL AND MINERALOGICAL PROCESSES ON PLANETARY SURFACES

The chemistry of planetary surface materials is exceedingly complicated. Even if no atmosphere or ocean is present, volcanic and metamorphic effects below the surface may produce continual changes in minerals on the surface. This

* Surface *structure* may be more important than *composition* in its effects on these measurements. For many years, while these were the best methods available, many observers tried to find a material that would match the reflecting properties of the lunar surface. In many respects, the best such material was reindeer moss, which simulated the porous, powder structures of the lunar soil's microsurface. Fortunately, no one was naive enough to propose that the moon's surface was composed of reindeer moss!

continues throughout the active period of the planet's thermal history. Only asteroidal-sized bodies less than 500 km in radius could have cooled from molten conditions to less than 300°K—cool enough to be "dead" at the present time (Anders, 1963)—and hence we expect active chemical processes on or inside all the planets. Atmospheres, hydrospheres, and the terrestrial biosphere complicate this still further, since they transport and alter rock particles. On the earth we recognize a geochemical cycle through which rocks are processed and reprocessed. We shall discuss this below and then give examples of processes applicable to Venus and Mars.

The Geochemical Cycle

The geochemical cycle is a useful schematic description of the complete history of rock material after it is injected into the planetary surface layers from deep-seated magma sources. The cycle is presented in diagrammatic form in Figure 11-9. In detail, each segment of the cycle is complex and worthy of continued major research.

Each planet has its own "geochemical" cycle. The details and the relative importance of even the major processes are different from planet to planet. On earth, sedimentation dominates in producing the observable rocks and may be fundamental in the development of the silieous continents. On planets with less extreme surface activity, direct production of igneous rocks from magma may be the principal process and only mechanical erosion may be important. Planetary rocks are altered by the metamorphic part of the cycle to a greater or lesser extent. For example, some meteorites and lunar rock samples show metamorphic effects of shock and other processes (see Chapter 9), but others appear to have crystallized in place with only minor shock metamorphism.

Meteoritic Material Added to Lunar Regolith

An example of chemical alteration of a planetary surface by cosmic material is the addition of meteoritic material to the lunar regolith. Two findings are of interest. First, the amount is small, only about 2 percent to the material in the surface soils. Second, the composition of this meteoritic material is close to that of carbonaceous chondrites but different from that of ordinary stone and iron meteorites (Anders et al., 1971).

Iron Minerals on Mars

For more than a decade it has been supposed that the reddish surface of Mars owes its color to iron oxide minerals, some of which are familiar as common

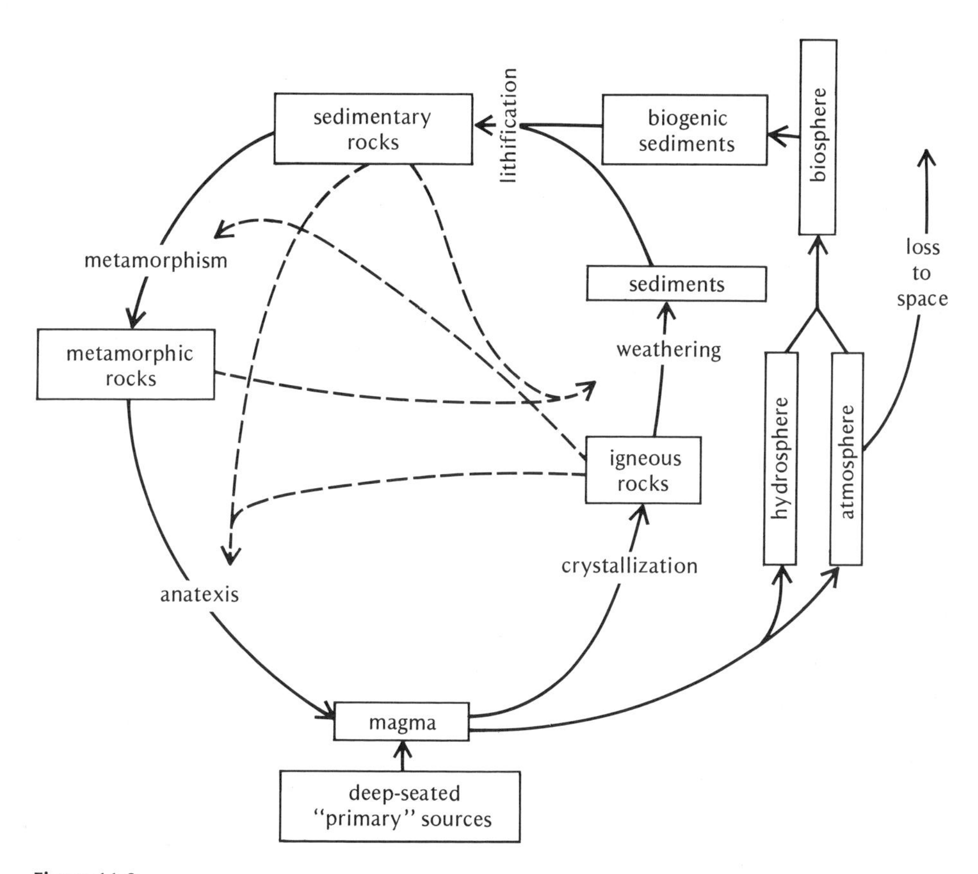

Figure 11-9

The geochemical cycle for terrestrial rock materials, showing various possible states of rock constituents (boxes) and alteration processes.

rust. Since 1960 much evidence has suggested oxidized basalt, especially the mineral limonite, a hydrated iron oxide of variable composition ($Fe_2O_3 \cdot nH_2O$), as the only material fitting polarimetric, visual colorimetric, and infrared colorimetric observations (Dollfus, 1961, p. 379; Binder and Cruikshank, 1966; Salisbury and Hunt, 1969; McCord and Adams, 1969). Binder and Cruikshank emphasize the occurrence of limonite as a surface stain on rocks and rock particles, produced during the weathering of the igneous rocks. They infer from this that the Martian limonite was produced by similar weathering in an earlier environment when the Martian atmosphere was denser than it is now. Very small particles of limonite or rock particles stained with limonite are compatible with observation of the Martian yellow clouds ("dust storms"). Estimates of the particle size in such dust clouds average around 1 micron (0.001 mm), with large particles ranging up to about 0.3 mm.

Carbon Dioxide and Carbonates

Urey (1952) pointed out the geochemical importance of the reaction between CO_2 gas and the silicate minerals to form carbonate minerals and silica:

$$MgSiO_3 + CO_2 \rightleftharpoons MgCO_3 + SiO_2.$$

High temperatures drive the reaction to the left, but the presence of water, to the right. Urey (1952) used this reaction to analyze Venus. He hypothesized that Venus initially had a composition similar to the earth's but lost its water by dissociation due to the high temperature on Venus. Urey theorized that in the absence of water and in the presence of volcanism, CO_2 gradually built up to its present very high abundance on Venus. On the earth, however, the oceans act as a CO_2 reservoir on the earth, producing carbonate-rich sediments and retaining most of the CO_2 that is not in rocks. To double the terrestrial atmospheric amount of CO_2, a large multiple of the present amount would have to be added, most of it going into the oceans (see Mason, 1966). Venus, which lacks oceans because of its high temperature, lacks this regulating effect of water and hence retains a tremendous amount of CO_2 in its atmosphere. The reaction discussed above is sometimes referred to as the *Urey reaction* and is just one example of the possible chemical complexities of planetary surfaces. Recent studies of Venus further complicate the picture.

The Chemistry of Venus

In 1967 Connes, Connes, Benedict, and Kaplan made the surprising discovery that hydrogen chloride (HCl) and hydrogen fluoride (HF) were trace constituents of the Venus atmosphere, in addition to CO_2, which was the main component.

Lewis (1968) investigated the physical chemistry of the Venus surface with a view to explain the halide constituents. By considering chemical equilibria between the extremely hot atmosphere (about 750°K) and the rocks, Lewis showed that the mineralogy and meteorology of Venus are very different from the earth's. Liquid water is absent; carbon dioxide dominates the atmosphere, and elements such as mercury and lead could be driven out of the crustal rocks and into the atmosphere as volatiles, with lead possibly condensing out at the poles. Thus high-temperature equilibrium between rocks and atmosphere predicts a very unearthly pattern of trace constituents.

RADAR, INFRARED, AND OTHER REMOTE SENSING OF PLANETARY SURFACES

Radar

Radar echoes were first received from another planetary body, the moon, in 1946 (Webb, 1946). In addition to determining distances—which are used in celestial mechanics and relativity theory—radar observations indicate certain surface properties. We shall not discuss the methods of radar observation but rather list some representative results. A good review of the field is given in the book *Radar Astronomy* (Evans and Hagfors, 1968).

In the interval 1957–1959 observations of the moon at different wavelengths (2 m to 10 cm) indicated that, contrary to the full moon's visual appearance, the radar reflection was much more nearly *specular*—like a

Figure 11-10

Radar map of the moon constructed from reflected radar signals of 70-cm wavelength. The bright craters Tycho (T), Copernicus (C), and Langrenus (L) are prominent (compare the infrared image, Figure 11-11, and the full-moon photo, Figure 11-12). These and other bright spots are associated with concentrations of rocks and boulders on the surface, causing enhanced radar reflectivity. (National Astronomy and Ionosphere Center.)

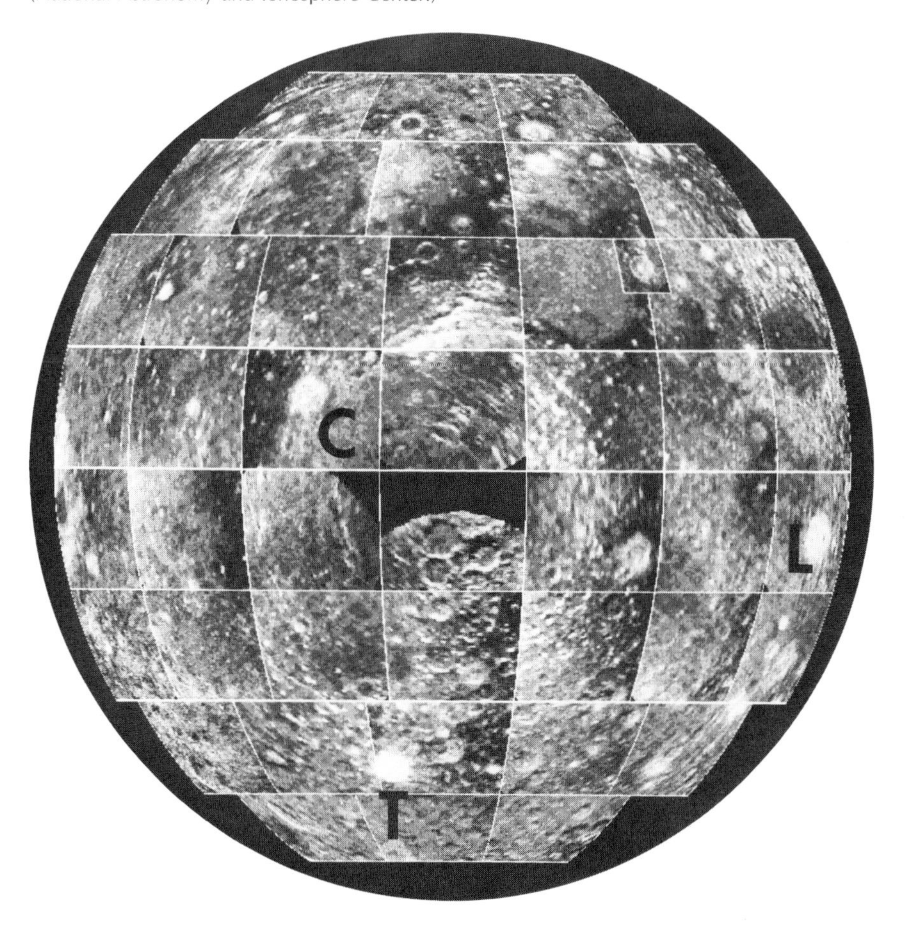

mirror—with a bright central "highlight." This suggested that the moon is relatively smooth at the scale of radar wavelengths, as was confirmed by Surveyor and Apollo photos of smoothly rolling landscapes. Viewed by radar, Mercury resembles the moon, while Venus appears significantly more smooth, on the average.

With certain assumptions, radar observations yield estimates of the dielectric constant of the reflecting material. Lunar observations yield unexpectedly low values attributed to the high porosity of the lunar surface layer. Observations of Venus yield values consistent with dry terrestrial rocks and appear to rule out significant amounts of liquid water.

Through combined measures of range and rotation-induced Doppler shift, it is possible to identify on the planet two restricted regions of the surface from which the return signal is being reflected. Images can thus be constructed, as shown in Figure 11-10. In this way localized rough areas, possibly mountain ranges, have been mapped on the surface of Venus (Carpenter, 1964), and certain craters on the moon, as shown by Figure 11-10, have been identified as sources of anomalous radar return, probably due to bare exposed rocks (e.g., Pettengill and Henry, 1962).

Infrared Studies

The anomalous lunar craters are further illumined by infrared observations. Shorthill and Saari (1965) constructed images of the moon in the far infrared (wavelength 10 to 12 microns), with resolution of a few kilometers. One such image is shown in Figure 11-11, which can be compared to the radar image in Figure 11-10. Infrared wavelengths reveal the moon by its own feeble thermal emissions, not by reflected sunlight. The infrared images indicate that during the sudden cooling produced when the moon enters eclipse many individual craters remain warm. These have been nicknamed *hot spots*. Many but not all of the hot spots are radar reflectors. The "hot" and radar-reflecting craters are typically young-looking, with sharp, bright rims and systems of bright rays, thought to be spurts of pulverized, glassy ejecta.

In these young craters, the low-density, finely fragmented, regolith has been blasted away. Thus a rough, thinly blanketed surface is exposed, enhancing the radar reflectivity and the absorption of solar radiation during the day. Such craters would stand out as hot spots during dark periods, as the stored heat is conducted back to the surface and radiated away. This explains many of the radar and infrared hot-spot observations (Salisbury and Hunt, 1967), but not all of them. For example, (1) the fairly young crater Eratosthenes shows enhanced radar return but is not a prominent "hot spot," and (2) "hot spots" are more common in lunar maria than in lunar uplands. Other

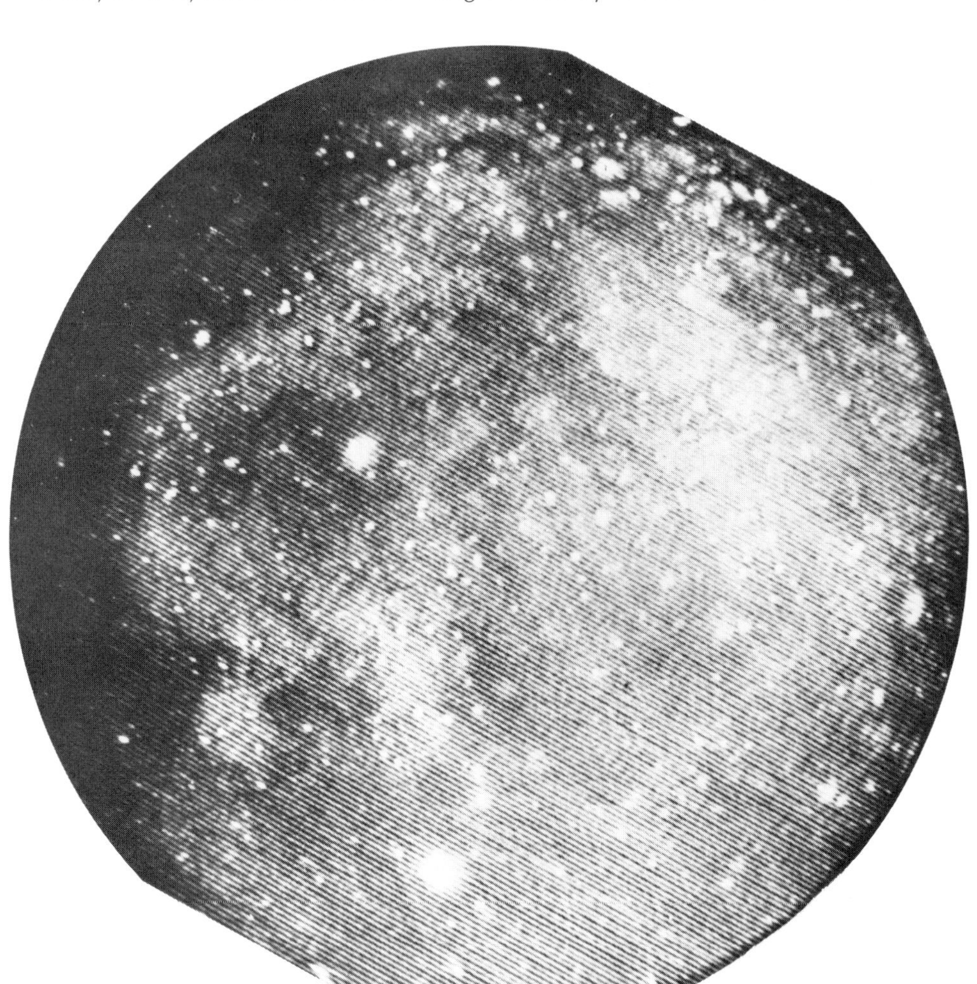

Figure 11-11

Infrared image of the moon during eclipse. Bright spots are radiating in the infrared, being warmer than their surroundings; they have been called "hot spots." Lunar hot spots are regions of thin regolith cover and greater-than-average exposure of bare rock. Because of thin insulation they have greater ability to absorb and reemit heat than their surroundings. Most hot spots coincide with areas of enhanced radar return (compare Figure 11-10). (Boeing.)

effects, such as nonuniform stratigraphy, differential erosion, and nonuniform local mineral properties, may be contributing factors.

PHOTOINTERPRETATION OF PLANETARY SURFACES

Methods of Analysis

More familiar than the remote sensing techniques described above is simple photography. There are two approaches one can take to photographs of a planet: quantitative and qualitative.

In quantitative photointerpretation, one tries to reduce the information content to numbers. For example, by carefully calibrating the film or electronics (depending on whether film or a TV system is being used) one can try to analyze the albedo and other reflecting properties of the surface. In addition, it is theoretically possible, by using the angle of solar illumination, to deduce slope angles on the surface photographed. In principle, if one knew enough about the surface's reflecting properties, one could compute a three-dimensional reconstruction of the surface. Severe limitations are met in practice because the exact reflecting properties of planetary surfaces are in general not known. An attempt at this kind of analysis is found in a study of the Mariner photography of Mars by Leighton et al. (1967).

The other approach, which has been more widely used, is direct interpretation of the photographs as pictorial documents. This is necessarily more subjective and qualitative and requires considerable familiarity with the appearance of various geological formations. Low lighting helps to emphasize relief, and high resolution permits identification of various types of detail as shown by comparing Figure 11-12 (at high lighting) with Figure 11-13 (low-lighting views, various resolutions). Information at other wavelengths, such as radar and infrared, can also be applied. (Compare Figures 11-10, 11-11, and 11-12.) The thrust of this work is to use geological information to interpret the history of the planetary surface and determine what forces have acted on it.

A useful and basic tool in these studies is *stratigraphy,* the science of identifying different rock units—lava flows, lake sediments, ejecta from craters, etc.—and determining their relative age. Generally, unless the surface has been greatly disturbed, the oldest units will be overlain by the youngest units. One thus speaks of a *stratigraphic column,* an imaginary cross section of the layers, showing the relative position of the various units. On the moon, for example, one can quickly see that parts of the ancient terrace, or upland surfaces, have been covered by relatively young lava flows, which in turn have been partly covered bright rays of ejecta from still younger craters. Thus we get an idea of the sequence of events on the moon. Once the sequence of *relative*

Figure 11-12

The full moon. The craters Tycho, Copernicus, and Langrenus are prominent bright spots and center of "ray" systems (see Figure 11-10 for identification). Dark areas are lunar maria, or "seas"; bright areas are lunar terra, or uplands. (Lunar and Planetary Laboratory.)

ages has been determined, it remains to determine the *absolute ages*. These can often be estimated from characteristics of the surface, such as the number of impact craters per square kilometer or the depth of the regolith, but the most accurate ages must come from radioisotopic dating of rock samples collected on the planet.

The best example of stratigraphic investigation is the work of the U.S. Geo-

C

Figure 11-13

Effects of increasing resolution in planetary photography. Each image has 10 times the resolution and scale of the preceding image. The lunar crater Tycho is shown. (A: Lunar and Planetary Laboratory; B and C: NASA.)

logical Survey in preparation for the lunar landing program. Over a period of years, Survey scientists observed regions of the moon in great detail and prepared stratigraphic maps. New types of features were noted, tests of theories of origin were generated, a relative chronology was derived, and detailed maps were prepared to guide the astronauts to landing sites. This work is summarized in detail and with many illustrations by Mutch (1970). Similar mapping is now planned for Mars.

Historical Achievements

Table 11-2 gives a listing of important milestones in the photography of planets and notes on programs presently in the planning stage. Some examples of interpretation and results of planetary photoanalysis will be found in the next section.

VOLCANISM AND TECTONICS

Probably the most important formative processes acting on planetary surfaces, with the possible exception of meteorite impact,* are volcanism and associated

* Meteorite impact craters and associated surface structures are discussed in Chapter 9.

Table 11-2

Achievements in planetary photography: historical events

First telescopic photograph of moon	Draper	ca. 1840
First photographs of moon's far side	Luna III (U.S.S.R.)	Oct. 7 1959
Meteorological surveillance of earth	Tiros (U.S.)	1960
First meter-resolution photos of moon	Ranger VII (U.S.)	July 31 1964
Discovery of Martian craters	Mariner 4 (U.S.)	July 15 1965
First millimeter resolution photos of moon	Luna 9 (U.S.S.R.)	Feb. 3 1966
First orbital reconnaissance of moon (95 percent complete)	Orbiter (U.S.)	1966–1967
First photograph resolving Martian satellite	Mariner 7 (U.S.)	1969
First orbital reconnaissance of Mars	Mariner 9 (U.S.)	1971
Planned missions[a]		
Closeup photography of Jupiter	Pioneer	1974–1975
Closeup photography of Mercury and Venus	Mariner	1974?
Surface photographs of Mars	Viking	1976?

[a] The U.S.S.R. also has a series of planetary probes planned, but the details are not publically available because of party reluctance to discuss missions other than *faits accompli*.

tectonic processes, such as mountain building. These processes are both driven by the internal thermal energy of the planet. They are often grouped under the name *endogenic* (internally caused) *processes,* distinguishing them from *exogenic processes,* such as meteorite impact, stream erosion, and aerial dust deposition.

Definitions

Volcanism includes all processes by which material—gaseous, liquid, or solid—is exhausted from the interior of a planet. Although many people associate volcanism only with eruptions of molten lava, volcanologists emphasize that many associated phenomena, such as hot springs and ash flows, must be considered if we are to understand volcanism. *Tectonics* includes all internal processes of distortion of the planetary surface. Tectonic processes produce folded mountains, faults, rifts, and other features.

Rocks Produced by Volcanism

Figure 11-1 showed a classification scheme for rocks. Extrusive (erupted) rocks are rocks formed by eruption from the subsurface. Extrusive rocks range in composition from siliceous rhyolites to basic basalts. In texture, they are usually fine grained or glassy, because once exposed at the surface they cool so fast that there is not enough time for large crystals to grow. This is not to say that extrusive rocks cannot be coarse grained. In a basaltic lava flow, for example, molten lava may be trapped and insulated in the interior of the flow, and may cool slowly enough to form a coarsely crystalline rock (gabbro) instead of a fine-grained rock of the same composition (basalt; see Figure 11-1). Classification by grain size was one of the first diagnostic tasks performed on lunar rocks after their arrival on earth.

As mentioned earlier, basalts have been regarded in the past as the most common of terrestrial extrusive rocks. However, there is a growing belief that eruptions of rhyolite, a light-colored, siliceous rock, are more frequent and important on the earth (in terms of rock volume) than flows of basalt. Study of terrestrial rhyolites has shown that many of them were not emplaced in the form of massive lava flows, as were basalts, but rather were erupted in the form of clouds of pulverized, red-hot rock fragments suspended in hot gasses. These are known as *fluidized eruptions.* There is no flow of liquid, but the cloud of gas and suspended particles is dense enough to flow along the ground like a swirling flash flood. Such a cloud is called a *nuée ardente* (glowing cloud) and is composed of small hot ash, cinder particles, or molten droplets, suspended in turbulent gas and smoke. A nuée ardente from Mt. Pelee raced across the city of St. Pierre, Martinique, May 8, 1902, at 150 m/sec, carrying along large boulders, leveling trees and buildings, and killing 29,000 inhabitants. There are terrestrial rhyolitic deposits less than 1 million years old, covering 26,000 km^2 (Rittmann, 1962). Nonfluidized ash and cinder deposits can also be destructive because of their sheer volume, as indicated by Figure 11-14.

The importance of terrestrial rhyolite deposits led to short-lived speculation that the lunar plains, or *maria,* might be composed of rhyolite sheets, rather than the basaltic lava flows proposed by Baldwin (1949). This hypothesis was rejected when the lava samples from the lunar maria were found to be basaltic crystalline rocks.

Lava flows can assume many forms, depending on their rate of flow and viscosity, which in turn may depend on the angle of slope, gas content, temperature, etc. Two common forms have Hawaiian names: Aa (pronounced áh-ah) is extremely rough and clinkery, being formed when solid blocks on the top of the flow are broken and upended (mnemonic: think of the pointed top of A) by the motion of molten material inside the flow; Pahoehoe (pronounced pa-hóy-hoy) is smooth and sometimes glassy-looking. Figure 11-15 shows a site in Hawaii where the two kinds of lava overlap.

A number of other volcanic products are common and of planetological interest. Very fine volcanic fragments—of millimeter scale—are called *ash.*

Figure 11-14

Effects of deposition of ash and cinders from a volcanic eruption. A highway has been buried and a forest devastated (background) by an eruption of the Hawaiian volcano Kilauea Iki.

Figure 11-15
A sea of overlapping lava flows. In this example, dark, rough *aa* lava has partly covered an earlier, lighter *pahoehoe* flow. Lunar maria also show flow units of different tone and may be analogous, except that the lunar surface has a regolith cover of pulverized rock (Mauna Loa volcano, Hawaii).

Volcanic fragments of the order of 10 mm in size are called *cinders*. *Tuff* is a type of rock formed by consolidation of ash particles, either by deposition in water or directly from the air. Tuff is sometimes classified as sedimentary on the grounds of its depositional character. *Welded tuff,* or *ignimbrite,* forms when hot, glassy ash particles weld to each other during deposition. Ignimbrites can form vast, blanketing deposits, usually of siliceous composition, as a result of nuée ardente eruptions.

So far we have considered only rocks extruded onto the surface, since all forms of *volcanism* involve magma gaining access to the surface. Some bodies of magma, however, may never gain access to the surface. Instead of erupting, they may solidify into large, underground rock masses, called *intrusions*. Although strictly not volcanic phenomena, since they are not eruptive, intrusions are common in areas of volcanism and mountain building. One area of current research involves a correlation between the places and times of intrusions and the subsequent development of mountain ranges in the same area. Granite intrusions are very common in these cases.

Siliceous igneous rocks, such as rhyolite flows, rhyolitic ignimbrites, and granite intrusions are relatively common on the earth's continents, but they may be rare on other planets, because other planets may not have gone through such extensive differentiation and weathering as the earth and may thus lack the siliceous continental crustal blocks where these rock types abound.

Structures Produced by Volcanism and Tectonic Processes

If we photograph, fly over, or land upon a planet, we shall see a great number of surface features. To an untrained observer, these may be bewildering in their variety. For example, on the earth they include ocean coastlines, mountain chains, and shifting desert sands; on the moon they include craters, mountain-rimmed basins, and canyon-like rilles; on Mars, closeup photographs reveal craters and "chaotic terrain." A major goal of planetary science is to sort out and identify the causes of these different features, and a major tool is knowledge of volcanic and tectonic landforms on earth. Such knowledge is the concern of the field known as *structural geology*.

Interpretations of planetary photographs show a detectable correlation between the familiarity of the interpreter with endogenic geological structures and the degree of success of his interpretation. Indeed, most investigators share an implicit faith that an imaginary ideal planetologist who had seen all the volcanic and tectonic structures of earth, and had total recall, would have a much better record of interpreting planetary surfaces than any real scientist has yet compiled.

Although identification and interpretation of planetary surface features requires extraordinary familiarity with the variety of volcanic and tectonic landforms, we can only give brief examples of some such landforms.

Volcanic landforms can have either positive or negative relief surrounding a central volcanic vent. If the vent expels viscous or nonfluid debris, the ejecta may pile up into a steep positive relief feature, such as a *cinder cone* (Figure 11-16), ranging up to hundreds of meters in dimension. Smaller examples are *spatter cones* or *hornitos,* where tacky lava is thrown out and piles up around the vent. On a much larger scale, *shield volcanoes* are built up by eruptions of fairly fluid lava which spreads out in all directions and builds up a gently sloping, dome-shaped mountain. An example is Mauna Kea in Hawaii, which reaches over 4200 m (13,796 ft).

Volcanic *craters* are negative relief features. Unfortunately, the word crater is such a general term that it has little specific meaning. "Crater" has been used to refer to almost any hole in the ground of whatever origin: explosion, gas coring, collapse, drainage, and so on, not to mention meteorite impact, which is a totally different phenomenon, and also the vent at the top of a positive feature such as a cinder cone. Thus the statement "There are craters on the moon" has no genetic significance. Some specific crater types must be discussed.

A *volcano-tectonic* sink is a very large, irregular depression caused by

Figure 11-16

A cinder cone, showing characteristic steep slope of loose material lying nearly at the angle of repose ("S.P." crater, near Flagstaff, Arizona).

Figure 11-17

The pit crater Halemaumau, occupying the central floor of the caldera Kilauea, in Hawaii. Diameter, 1 km. Successive generations of layered basaltic lava flows are exposed in the walls.

collapse or subsidence as lava is erupted from underground sources. The lava may either flood into the sink or erupt outside. Volcano-tectonic sinks may range up to tens of kilometers in size.

A *caldera* is a smaller, more circular collapse or subsidence crater. The diagnostic feature of a caldera is that the depression is created predominantly by material moving downward, usually into the void created by the simultaneous or prior eruption of magma from underground chambers (Figure 11-17). Most calderas range from about 1 to 30 km in diameter.

Drainage craters occur by the draining of loose surface materials into underground cavities that are common in lava flows. The cavities are places evacuated by the lava. Drainage craters have dimensions of several meters and larger (Figure 11-18). Kuiper, Strom, and Le Poole (1966) gave evidence

Figure 11-18

Examples of volcanic landforms, correlated according to lava type and quantity. (After Rittmann, 1962, and others.)

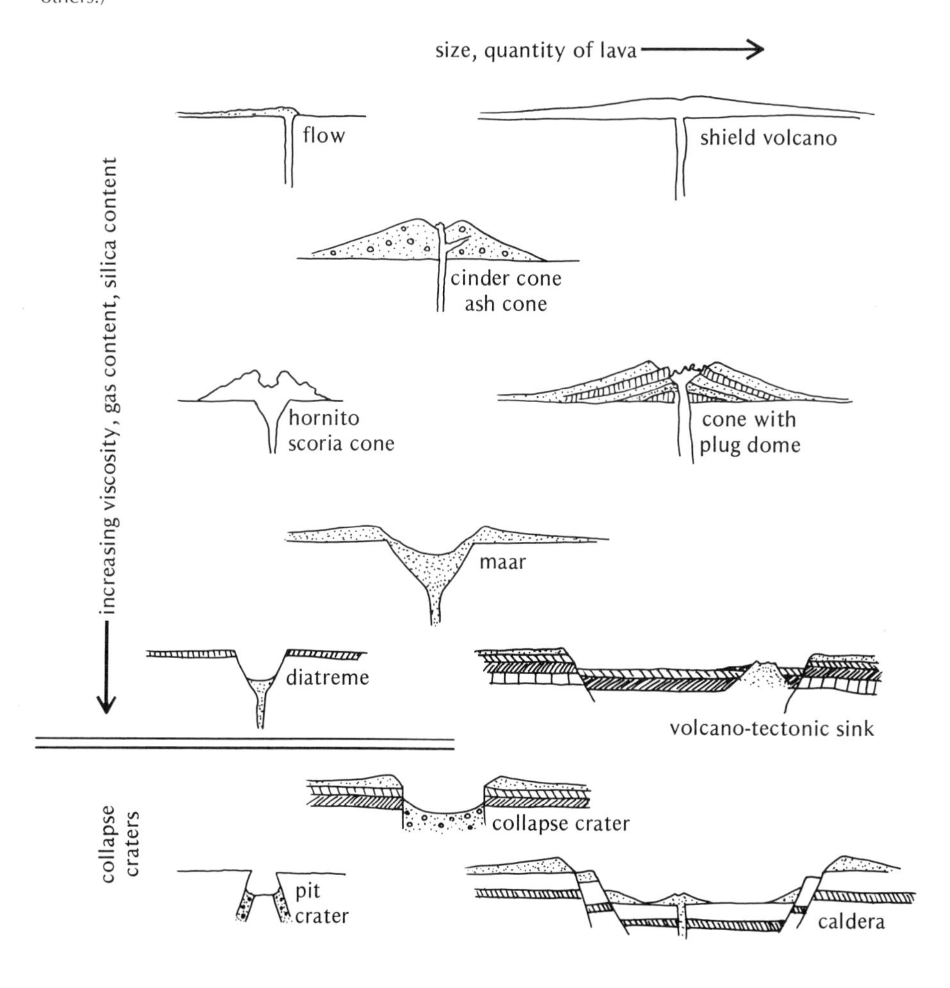

that many of the shallow craters in lunar lava fields, ranging up to several hundred meters in diameter, are drainage craters.

Many other kinds of craters are formed primarily by blowing material upward out of volcanic vents, rather than by downward withdrawal of material. Ordinary craters at the tops of cinder cones and larger volcanic mountains are often of this kind.

Diatremes are craters in which the expelled material is gas rich. The gas carries away bits of magma and rocks broken off the walls of the vent. Thus the vent is enlarged and a large crater with a rim can result. *Maars* are another name for diatreme craters; the name originated from a number of examples in Germany that were filled with water, forming small round lakes. Diatremes and maars are typically about 1 km in scale. Certain lunar craters of about this size, known as *dark halo craters,* are surrounded by dark, smooth deposits and often occur along apparent fractures. Figure 11-19 shows an example. It is thought that they may be diatremes, with the "halos" being erupted material.

Ash and tuff rings are often built up around vents with both gas and ash ejecta. Diamond Head, a famous example in Hawaii, is shown in Figure 11-20.

Tectonic activity produces a variety of features of different forms, caused either by compressional or by stretching forces. Gradual buildup of compressional forces may result in buckling of the surface layers, producing folded

Figure 11-19

Lunar dark halo craters, possible examples of lunar diatremes. A: Three dark halo craters associated with rilles in the crater Alphonsus. B: Detail of A, showing a dark halo crater that has interrupted a rille.

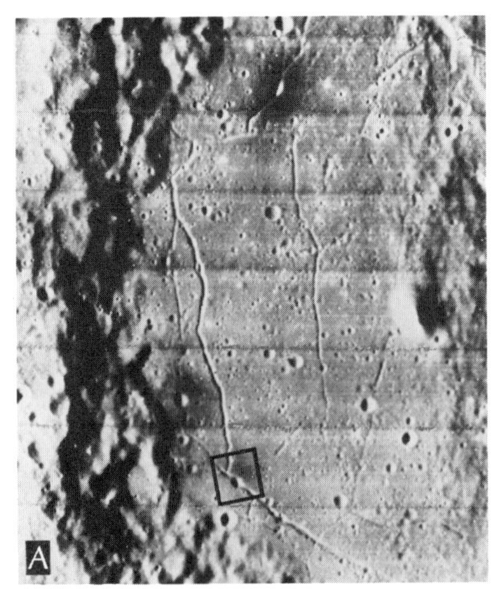

Figure 11-20
A crater within a city. The famous Hawaiian landmark Diamond Head is an ash-tuff cone that originally formed about 150,000 years ago.

mountains, composed of alternating synclines and anticlines, discussed in Chapter 10.

A different family of features is produced if the stresses are relieved not by plastic deformation but suddenly by fracturing, as in earthquakes. Such fractures may create remarkable surface structures on planets. An example of this is the *lunar grid system,* which is a latticework of *lineaments* or linear features such as those shown in Figure 11-21. Lineaments are valleys, ridges, and cliffs, probably developed over subsurface fractures. The preferential direction of the lineaments gives some clue as to the nature of the stresses that produced the fractures. In the lunar case the grid system is strongly developed along the NW-SE and NE-SE directions (Fielder, 1961; Strom, 1964), and there is some evidence of a similarly oriented grid on Mars (Binder, 1966). These lineament systems suggest early, global stresses in the surface layers of these planets.

Fractures along which there has been relative motion are called *faults.* Faults are common on the earth and in the lunar grid system. There are several varieties of faults, depending on the directions of relative motion, as illustrated in Figure 11-22.

Causes of Volcanism

The basic causes of volcanic activity are still shrouded in mystery. In the first place, the mechanisms that create magma are uncertain. On the moon, the

Figure 11-21
Tension-produced graben in Iceland. Layered basalt flows can be seen in the far wall. The whole island of Iceland is being rifted apart by the outward motion of material from the mid-Atlantic ridge, which passes through Iceland. Graben width, approx. 100 meters.

major lava flows are about 3 to 4 billion years old, and it appears that most of the lava was generated in an early melting of the lunar interior. This early melting, which seems to be characteristic of planetary bodies, may have been caused by heat produced by the short-lived isotopes we have discussed before.

In the case of the earth, however, volcanism continues to the present day. What causes continued production of molten rock near the surface? Possibly convection in the mantle brings up enough hot material to create molten pockets just below the crust. Also, radioactive elements are concentrated in the continental sediments which accumulate in geosynclines (see Chapter 10), and may produce heat that aids in creating magma. The volcanologist Eaton (1962) used seismic data to show that some magmas originate in the upper mantle, some 50 to 60 km beneath the Hawaiian volcano Kilauea.

What drives the magma upward? The magma is lighter than the surrounding rocks, and therefore hydrostatic pressure forces it to rise. The lower density of the magma results from its containing gases in solution. Since the weight of the magma column is less than the weight of the surrounding rocks, the magma can be driven upward several kilometers above mean surrounding countryside, even to the top of a high volcanic mountain such as Mauna Kea.

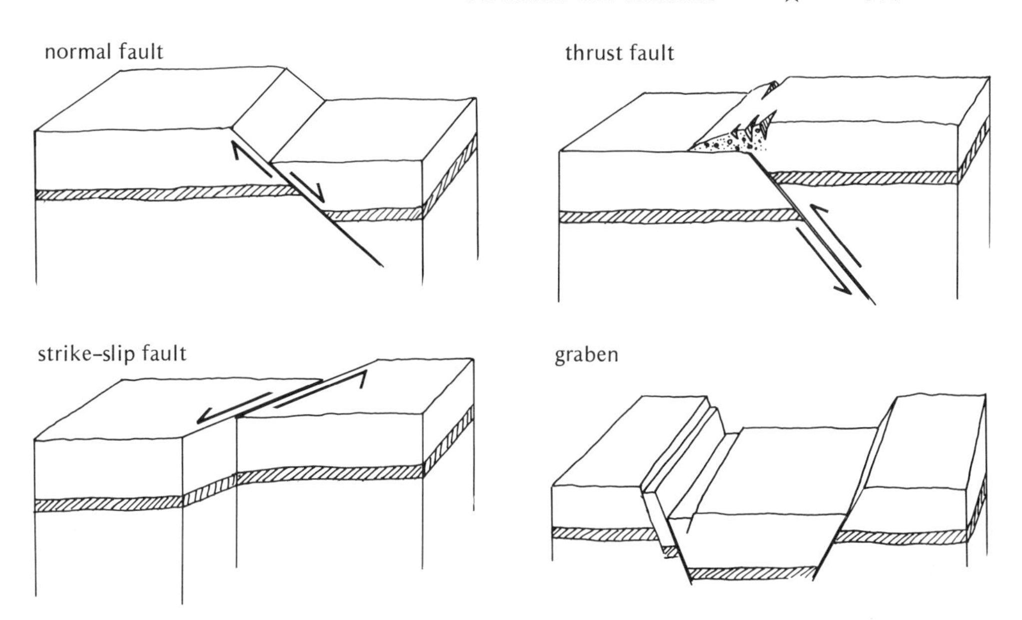

Figure 11-22
Nomenclature for various fault types.

As the magma ascends, it may temporarily collect in various conduits and magma chambers. Eventually, when it nears the surface, the pressure is sufficiently low that the gases begin to be released and the lava may become very frothy, just as a soft drink froths when the pressure is reduced by opening the bottle. This frothing action may cause spectacular activity at the mouth of the vent when the lava is released. The lava may retain the bubbles after it flows onto the surface. The bubbles are called *vesicles* and such lava is called *vesicular lava*. Vesicular fragments are known as *scoria*. Figure 11-23 shows

Figure 11-23
Vesicular lava, showing frothy structure produced by degassing of volatiles in the magma as the pressure is reduced during eruption of magma at the surface. Specimen length approx. 10 cm.

an example. An extreme case is *pumice,* which has so many bubbles that its bulk density is less than 1.0, allowing it to float on water.

Volcanic Activity on the Moon

When Galileo first looked at the moon through a telescope about 1610, he could see mountains, craters, and vast flat gray areas that he interpreted as oceans. These were given the name *mare* (pronounced mah'-ray; pl. maria), the Latin word for sea. They are of course, not oceans but dusty, rock-littered plains.

We have already mentioned some of the theories that were put forward about the nature of the lunar maria. They have been variously attributed to lava flows, ash flows, dust deposits, and water-laid sediments. We have also noted that the first Apollo missions brought back mare rocks that were pieces of lava flows. Specifically, most rocks were crystalline, moderately coarse grained gabbros (see Figure 11-1) with some vesicles. Analysis indicated that there were slightly different compositions at each mare site, probably corresponding to individual lava flows. The Apollo 15 astronauts observed layering in the walls of the Hadley rille, another indication of individual flows.

These findings show that there were definitely lava flows on the moon. Apparently, the maria were produced by vast outpourings of lava on many parts of the moon about 4 to 3 billion years ago (the solidification ages found for the lunar lava samples). Figure 11-24 shows an example of a probable lava flow photographed by the Lunar Orbiter spacecraft in one of the lunar maria.

A question raised by these findings concerns the total amount and importance of volcanic activity on the moon and, by inference, other planets. Was volcanic activity responsible for most of the moon's features?

A controversy raged for more than a century with participants taking various sides of the question. Some invoked the fact that many terrestrial volcanic craters resembled those on the moon and pointed to other lunar structures, such as long faults, to show that the moon had been sculpted almost entirely by volcanic and tectonic forces. Others showed that lunar craters resembled bomb explosion and meteorite-impact craters, and could be simulated by dropping projectiles into plaster or powder; they argued that impacts must be the dominant surface process. Interesting from the point of view of the history of science was a tendency for astronomers to favor volcanic mechanisms, while geologists argued an impact origin, as if each discipline wished to pass the hot potato to the other! The inference to be drawn from this is that the moon is complex, and each traditional discipline alone was hard put to find suitable explanations of its structures. Baldwin's (1949) study of the formation of the lunar surface was the closest to the truth at its time, as we can now judge in retrospect. Baldwin argued that the maria were overlapping flows of very fluid basalt, a conclusion basically correct, since the rocks actually brought back from the moon are basalts and gabbros, the coarse-grain equivalent of

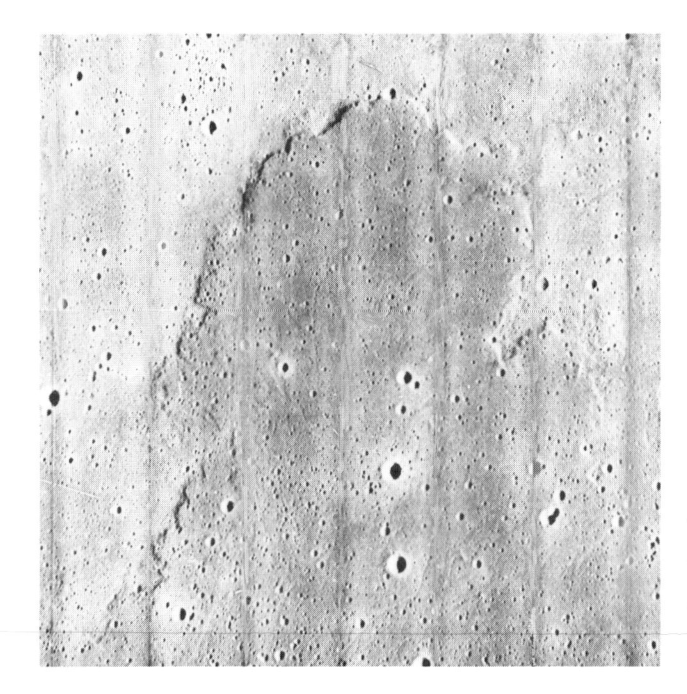

Figure 11-24
Tip of a lunar lava flow in Mare Imbrium, photographed by Lunar Orbiter V. Width of lobe at top approx. 15 km. (NASA.)

basalt (see Figure 11-1). Baldwin correctly surmised that the lavas had been very fluid, in order to flow the long distances and form the smooth surfaces we observe. Studies of Apollo rocks indicate that they were erupted at unusually high temperatures, about 1400 to 1600°K (Epstein and Taylor, 1970), which increased their fluidity.

Turning to the craters, Baldwin plotted a number of relationships among crater parameters, such as diameter and depth, and showed that they fitted neatly on curves defined by impact craters rather than volcanic craters. Baldwin's evidence was much more complete and more systematically presented than any earlier evidence, and he persuaded many researchers that most lunar craters were produced by meteorite impacts. He also noted that many of the maria, or lava plains, which were much larger in dimension than the craters, were located in huge circular basins, and he proposed that these circular basins were also the sites of impacts, on a larger scale than those that formed most craters. After the basin-forming impacts occurred, volcanic activity flooded the basins with lava. Baldwin's general picture is accepted today.

In spite of the abundance of impact craters, a large number of nonimpact craters of volcanic origin certainly appear on the moon. Among these are *crater chains,* which probably mark subsurface faults, and *dark-halo craters,* which may mark diatremes (Figure 11-19). Whether some of the other large craters may be calderas is an open question.

To complicate matters, faulting and lava flooding have modified a number of the large impact features, making them hybrid craters with features of both impact and volcanic craters.

The moon lacks some of the larger, positive-relief volcanic features characteristic of the earth, such as large volcanic mountains like Mt. Fuji. Possibly this is due to the greater fluidity of the lunar lavas, which allowed them to spread laterally instead of piling up into mountains.

Features related to lunar volcanism include the following. *Domes* are low, rounded hills, sometimes with central summit craters. They are a few kilometers across and may be lunar volcanoes, the equivalents of large cinder cones or small shield volcanoes. Figure 11-25 shows a major grouping of these—the Marius Hills, located in the Ocean of Storms. Running through the middle of the Marius Hills is a *wrinkle ridge,* a common type of lunar mare ridge from which lava flows emanate. The Marius Hill complex is probably an example of late lunar volcanism.

Rilles are long, canyon-like depressions. There are two types, which are

Figure 11-25

The Marius Hills region of the moon. The various rounded hills, or domes, are thought to be volcanic features. A prominent wrinkle ridge extends north and south through the region to the east (right) of center. Typical domes are about 8 km in diameter. (NASA.)

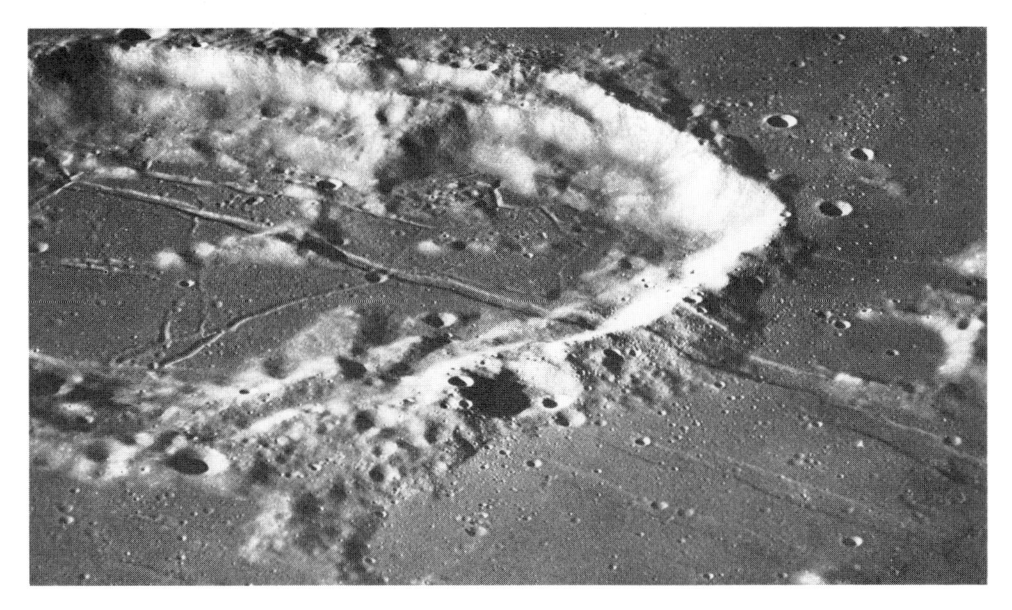

Figure 11-26
Strong linear trends on the moon. In this area linear rilles (graben) cut mare, crater rim, crater floor, and crater central peak surfaces. (NASA.)

probably distinct genetically. *Linear rilles* are probably examples of graben (grah'-ben), valleys produced by parallel faulting (Figure 11-26). *Sinuous rilles* wind irregularly and may be examples of large flow channels produced when very fluid lava cut channels through the loosely consolidated regolith. Figure 11-27 is an Apollo 15 astronaut's photograph of one prominent example—the Hadley rille. Some sinuous rilles meander back and forth in channels reminiscent of the Mississippi and other old, meandering rivers. Other sinuous rilles evidently are caused by the collapse of underground *lava tubes*—conduits through which lava once flowed but later exited. Figure 11-28 compares a terrestrial and probable lunar example of collapsed tubes.

Are volcanoes still active on the moon today? For a long time the moon was regarded as a dead world. Indeed, this accounted for a certain lack of interest in the moon on the part of astronomers until the middle 1900s. A few amateur astronomers had reported changes in certain lunar structures, but none of the reports was well enough documented to be taken seriously by most professional astronomers. However, on November 19, 1958, the Soviet astronomer Kozyrev obtained a photographic spectrum indicating an emission of gas in the lunar crater Alphonsus (Kozyrev, 1959). Next, on two occasions in 1963, a number of observers at Lowell Observatory saw red glowing spots in and near the crater Aristarchus. Although all observers agreed that lunar vol-

Figure 11-27

Hadley rille—a lunar sinuous valley. Apollo 15 astronauts explored the margins of this rille and noted rock layers outcropping in the valley wall (inset). This observation affirms that the lunar maria consist of layered lava flows and suggests that sinuous rilles are eroded flow channels or collapsed lava tubes. (NASA.)

canism was at best much rarer than terrestrial volcanism, these observations spurred a renewed interest in present-day lunar activity. A statistical study of many rather poorly documented reports of lunar transient phenomena showed that they were most frequent when the moon passed through its perigee point, when tidal forces on the moon were largest (Middlehurst and Chapman, 1968).

During the first lunar landings, astronauts set up seismometers to search for lunar volcanic activity. Two results were found by analysis of the seismic records. First, the moon was seismically very quiet, confirming that lunar volcanism and tectonic activity must be very rare (see Chapter 10). Second, a particular type of "moonquake" occurred every time the moon passed its perigee point, indicating that strains are released when the moon experiences maximum flexing during its closest approach to the earth.

In summary, it appears that although the moon once had an exceedingly

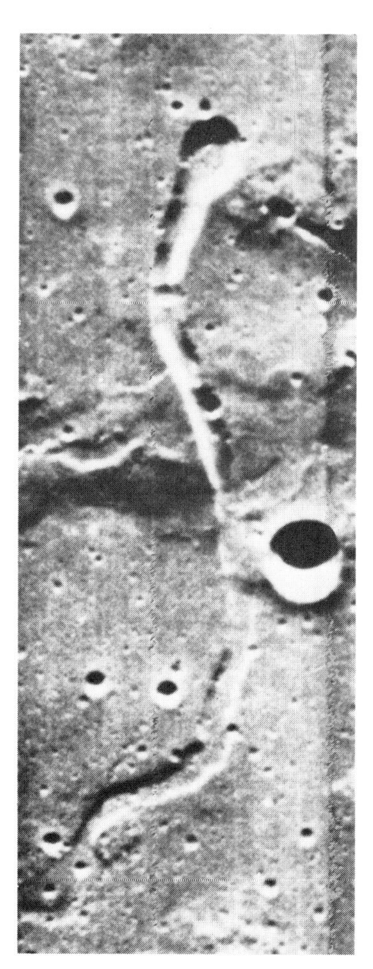

Figure 11-28

Comparison of a terrestrial flow channel in Iceland (A) with a probable lunar flow channel, or sinuous rille (B). Note collapsed craters along the terrestrial example. Although instructive, such comparisons are qualified by differences in scale, the lunar features usually being larger than comparable terrestrial features. (Lunar and Planetary Laboratory; NASA.)

active volcanic and tectonic phase in its history, it is now marked by only sporadic, minor volcanic and tectonic activity, much of which is induced by periodic tidal strains in the body of the moon.

Tectonic Modification of Impact Structures—Lunar Basins

In Chapter 9 we talked of impact structures, and above we talked of volcanism. We have only hinted at the complexities that arise when these two land-

Figure 11-29
The Orientale basin system, showing concentric ring structure, central lava flooding, and striated ejecta blanket. Map at lower left indicates scale. (NASA.)

forming agents are mixed. The huge lunar circular basins provide an example and are of interest in their own right as well.

The basins, which are typically several hundred kilometers in diameter, show some features of ordinary craters, such as a raised rim and depressed interior. However, they show a number of additional features. Baldwin (1949) pointed out the existence of outer, concentric cliffs surrounding several of the mare-filled basins, such as Mare Nectaris (Sea of Nectars), which is surrounded by the Altai scarp, or Mare Imbrium (Sea of Rains), which is surrounded by the lunar Alps, Apennine, and Carpathian Mountains. Hartmann and Kuiper (1962) made a systematic survey and discovered a number of new, concentric multiring basin systems, including the Orientale "bulls-eye" system on the extreme limb of the moon, barely visible from earth. As shown in Figure 11-29, the basin system surrounding Mare Orientale was later photographed in a series of remarkable photographs by the Lunar Orbiter spacecraft.

In addition to a series of concentric cliff-like rims, the large basins are surrounded by radiating lineaments, often in the form of long valleys, crater chains several kilometers in width, and fine striations as small as a hundred meters across. Furthermore, the basins are surrounded by *ejecta blankets,* layers of blasted-out debris that cover and partially obscure preexisting terrain.

How did these complex features form? Let us take Mare Orientale and its basin system as an example. The ejecta blanket and the symmetric system of radial and concentric structures indicate that the central basin was the site of a tremendous explosion. Only the impact of an asteroidal-sized projectile could reasonably provide enough energy for such an event.

During the impact, the ground immediately outside the impact zone was apparently scoured by an outrush of material that led to deposition of the ejecta blanket. This may account for the peculiar, grooved terrain surrounding the basin, shown in detail in Figure 11-30.

The surrounding cliffs, on the other hand, evidently were produced by faulting—by the ground fracturing and dropping on the inner sides. Figure 11-31 shows a theoretical cross section of such a fault-ring system. There is some uncertainty as to whether the faulting occurred during the impact event as part of the dissipation of the strong shock wave, or after the event, as part of the adjustment of the lunar crust to internal melting and volcanic activity, but in any case the rings seem to be examples of tectonic adjustment.

The outlying radial valleys and ridges may also be examples of tectonic adjustment, and the radial crater chains must have involved volcanic activity along radiating fractures caused by the initial impact.

Some time after the impact occurred, lava worked its way to the surface and flooded the interior of the basin. Probably the inner regions of all the basins were most accessible to lava flooding because that is where the impact produced the deepest fractures, allowing lava to gain access to the surface. Another clue that lava came up along fractures is that some of the lava lies not

Figure 11-30

Detail of striations in the Orientale ejecta blanket. Striations lie radial to the basin center and have characteristic width about 1 km. (NASA.)

Figure 11-31

Schematic cross section of a major lunar basin system (compare Orientale basin, Figure 11-30). The diagram shows the highly fractured subsurface produced by impact, ejecta in the bottom of the basin and in a surrounding ejecta blanket, concentric fracture rings, and lava flooding.

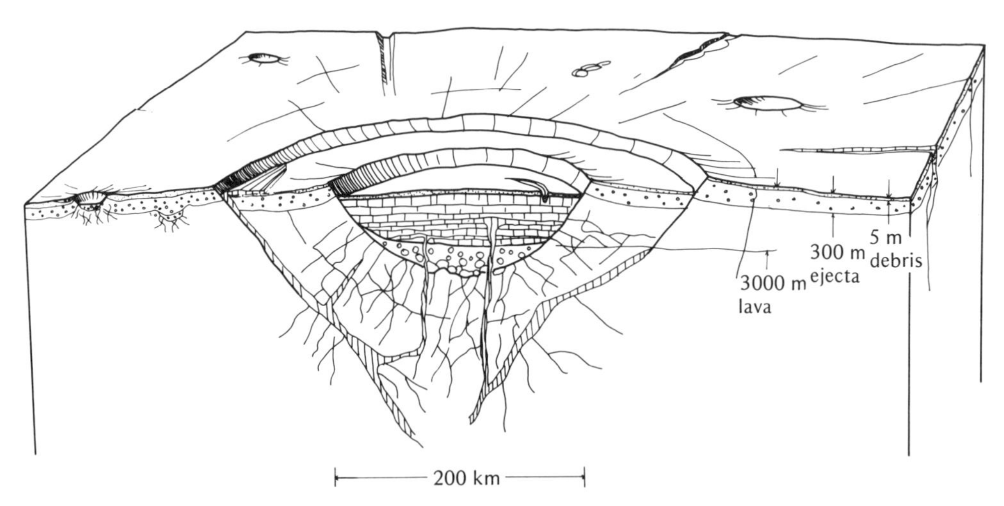

in the center but in arcuate strips along the base of some of the concentric fault cliffs. Ultimately, the interiors of most large basins were flooded by lava.

The ground must have settled and cracked as volumes of lava came up from below and settled on the surface. Rilles and other complex evidence of tectonic adjustment occur on the surface of the mare, as shown by Figure 11-32.

Figure 11-32

Terra and mare structure in one quadrant of the inner Orientale basin (compare Figure 11-29). Arcuate cliffs and fractures are prominent. (NASA.)

The exact dates of formation of the basin systems are not known, but all of them formed before most maria formed. They probably date back to a period of early intense bombardment, when the last planetesimals were colliding with the planets. Since the surface lavas in the maria are about 3 to 4 billion years old, most basins must have formed approximately in the first 0.5 billion years of lunar history. Thus they have been exposed for a long time to the many modification processes of the lunar surface. A schematic diagram of the history of a volcanically modified impact basin is shown in Figure 11-33.

PLANETARY SURFACE TEMPERATURES

Some considerations in measuring planetary temperatures were mentioned in Chapter 10. Since the radiation from a solid body varies with wavelength in a well-known fashion, called the *Planck radiation* law, a planet's temperature can be determined by measuring the amount of radiation at different wavelengths in the infrared, where the radiation is concentrated. Hence we can measure planetary surface temperatures without actually visiting the planets.

A difficulty arises if the planet has an atmosphere, because we must then know if the radiation we see comes from the top of the atmosphere or the surface. If we measure at a wavelength that is absorbed by the planet's atmosphere, the atmosphere will appear opaque to our instrument and we will measure upper-atmosphere temperatures. If we observe at a wavelength at which the atmosphere does not absorb, we will see through the atmosphere to the ground level. An example of this is Venus, for which radiation of wavelength shorter than 1 cm gives a temperature of about 240°K, for the top of the atmosphere, while radiation of wavelength longer than a few centimeters gives a temperature near 700°K for the surface.

The following summary includes results of temperature measurements.

CONCLUSION: SURFACE CONDITIONS ON THE PLANETS

We shall now give a brief sketch of the conditions to be encountered on the surface of each planet, as well as can be estimated by putting together all the types of evidence discussed in this chapter.

Mercury

Mercury's surface is rather like that of the moon, a finding based on photometry, colorimetry, polarimetry, and radar. When the sun is up—a period that lasts over 27 days—the surface temperatures probably reach as high as

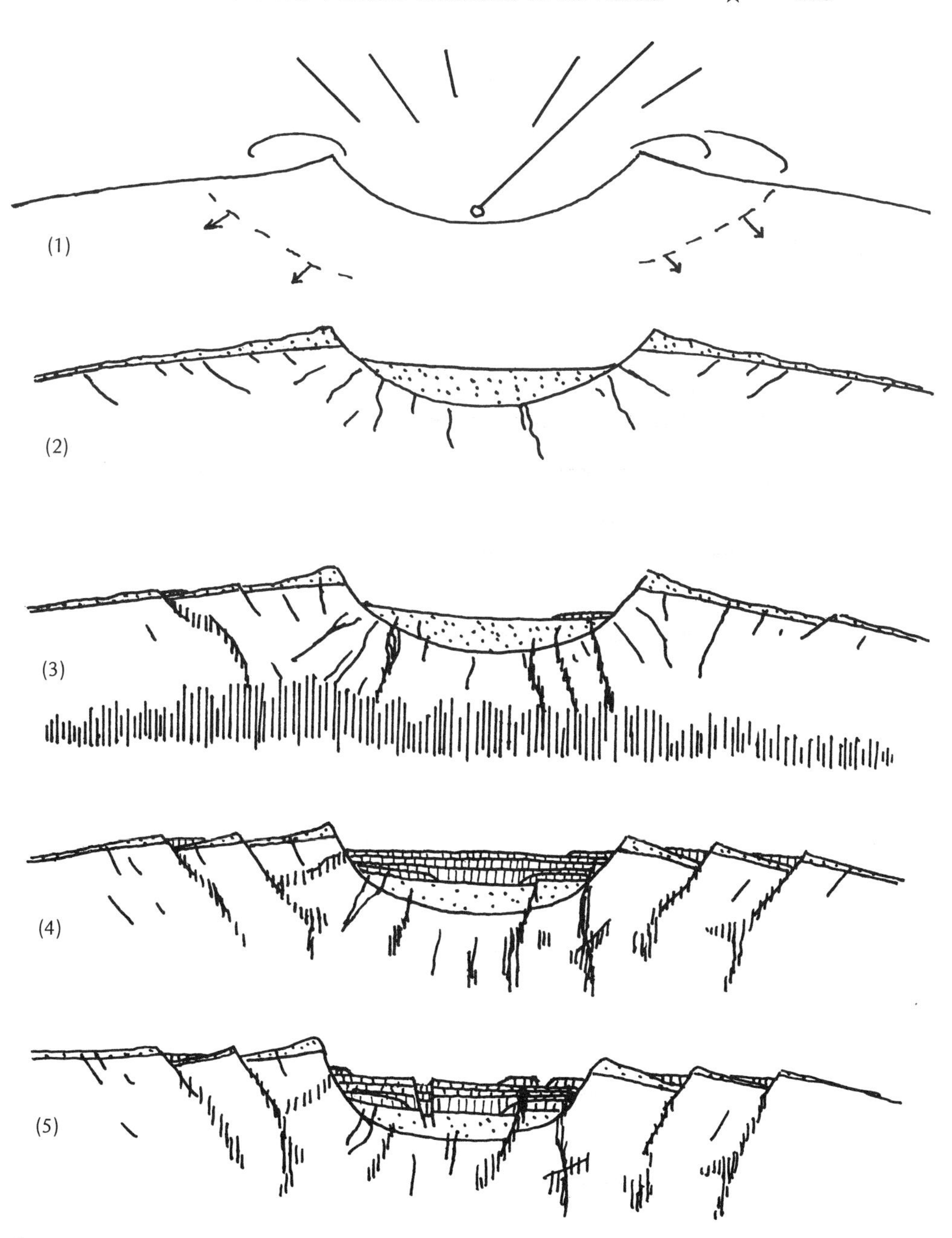

Figure 11-33

Schematic diagram of evolutionary history of a major lunar basin system: 1, impact and expanding shock wave; 2, impact crater and ejecta immediately after formation; 3, intrusion of subsurface magma, faulting; 4, extrusion of lava flows and major faulting, producing concentric cliffs; 5, present state of basin after additional lava extrusion and faulting.

420°K (297°F). For certain models of the surface soil, temperatures could exceed 500°K. This compares to the melting point of tin (505°K) or lead (600°K). The pre-dawn equatorial temperature is estimated to be 100 ± 15°K (−279°F) (Morrison, 1970).

Morrison's data suggest a surface layer of density only 1.5 ± 0.4 g/cm^3, which may resemble the lunar regolith. Indeed, Morrison concludes that Mercury looks very much like the moon moved nearer to the sun.

The dark markings of Mercury may be vast, ancient lava flows like the lunar maria. The number of craters caused by impacts of asteroidal meteorites will probably be fewer than on the moon, because Mercury is so far removed from the asteroid belt. However, the number of craters caused by impacts of comet nuclei may be significant and similar to that on the moon, since the density of cometary nuclei among the terrestrial planets is more or less constant. Mercury has no appreciable atmosphere.

Venus

The surface of Venus must be forbidding. There is no liquid surface water, and typical daytime temperatures are near 750°K (890°F). The atmospheric pressure is enormous, about 90 times that on the earth, according to interpretations of measures made by probes parachuted into the atmosphere. Venus is so densely obscured by clouds that it is uncertain how much direct sunlight gets through to the surface. The surface may be perpetually fogged by mist or blowing dust, and the hot, dense atmosphere if not opaque would cause extreme optical distortions, beyond the terrestrially familiar flattening of the setting sun, caused by the earth's atmosphere. Geochemistry is dominated by high-temperature equilibria between the rocks and atmosphere. Some mountain ranges, suggested by radar, may lie shrouded in the clouds.

The Earth

If an alien artist were commissioned to make a painting of the earth's surface, and if he had a good knowledge of the earth, what would he paint? What would be representative? Should he show a deciduous forest, the frozen wastes of the Antarctic, the Sahara, or a volcano? Should he show sunshine, rain, fog, or smog? He could argue that the most representative picture would be the view shown in Figure 11-34, since the earth is mostly covered by oceans. Or he could paint the bottom of the sea. This dilemma reminds us of the difficulties in giving thumbnail sketches of planetary surfaces, as we are attempting to do.

Salient planetological features of the earth include the following points. Earth is a remarkable planet in that it maintains a surface temperature and pressure regime allowing large amounts of liquid water. Fully 70.8 percent of its

Figure 11-34
Representative view of the surface of the earth.

surface is covered by liquid water, while abundant white clouds of liquid water droplets and ice crystals obscure the surface. Large polar ice caps contain more water. Liquid and solid water frequently precipitate onto the surface and cause substantial erosion of the silica-rich soils and rocks. The interior is active, causing continued mountain building and volcanic activity at the present time. Man appeared in the last 0.1 percent of the earth's history and is biologically adapted to this planet—it is the only known planet on whose surface he can stand naked and remain alive for more than a few minutes. The delicate geochemical and thermal environment could be slightly altered (i.e., rendered uninhabitable) if man pursues certain of his current pursuits with his present abandon.

The Moon

The surface of the moon is dry and desolate, and yet in the opinion of the few men who have been there, dramatic and beautiful. Scattered boulders of gabbroic and anorthositic igneous rock are strewn about, but the landscape is softened by a pulverized layer of surface debris, or regolith. There is no air or water, and even the rocks contain unusually small amounts of water by terrestrial standards. To an observer on the front side, the earth remains fixed in the sky because the moon keeps the same side earthward. The sun rises and sets 2 weeks later, but there is virtually no other change. At local noon, the rock temperatures rise as high as 407°K (273°F), but the dusty surface material insulates the subsurface and prevents absorption of much thermal energy. Consequently, as soon as the sun goes down or shadows fall on a local area, the rock temperatures fall rapidly, reaching 120°K (−243°F) during the lunar night. In the previous section we gave a detailed account of various types of lunar features. A point to be emphasized here is that to understand these features, we must order them into some kind of chronological sequence. Which formed first and which most recently? Did most of them form at the beginning, or has the development of the lunar surface been uniform over time? These questions could apply to any planetary surface. In the case of the moon a clue has been provided by dates determined from rock samples. These dates allow us to reconstruct the lava flow and meteoritic bombardment history. Figure 11-35, based on rock dates and crater counts, shows that the meteoritic impact rate was highest in the first few hundred million years. This accords with the evidence cited in the last section that the basins formed also in the moon's early history, and is in accord with Chapter 6.

Figure 11-35

Schematic history of lunar cratering, reconstructed from crater counts and dates of Apollo lunar samples. Most maria date from the shaded zone. Pre-mare cratering was much more intense than post-mare cratering. Cratering rate may have had small-scale fluctuations resulting from discrete fragmentation events among meteorite parent bodies. Proposed ages for the craters Copernicus and Tycho (see Figure 11-10) are shown.

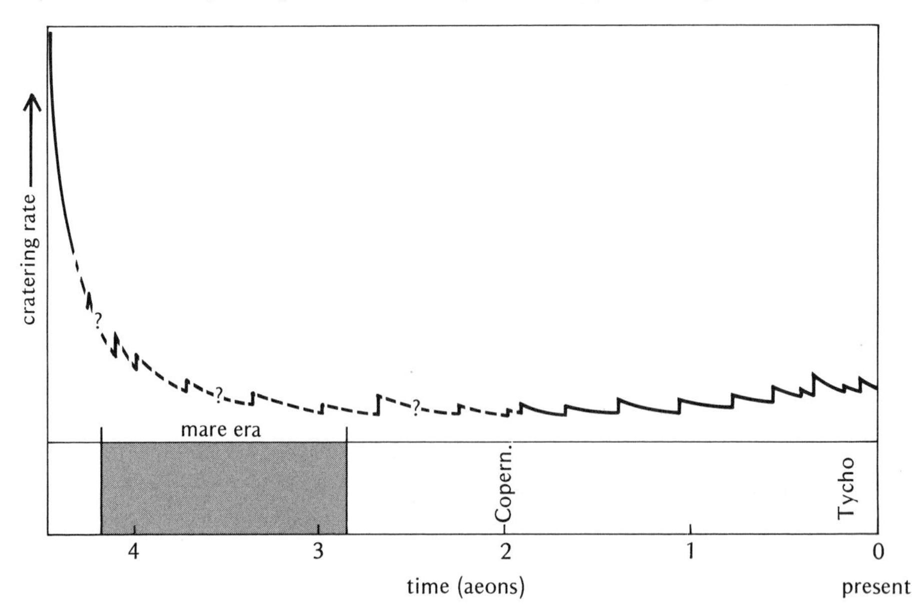

A crude breakdown of the lunar time scale has three parts: pre-mare time, the mare-forming interval, and post-mare time. Pre-mare time involves the first 0.5 billion years, judging from the dated Apollo samples of lunar lavas. In pre-mare time the impact rate was very high. The mare-forming interval occurred about 3 to 4 billion years ago. During this period great amounts of lava flooded the surface, especially the low-lying basins. For some reason not yet known, this lava extruded primarily on the side of the moon facing the earth. This may reflect some early asymmetry in the interior of the moon. Post-mare time has been relatively quiescent, with sporadic meteorite impacts.

Mars

The surface of Mars is probably dusty, with scattered rock outcrops. There is very little water. The atmospheric pressure is very low, only about 0.008 that of the earth, or similar to that about 20 km (65,000 ft) above the earth's surface. As a result of the thin air, the sky is almost black, but it is blue toward the horizon. The sun is noticeably smaller and casts fainter light than we on earth are used to. A imaginary Martian landscape is shown in Figure 11-36. There are

Figure 11-36

Imaginary view of the surface of Mars. Relief is probably generally subdued, with dusty deserts dominating the landscape. The sun appears smaller than from the earth.

occasional clouds and haze, especially at dawn and dusk and near the poles. Frozen carbon dioxide, possibly with some frozen water, constitutes the polar caps, each of which disappears or nearly disappears in summer. Summer equatorial temperatures can occasionally exceed 300°K (81°F), but pre-dawn equatorial temperatures can drop to 170°K (−153°F). Generally, Martian temperatures run about 50°K (90°F) cooler than the earth's.

Martian craters have a flattened appearance, suggesting effective erosion and deposition processes.

A major question concerns life, either present or past. There is no proof that the surface markings are biogenic, in spite of their seasonal and non-seasonal changes. The famous "canals" have been found to be merely rough alignments of patchy markings, systems of linear faults, and borders between different provinces. Nonetheless, Mariner 9 photography has revealed seeming dendritic erosion patterns such as would be produced by flowing liquid surface water. Since no pools of surface water exist now, the photographs raise the possibility of ancient or episodic concentrations of volatiles (see p. 293). Could these have supported simple Martian life in the past? Clearly an exciting search is underway and our conception of Mars will evolve in the coming years.

The Giant Planets

The nature of the surfaces of the giant planets is not known, except that they are probably dark, with high gravity and very high atmospheric pressure.

Satellites of the Giant Planets

The satellites are probably covered with, or composed in part, of snow and ice, as suggested by Figure 11-37. Some of Saturn's satellites have bulk density as low as ice, and some satellites, such as Jupiter's Io and Europa, have quite high albedos, also suggesting icy or snowy surfaces. Saturn's sixth satellite, Titan (largest satellite in the solar system), has a spectrum indicating a methane-rich atmosphere, indicated in Figure 11-38. The Galilean satellites of Jupiter may also have thin atmospheres (see Chapter 12). Reports of dusky markings on these satellites suggest that they are not uniformly snow-covered, and reports of irregular brightness changes suggest that the snow deposits may be variable (see Chapter 12). Johnson and McCord (1970), in a study of visual and infrared reflectivities of the Galilean satellites, find that they have surfaces different from the moon and Mars but consistent with some type of ferrosilicate rocks with a possible partial frost or snow cover (McCord, Johnson, and Elias, 1971). The Saturn satellites, with the exception of Titan, have surfaces still more strongly indicative of ice or snow cover, with rather flat and featureless spectral reflectivities. Polarization studies by Veverka (1971) are consistent, indicating

Figure 11-37
Imaginary view of Jupiter as seen from a snowfield on one of its Galilean satellites.

that the Galilean satellites Io, Ganymede, and Europa are covered by a highly reflective, transparent material, probably frost. Callisto, however, is darker and more like the moon, suggesting that at least part of the surface is bare rock. In this context it is interesting that Callisto is significantly less dense than the other three Galilean satellites, an unexpected characteristic in view of the evidence for a more rocky surface than the other satellites. On the basis of spectroscopic and colorimetric evidence, Veverka and others have suggested that surface material of Io, Ganymede, and Europa is H_2O frost with an admixture of impurities such as CH_4, NH_3, NH, CH, or more complex organic compounds.

Many of these satellites keep one face constantly toward the primary planet, just as the moon does. This is known as *synchronous rotation,* since the rotation and revolution rates are synchronized, and is a result of tidal interaction. In most cases either the leading or the trailing side (90° around from the primary) is brightest. The very unusual satellite Iapetus, eighth satellite of Saturn, has a trailing side roughly six times brighter than its leading side. Cassini first noted this in 1672, and it is said that many early observers could see

Figure 11-38
Imaginary view of the surface of Saturn's sixth satellite, Titan, the only satellite known to have an atmosphere.

Iapetus only at western elongation. Possibly a frost or snow mantle has been removed from the leading side of the satellite by micrometeorite erosion or a major impact event, so that the leading side has a lunar or rock-like albedo of perhaps 7 to 10 percent, while the trailing side has the albedo of a snowfield.

References

Anders, E. (1963) "Meteorite Ages," in *The Moon, Meteorites, and Comets,* B. M. Middlehurst and G. P. Kuiper, eds. (Chicago: University of Chicago Press).

______ J. Laul, R. Keays, R. Ganapathy, and J. Morgan (1971) "Elements Depleted on the Lunar Surface: Implications for Origin of Moon and Meteorite Influx Rate," presented at 1971 Lunar Science Conference, Houston.

Bagnold, R. A. (1941) *The Physics of Blown Sand and Desert Dunes* (London: Methuen & Co. Ltd.).

Baldwin, R. B. (1949) *The Face of the Moon* (Chicago: University of Chicago Press).

Bauer, C. A. (1947) "Production of Helium in Meteorites by Cosmic Radiation," *Phys. Rev., 72*, 354.

Binder, A. B. (1966) "Mariner IV: Analysis of Preliminary Photographs," *Science, 152,* 1053.

——— and D. P. Cruikshank (1966) "Lithological and Mineralogical Investigation of the Surface of Mars," *Icarus, 5,* 521.

Bowen, N. L. (1928) *The Evolution of the Igneous Rocks* (Princeton, N.J.: Princeton University Press; reprinted Dover Publications, Inc., New York, 1956).

Carpenter, R. L. (1964) "Study of Venus by Continuous Wave Radar—Results of the 1964 Conjunction," *Astron. J., 70,* 134.

Coffeen, D. (1969) "Wavelength Dependence of Polarization. XVI. Atmosphere of Venus," *Astron. J., 74,* 466.

Connes, P., J. Connes, W. Benedict, and L. Kaplan (1967) "Traces of HCl and HF in the Atmosphere of Venus," *Astrophys. J., 147,* 1230.

Daly, R. A. (1933) *Igneous Rocks and the Depths of the Earth* (New York: McGraw-Hill, Inc.).

Dollfus, A. (1961) "Polarization Studies of Planets," in *Planets and Satellites,* G. P. Kuiper and B. M. Middlehurst, eds. (Chicago: University of Chicago Press).

Duke, M., C. Woo, M. Bird, G. Sellers, and R. Finkelman (1970) "Lunar Soil: Size Distribution and Mineralogical Constituents," *Science, 167,* 648.

Dyce, R. B. (1970) "Radar Studies of the Planets," in *Surfaces and Interiors of Planets and Satellites,* A. Dollfus, ed. (New York: Academic Press, Inc.).

Eaton, J. P. (1962) *Crustal Structure and Volcanism in Hawaii, Geophys. Monograph 6,* p. 13 (Washington, D.C.: American Geophysical Union).

Epstein, S., and H. P. Taylor, Jr. (1970) "$^{18}O/^{16}O$, $^{30}Si/^{28}Si$, D/H, and $^{13}C/^{12}C$ Studies of Lunar Rocks and Minerals," *Science, 167,* 533.

Evans, J. V., and T. Hagfors, eds. (1968) *Radar Astronomy* (New York: McGraw-Hill, Inc.).

Fielder, G. (1961) *Structure of the Moon's Surface* (Elmsford, N.Y.: Pergamon Press, Inc.).

Fireman, E. L. (1959) "The Distribution of Helium 3 in the Grant Meteorite and a Determination of the Original Mass," *Planet. Space Sci., 1,* 66.

Fowler, W. A., J. Greenstein, and F. Hoyle (1962) "Nucleosynthesis During the Early History of the Solar System," *Geophys. J. Roy. Astron. Soc., 6,* 148.

Gault, D. E., E. D. Heitowit, and H. J. Moore (1964) "Some Observations of Hypervelocity Impacts with Porous Media," in *The Lunar Surface Layer,* J. W. Salisbury and P. E. Glaser, eds., p. 151 (New York: Academic Press, Inc.).

Gehrels, T., D. Coffeen, and D. Owings (1964) "Wavelength Dependence of Polarization. III. The Lunar Surface," *Astron. J., 69,* 826.

Geikie, A. (1905) *The Founders of Geology,* 2nd ed. (New York: The Macmillan Company; reprinted by Dover Publications, Inc.).

Gold, T. (1955) "The Lunar Surface," *Monthly Notices Roy. Astron. Soc., 115,* 585.

Hapke, B. (1965) "Effects of a Simulated Solar Wind on the Photometric Properties of Rocks and Powders," *Ann. N.Y. Acad. Sci., 123,* 711.

——— (1967) "A Readily Available Material for the Simulation of Lunar Optical Properties," *Icarus, 6,* 277.

Harris, D. L. (1961) "Photometry and Colorimetry of Planets and Satellites," in *Planets and Satellites,* G. P. Kuiper and B. M. Middlehurst, eds. (Chicago: University of Chicago Press).

Hartmann, W. K. (1966) "Martian Cratering," *Icarus, 5,* 565.

______ and G. P. Kuiper (1962) "Concentric Structures Surrounding Lunar Basins," *Comm. Lunar Planet. Lab., 1,* 51.

Howell, J. V., ed. (1960) *AGI Glossary of Geology* (Washington, D.C.: American Geological Institute).

Jacobs, J. A., R. D. Russell, and J. T. Wilson (1959) *Physics and Geology* (New York: McGraw-Hill, Inc.).

Johnson, T. V., and T. B. McCord (1970) "Galilean Satellites: The Spectral Reflectivity 0.30–1.0 Microns," *Icarus, 13,* 37.

KenKnight, C., D. Rosenberg, and G. Wehner (1967) "Parameters of the Optical Properties of the Lunar Surface Powder in Relation to Solar Wind Bombardment," *J. Geophys. Res., 72,* 3105.

Kozyrev, N. (1959) "Observation of a Volcanic Process on the Moon," *Sky and Telescope, 18,* 184.

Kuiper, G. P., R. G. Strom, and R. S. Le Poole (1966) "Interpretation of the Ranger Records," in *Ranger VIII and IX. Part II. Experimenters' Analyses and Interpretations, JPL TR 32-800,* p. 35 (Pasadena: Jet Propulsion Laboratory).

Lebofsky, L., T. Johnson, and T. McCord (1970) "Saturn's Rings: Spectral Reflectivity and Compositional Implications," *Icarus, 13,* 226.

Leighton, R. B., et al. (1967) *Mariner IV Pictures of Mars, JPL TR 32-884* (Pasadena: Jet Propulsion Laboratory).

Lewis, J. S. (1968) "An Estimate of the Surface Conditions of Venus," *Icarus, 8,* 434.

Mason, B. (1966) *Principles of Geochemistry* (New York: John Wiley & Sons, Inc.).

______ and W. G. Melson (1970) *The Lunar Rocks* (New York: John Wiley & Sons, Inc.).

McCord, T. B., and J. Adams (1969) "Spectral Reflectivity of Mars," *Science, 163,* 1058.

______ J. Adams, and T. V. Johnson (1970) "Asteroid Vesta: Spectral Reflectivity and Compositional Implications," *Science, 168,* 1445.

______ T. V. Johnson, and J. Elias (1971) "Saturn and Its Satellites: Narrow-Band Spectrophotometry (0.3–1.1 μ), *Astrophys. J., 165,* 413.

McCracken, C. W., and M. Dubin (1964) "Dust Bombardment on the Lunar Surface," in *The Lunar Surface Layer,* J. Salisbury and P. Glaser, eds. (New York: Academic Press, Inc.).

Middlehurst, B. M., and W. Chapman (1968) "Tidal Cycles and Lunar Event Mechanisms," *Astron. J., 73,* 192.

Milton, D. J. (1967) "Slopes on the Moon," *Science, 156,* 1135.

Minnaert, M. (1961) "Photometry of the Moon," in *Planets and Satellites,* G. P. Kuiper and B. M. Middlehurst, eds. (Chicago: University of Chicago Press).

Morrison, D. (1970) "The Thermophysics of Mercury," *Space Sci. Rev., 11,* 271.

Mutch, T. A. (1970) *Geology of the Moon* (Princeton, N.J.: Princeton University Press).

Öpik, E. J. (1965) "Mariner IV and Craters on Mars," *Irish Astron. J., 7,* 92.

Pettengill, G., and J. Henry (1962) "Enhancement of Radar Reflectivity Associated with the Lunar Crater Tycho," *J. Geophys. Res., 67,* 4881.

Rittmann, A. (1962) *Volcanoes and Their Activity* (New York: John Wiley & Sons, Inc.).

Sagan, C., and J. B. Pollack (1969) "Windblown Dust on Mars," *Nature, 223,* 791.

Salisbury, J. W., and G. R. Hunt (1967) "Infrared Images: Implication for the Lunar Surface," *Icarus, 7,* 47.

______ and G. R. Hunt (1969) "The Compositional Implications of the Spectral Behavior of the Martian Surface," *Nature, 222,* 132.

______ and V. G. Smalley (1964) "The Lunar Surface Layer," in *The Lunar Surface Layer,* J. Salisbury and P. Glaser, eds. (New York: Academic Press, Inc.).

Shoemaker, E. M. (1965) "Preliminary Analysis of the Fine Structure of the Lunar Surface," in *Ranger VII. Part II. Experimenters' Analyses and Interpretations, JPL TR 32-700,* p. 75 (Pasadena: Jet Propulsion Laboratory).

______ et al. (1970) "Lunar Regolith at Tranquillity Base," *Science, 167,* 452.

Shorthill, R., and J. Saari (1965) "Nonuniform Cooling of the Eclipsed Moon: A Listing of Thirty Prominent Anomalies," *Science, 150,* 210.

Strom, R. G. (1964) "Tectonic Maps of the Moon, I," *Comm. Lunar Planet. Lab., 2,* 205.

Sytinskaya, N. N. (1957) *Uch. Zap. Leningrad Univ., 190,* 74.

Urey, H. C. (1952) *The Planets: Their Origin and Development* (New Haven, Conn.: Yale University Press).

Veverka, J. (1971) "Polarization Measurements of the Galilean Satellites of Jupiter," *Icarus, 14,* 355.

Webb, H. D. (1946) "Project Diana," *Sky and Telescope, 5,* 3.

Wehner, G. K. (1964) "Sputtering Effects on the Lunar Surface," in *The Lunar Surface Layer,* J. Salisbury and P. Glaser, eds. (New York: Academic Press, Inc.).

Whipple, F. L. (1959) "Solid Particles in the Solar System," *J. Geophys. Res., 64,* 1653.

PLANETARY ATMOSPHERES

THE PURPOSE OF STUDYING planetary atmospheres is to learn their origin and evolution, the principles that determine their structure, and their meteorological conditions.

ORIGIN OF PLANETARY ATMOSPHERES

Primitive Atmospheres

As we discussed in Chapter 6, the planets were immersed in a gaseous medium when they formed. This medium was rich in hydrogen and had a pressure that can be easily calculated from the universal gas law, which relates pressure, density, and temperature of any gas. The density of the gas was probably not more than 10^{-6} g/cm^3, and the temperature not more than a few hundred degrees near the surfaces of the accumulating planetesimals. The pressure, calculated from the gas law, is thus found to be less than 2 percent of that in the atmosphere at the earth's surface.

As the planets grew, they attracted gas, and the planets' gravitational fields could compress this gas and hold it in the form of an atmosphere blanketing the planetary surfaces. These *primitive atmospheres* may have been quite dense. They were very different from the earth's present atmosphere because they formed from gas of cosmic composition and were rich in hydrogen compounds such as molecular hydrogen (H_2), methane (CH_4), ammonia (NH_3), and water (H_2O). In this sense they were rather like the present-day atmospheres of the giant planets.* In chemical terms, they were not *oxidizing* atmospheres, like the earth's is now, but *reducing* atmospheres (i.e., hydrogen-rich atmospheres).

Secondary Atmospheres on the Earth, Mars, and Venus

Secondary atmospheres are atmospheres that have been produced or significantly altered by gases exhausted from the planetary interior. Although large

* The composition of such an atmosphere will be compared with observed planetary atmospheres in Table 12-2.

planets, such as Jupiter, have gravitational fields strong enough to retain their primitive atmospheres, smaller planets lost much of their primitive atmospheres through a process called thermal escape, which will be discussed in the final section of this chapter. At the same time, new gases were being added from the interiors of the planets. The new gases arose when internal energy sources heated the planetary interiors. This created large quantities of molten magma saturated with dissolved gases that eventually escaped into the atmosphere, either by eruption or slow fumarole activity (Brown, 1948).

What kinds of gases would be added to planetary atmospheres by such activity? This problem was analyzed by Rubey (1951). Rubey's classic paper showed that most gases now in the earth's atmosphere neither are original nor came from weathering of surface rocks but came from outgassing accompanying volcanism. A major extension of this work was a study by Holland (1962), who showed that the gases that came out of the *earliest* terrestrial volcanoes were not the same as those emitted today. This finding was based on physical chemistry. The metallic iron in the early, undifferentiated earth would combine with and remove oxygen from the early volcanic gases. Thus, instead of creating an oxidizing atmosphere, the early volcanic gases produced a reducing environment. These gases were probably rich in molecular hydrogen (H_2), water (H_2O), carbon monoxide (CO), and hydrogen sulfide (H_2S).

By the time the earth's interior had melted and iron had drained toward the center of the earth, the escaping volcanic gases resembled those emitted today, because the escaping gases were then in chemical equilibrium with iron-poor minerals instead of iron-rich minerals. Table 12-1 compares the gases emitted by present-day volcanoes with the gases, which, according to Rubey's calculation, must have been added in the last few billion years. It can be seen that the agreement is good, in the sense that a mixture of mantle-derived gas from oceanic lavas and gases from the more siliceous continental magmas would go a long way toward accounting for the observed abundances in our atmosphere.

One peculiarity of the earth's atmosphere not explained by this scheme is the fact that the inert gases (helium, neon, argon, krypton, and xenon) are now present in only the minutest fraction of their original cosmic abundance. Relative to cosmic abundances, the earth's atmosphere is depleted in neon, for example, by more than a factor of 10^{10}, as was pointed out by Brown (1948) and Suess (1949). The probable explanation invokes the fact that inert gases are chemically nonreactive (i.e., don't combine into compounds with other chemicals). They were thus not incorporated into the early planetesimals and hence were never incorporated into the planets. The inert gases probably were carried into interstellar space when the solar nebula dispersed. The depletion of the inert gases in the earth's atmosphere is an evidence that (1) the earth's atmosphere is not a residue of cosmic gas, and (2) the earth accreted from small particles that never contained inert gases. If a body the size of the earth

Table 12-1

Gases added to the atmosphere by volcanic outgassing of the earth (percent composition by weight)

	Observed from eruptions that may tap the mantle (Hawaiian volcanoes)[a]	Observed from continental fumaroles, geysers, etc.	Calculated by Rubey (1951)
H_2O	57.8	99.4	92.8
CO_2[b]	23.5	0.33	5.1
Cl_2	0.1	0.12	1.7
N_2	5.7	0.05	0.24
S_2	12.6	0.03	0.13
Others	<1	<1	<1
	100	100	100

[a] This represents the most primitive, deep-seated magma available. Hawaiian volcanoes lie above a thin oceanic crust, and the magma tapped by them taps the upper mantle. Continental fumaroles, on the other hand, represent material that has been chemically reprocessed and contaminated with groundwater, etc.

[b] Plus small amounts of CO.

had ever accreted the heaviest inert gases, those gases would for the most part still be present, because they are too heavy ever to have escaped the earth's gravitational field (see the last section of this chapter).

Although the primitive reducing atmosphere of the earth had little oxygen, the oxygen content increased by two processes. The first was dissociation of water molecules into hydrogen and oxygen, which occurs in the upper atmosphere when water molecules are struck and split by energetic ultraviolet light rays from the sun. The second process was photosynthesis, which began after green plants evolved. The history of oxygen in the earth's atmosphere was discussed by Kuiper (1952) and Berkner and Marshall (1965) (see also Chapter 13, where we discuss the origin of life).

Mars and Venus also appear to have at least partially secondary atmospheres. In both cases their present atmospheres are dominated by carbon dioxide (CO_2). A clue as to whether Venus has a secondary atmosphere would be given if we could determine the abundance of the next most important gases after CO_2. Is nitrogen (N_2) the second most abundant constituent? If Venus has a secondary atmosphere, predictions are that N_2 should be the secondary constituent (Lewis, 1968). Cameron (1963), however, suggested that the Venus atmosphere is largely a remnant of a primordial hydrogen-rich atmosphere and *not* a secondary atmosphere. In this case, neon would be the second most abundant constituent. Lewis (1968, p. 450), who discusses this problem, finds that neon may be present, although the abundance is uncertain. Neither nitrogen nor neon has easily observable spectral absorptions, at visible wavelengths, and their abundance has not been established from earth by present techniques. The four Soviet spacecraft parachuted into the Venus atmosphere have not firmly distinguished among nitrogen, neon, and other inert gases, and so future determinations of their abundance will be of interest to settle whether Venus has a secondary atmosphere, as is usually assumed.

Mars is only about half the size of the earth and Venus; hence it lacks the gravity to retain much of an atmosphere. Mars has certainly lost most of its primitive atmosphere, and the thin one that exists may be a remnant of a secondary atmosphere. On the other hand, Mars may not have had sufficient internal thermal activity ever to have exhausted a substantial secondary atmosphere.

The Martian atmosphere has a high percentage of CO_2. One may ask why, if Mars's atmosphere is secondary, there is not a large percentage of nitrogen, as on the earth. We can give a probable answer for this, even though the Martian N_2 content is not yet accurately known. In comparing the N_2/CO_2 ratio of either Mars or Venus with that of the earth, we must recall the earth's geochemical cycle, discussed in Chapter 11. Because of the presence of water on the earth, vast amounts of terrestrial CO_2 have been locked up in sedimentary carbonate rocks by the Urey reaction. Rasool and de Bergh (1970) show that the total CO_2 locked in the earth's crust is comparable to that in the Venus atmosphere. Thus we should say that the earth's atmosphere is CO_2-poor, rather than saying that Mars or Venus is N_2-poor. Mars and Venus have "normal" CO_2-rich atmospheres, whereas the CO_2 on earth has been locked in rocks.

STRUCTURE OF PLANETARY ATMOSPHERES

Hydrostatics

The same principles that were discussed in the beginning of Chapter 10 can be used to compute the structure of a planetary atmosphere. The hydrostatic equation $dP = -\rho g\, dz$ applies to both planetary atmospheres and interiors because both behave like fluid systems. (The minus sign appears here since z is chosen to increase upward as P decreases.)

★ PROBLEM ★

Use the hydrostatic equation to estimate the order of magnitude of the thickness of the earth's atmosphere.

Answer

Laboratory experiments show the pressure at sea level (the bottom of the atmosphere) to be about 1.0×10^6 dynes/cm^2 and the density to be 1.3×10^{-3} g/cm^3. Gravitational acceleration at the earth's surface is about 1×10^3 cm/sec^2. If we assume that the *mean* density of the atmosphere is about half that at the surface, and then substitute the values for dP, ρ, and g, respectively, and then solve for the height dz, we get an answer of the order 15 km. This rough answer (about 9 miles) shows that the atmosphere is remarkably thin, compared to the size of the planet. This statement applies to all planets. In reality the density has dropped to about 10 percent of its surface value at this altitude. In a sense, we all live only a few dozen miles from interplanetary space.

The calculation above was very crude because it assumed a linearly varying pressure and density. To compute the exact height and vertical structure of the atmosphere, we would have to know the exact pressure and density at each point. This, in turn, raises the question of temperature at each point, which we discuss next.

Temperature Structure

In Chapter 10, when we tried to compute the internal structure of a planet, we saw that the equation of state became important. The same applies to atmospheres, except that in this case the equation of state is much simpler, namely the *ideal gas law,* which applies to any gas. The ideal gas law relates the pressure P, to the density ρ, the temperature T, and the composition, which in the case of gases can be indicated simply by the mean molecular weight μ. The ideal gas law reads

$$P = \frac{\rho}{\mu M_{\mathrm{H}}} kT,$$

where M_{H} is the mass of a hydrogen atom and k the Boltzmann constant (Table 2-2).

Suppose that we now start trying to compute a crude model of the atmosphere, using the hydrostatic equation and measurements of the surface P, ρ, and T. Following the same scheme as in Chapter 10, we could divide the atmosphere into (say) 100 layers, each 1 km thick, and try to compute the pressure and other conditions at each layer. The atmosphere is so thin that we can treat g as a constant, to a first approximation.

Using the hydrostatic equation, we insert the known values of ρ, g, and $dz = 1$ km. We compute dP, the change in pressure between the ground and the top of the first layer. We now have P_1, the pressure 1 km above the ground.

Now we are stuck, because we have no way of getting the density at that level to use in our next calculation, for the second kilometer. If we had measures of ρ at all levels we could proceed. Or, if we had measures of T at all levels, we could proceed by using the ideal gas law to compute ρ, since we would then know P, T, and the composition (assumed uniform).

In other words, to compute a reliable model of an atmosphere, given the conditions at any one level, we have two alternatives: either we must (1) have a set of measures of P, ρ, *or* T, or (2) make some simplifying assumption about the behavior of one of these.

One of the simplest and conceptually most useful assumptions is that the temperature is constant at all levels (i.e., that we have an *isothermal atmosphere*). At first sight this assumption seems a gross mistake, since we know that it gets noticeably cooler as we ascend even a modest mountain. But the changes in atmospheric temperature are only some tens of degrees out of about 300°K (i.e., only a percentage change rather than a change by a substantial factor). Therefore, a simple isothermal atmosphere bears some relevance to the real physical world.

★ MATHEMATICAL THEORY ★

Our problem now is to derive the pressure and density structure in an isothermal atmosphere of uniform composition. By eliminating ρ in the hydrostatic equation and the ideal gas law, we can get an equation with only the single remaining unknown, P. The student should verify that

$$dP = \frac{-\mu M_{\mathrm{H}} P g}{kT}\, dz.$$

This differential equation is readily solved. Dividing by P, we have on the left dP/P, which coincidentally is identical to $d \ln P$. Integrating this logarithmic form from the surface ($z = 0$) to any arbitrary height, we find that

$$P = P_0 \exp\left(\frac{-\mu M_{\mathrm{H}} g}{kT}\, z\right),$$

where P_0 is the pressure at the ground level. We thus have solved for the pressure structure P as a function of z.

Two concepts are immediately evident. First, the density structure in an isothermal atmosphere is just the same as the pressure structure, since, by the ideal gas law, with T constant, we have $P/P_0 = \rho/\rho_0$. The second concept is the concept of *scale height*. If we define

$$\text{scale height} = H = \frac{kT}{\mu M_{\mathrm{H}} g},$$

substitution into the equation in the last paragraph shows that the density or the pressure will decrease by a factor $1/e$ if we ascend to a distance of one scale height (i.e., at $z = H$). The student should verify this.

The concept of *scale height,* mathematically derived above, expresses the vertical distance over which the density of the atmosphere decreases by a factor of $1/e$, or 0.368. Sometimes the *decimal scale height* is used instead; this gives the height at which the density falls by a factor $\frac{1}{10}$. The scale height is a useful indicator of the effective "thickness" of a planetary atmosphere, which is highly concentrated toward the planetary surface but grades off indefinitely into interplanetary space.

Although we derived the definition of scale height for an isothermal atmosphere, the definition can be applied to any real atmosphere, as long as we state the level—hence temperature—of which we are speaking (surface, cloud tops, etc.).

Real atmospheres are not isothermal. In fact, if we could somehow manage to create an isothermal atmosphere and then let nature take its course, the atmosphere would soon develop its own temperature structure. The two major sources of heat energy which cause this new distribution of temperature are (1) diurnally (day–night) varying inputs of radiation from the sun (visible radiation) and from the planet (infrared radiation); (2) condensation or evaporation of certain atmospheric constituents (heat input governed by latent heat of condensation). In order to see how these affect atmospheric structure, we must first discuss how energy is distributed through the atmosphere.

Radiative Transfer of Heat Energy

The theoretical treatment of radiative transfer of heat is exceedingly complex and is the basis of much of astrophysics. We can only touch on it here. (It was mentioned briefly in Chapter 10, with regard to heat transfer in planetary interiors.)

The keys to the theory of radiative transfer are the concepts of opacity and optical depth. *Opacity* is a measure of how much radiation is absorbed* over a given distance through the atmosphere. In a dense fog on the earth, the opacity is very high and we can see perhaps 10 m. On a clear day we can see 100 km. The *optical depth* helps to express the effect of opacity. The optical depth is a dimensionless number that expresses the amount of radiation lost

* Or scattered. Scattering and absorption are two distinct mechanisms for removing energy from a beam of light. For simplicity we shall minimize the difference between these mechanisms, although the difference is crucial in advanced radiation transfer theory.

from a beam of light. The optical depth is defined so that at optical depth 1, most of the light has been lost out of the original beam. Consider a beam of light approaching the top of an atmosphere from space. At the top the optical depth is zero. If the atmosphere is cloudy, like that of Venus, optical depth 1 may occur at the cloud tops. But if the atmosphere is clear, like much of the earth's, then we may have an optical depth of only 0.5 or some other fraction at the ground. At optical depth 10, virtually no radiation penetrates.

The subtle point about opacity and optical depth is that they change with the wavelength (color) of the radiation being considered. If we happen to pick a wavelength where some atmospheric constituent, say CO_2, strongly absorbs, the opacity will be very high at that precise wavelength even though it may be nearly zero at nearby wavelengths. To say it another way, if we looked through a filter of precisely that color, the atmosphere would look entirely opaque, but if we looked through a filter of a slightly different color, the atmosphere would appear transparent. To say it still another way, optical depth 1 would occur high in the atmosphere at the absorbing wavelength but low in the atmosphere (or not at all) at the nearby wavelength.

To give another example, an atmosphere could be quite transparent in the visible wavelengths but quite opaque at infrared wavelengths.

★ QUESTIONS ★

What effects would this situation have on atmospheric temperatures? On observations of planets from the earth's surface? These questions will be taken up later in this chapter.

To see why an atmosphere develops an irregular temperature profile, suppose that we could somehow create a perfectly isothermal atmosphere and then examine the consequences from the point of view of radiative transfer theory. Consider two cases. In the first case, suppose that the atmosphere was absolutely transparent at all wavelengths. Then the radiation from the sun would come through the atmosphere without any absorption. It would strike and heat the ground and cause the ground to radiate in the infrared, as in Figure 12-1(A) (see the discussion in Chapter 10 on the heat balance of planets). The infrared radiation would also escape into space without affecting the atmosphere, since the atmosphere has no opacity at infrared wavelengths. Thus we see that there would be no radiative influence of the sunlight on the atmosphere. No radiative energy would be absorbed at any wavelength or at any level in the atmosphere, and so the atmosphere would remain isothermal (neglecting conduction effects where the atmosphere contacts the heated ground).

For our second case, suppose there is a small amount of opacity at various

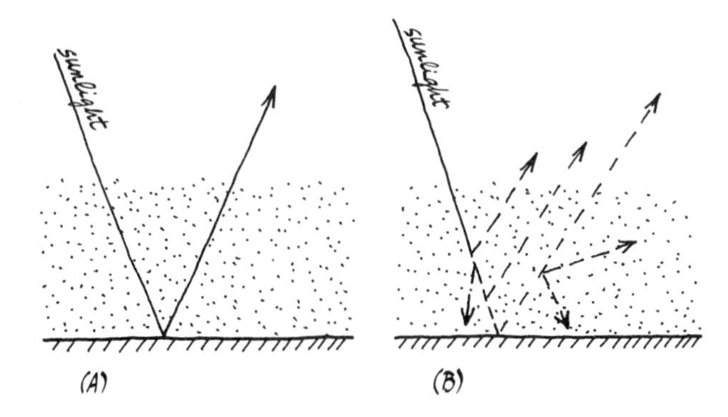

Figure 12-1
Transmission of sunlight through (A) a hypothetical transparent atmosphere, and (B) a more realistic atmosphere with absorption and reemission of radiation.

wavelengths. This is a more realistic model, whereas the first case was unrealistic. Some incoming sunlight and some outgoing infrared radiation would be absorbed. Thus the sunlight would cause a heating of the atmosphere. Sunlight can cause heating only insofar as there is some opacity causing radiative energy to be absorbed [Figure 12-1(B)].

To see one reason why the temperature structure of an atmosphere can be quite irregular, let us consider a still more realistic example. Suppose that the various absorbing constituents are arranged in layers. Take the earth as an example. Ozone (O_3), which strongly absorbs ultraviolet sunlight, is created and concentrated by photochemical reactions only in a layer at high altitudes. Water vapor, which is a very strong absorber of infrared, is created primarily at low levels by evaporation of surface water. If these two constituents were the main absorbers, some of the incoming sunlight would be absorbed at high altitudes, heating the upper atmosphere, and most of the outgoing planetary infrared radiation would be absorbed at low levels, causing the air temperature near the surface to be higher than that on mountain tops. The reader may recognize that this resembles the real situation on the earth.

Greenhouse Effect

The greenhouse effect is a strong heating of the lower atmosphere of a planet due to selective absorptions in the infrared. We can easily understand it by reviewing the above model. Suppose, as is realistic, that the atmosphere is *relatively* transparent in the visible wavelengths but has some strong absorbing medium in the infrared, such as water vapor (H_2O) or carbon dioxide (CO_2). The surface of the planet is heated to a few hundred degrees Kelvin by whatever solar radiation gets through. Any surface heated to such temperatures will radiate in the infrared. This outgoing infrared radiation gets absorbed by the atmosphere. What happens to that energy? The atmosphere cannot keep absorbing infrared energy indefinitely without getting warmer, so the atmosphere heats up and itself radiates in all directions. The lower atmosphere and the ground heat up, and radiate still more infrared until the amount of infrared energy escaping from the top of the atmosphere is equal to the amount of visible solar energy coming in. Only then is equilibrium achieved.

But the new equilibrium may leave the ground level much hotter than the few hundred degrees Kelvin that we initially expected from solar input. The greenhouse effect is named after the same phenomenon that occurs in a greenhouse, where the glass panes are analogous to our infrared absorber. Glass lets in sunlight, but it blocks the outgoing infrared and traps the warm air, causing the greenhouse to get warmer until enough infrared escapes to reach equilibrium.

The greenhouse effect is familiar on the earth. For example, during a cloudy night, when there is much water vapor in the air, the lower atmosphere absorbs heat and the air is likely to stay warm, but during a sparkling clear, dry night, the temperature may drop rapidly because the earth can radiate in the infrared the energy it gained from the sun in the daytime. Cool nights are common in dry deserts.

The greenhouse effect is probably responsible or partly responsible for the 750°K surface temperatures measured on Venus (Sagan, 1962; Rasool and de Bergh, 1970). Rasool and de Bergh reconstruct the history of the Venus atmosphere in contrast to the earth's and find that it is the greenhouse effect that is crucial in explaining the difference. When the secondary atmosphere of CO_2 and H_2O began to accumulate on each planet, a greenhouse effect began to occur. In the case of the earth, the presence of liquid water allowed the Urey reaction to drive most of the CO_2 into the rocks in the form of carbonates, but on Venus, a CO_2-caused "runaway greenhouse effect" occurred. Being closer to the sun, Venus began with higher temperatures, and the atmosphere was never cool enough to allow liquid water. What water was present remained gaseous, and dissociated because of solar radiation. The hydrogen escaped. The oxygen may have oxidized the surface rocks or partially escaped. Without abundant water, the rocks could not effectively trap CO_2. Consequently, CO_2 continued to build up, causing still greater greenhouse effects, until the high-temperature, CO_2-rich, present-day Venus atmosphere was created. In this view, if the earth had been only 5 percent closer to the sun, it too, would have suffered Venus's fate, and life would not have evolved.

Condensable Substances— "Moist Atmospheres"

Meteorologists speak of an atmosphere as moist when it contains substances that can condense into liquid or solid form. On the earth, water vapor is the familiar example, condensing to form cloud layers composed of water droplets or ice crystals. Such cloud layers may affect planetary atmospheric temperatures in two ways: (1) Clouds interrupt the flow of radiation with an abrupt layer of high opacity, and (2) the condensation of some substances is accompanied by the release of enough heat to affect the atmosphere structure. This is the case with water vapor on the earth and, indeed, water vapor is so important that a model of the lower earth atmosphere will be substantially wrong without taking it into account, as follows.

Suppose an imaginary "parcel" of air moves upward in the atmosphere, as happens in a *cumulus cloud*. It rises to a region of lower pressure and cooler air. It expands immediately* to adjust to the new pressure, and it also is cooled by the surrounding air. If there were no water vapor, it would tend toward a certain equilibrium temperature and pressure. If it were moist to start with, the cooling might make water vapor condense, liberating heat and causing the parcel to tend toward a different equilibrium temperature than a dry parcel would have had. Thus condensable substances change the temperature structure of atmospheres.

The temperature structure is often expressed in terms of the *temperature gradient,* which is the change in temperature per kilometer as we ascend in the atmosphere. Since this quantity is negative near the earth's surface, meteorologists often use for convenience the *lapse rate,* which is simply the negative of the temperature gradient. Thus, on the earth, the typical near-surface temperature gradient is −6.5°K/km, while the lapse rate is +6.5°K/km. For comparison, most estimated values of the Martian surface lapse rate range from 4 to 6°K/km.

DYNAMICS OF PLANETARY ATMOSPHERES

In the preceding sections we considered the statics of atmospheres. We presented simple models of pressure and density structure but did not consider atmospheric motions.

★ QUESTIONS ★

Why do the winds blow? Why should Jupiter have tremendous, chaotic cloud motions, and why shouldn't the winds on the earth blow themselves out and remain calm forever more?

In this section we will show that winds are a necessary consequence of the input of solar radiation.

* This adjustment is much more rapid than the thermal adjustment, since pressure differences are transmitted at the speed of sound.

Vertical Mixing by Convection

First, let us consider the generation of winds on a very local scale. Recall the ascending parcel of air that we described two paragraphs back. As it ascends, it immediately adjusts its pressure to the surrounding pressure, which is lower. Therefore, it is expanding and growing less dense as it rises. Since any heat transfer between the parcel and its new surroundings takes much longer than the pressure adjustment, the pressure adjustment takes place with essentially no energy input. It is what is called an *adiabatic change*. This adiabatic change in pressure will create a new density and temperature in the parcel. *If the new density is less than that of the surroundings, the parcel will be buoyed up and will continue to rise automatically.* The resulting motion of the air is called *convective motion*.

A familiar example of atmospheric convective motion is the vertical buildup of a towering cumulus cloud, or thunderhead. Airplane travelers are familiar with the fact that passage through a cumulus cloud is bumpy. This happens because the air is unstable with ascending *convection currents*. The reason such convective cells of air are visible as clouds is that any moisture usually condenses out into water droplets at the cooler upper-air temperatures.

Convection will begin if the actual lapse rate in the air is greater than the theoretical adiabatic lapse rate. The adiabatic lapse rate is that experienced by an ascending parcel of air that rises adiabatically (i.e., without energy input). Another way of saying the same thing is that convection sets in if the air is *superadiabatic*.

Thus we see one way that winds could start. Suppose we magically calmed all the winds on a planet. Sunlight strikes the planet. Perhaps a certain patch of dark ground gets quite hot and heats the parcel of air above it by conduction. That parcel expands and rises. New air has to rush in horizontally to fill the potential vacuum. Already this is a local wind. If the temperature gradient happens to be superadiabatic, the process is self-regenerating and a substantial wind current may start. Next we shall see how larger-scale winds can start.

Global Circulation of the Planetary Atmospheres

Let us now consider the whole globe instead of concentrating on a local area. Suppose again that we could magically stop all winds. Stagnant air at the equator would then be heated and tend to rise, while cold air at the poles would tend to sink. This would set up a circulation pattern with air moving down at the poles, then toward the equator, and then upward in the equatorial regions. This very simple model does not accurately predict planetary circulation patterns (although southward-moving weather patterns are familiar in the United States, especially during winter), but what it does show is that the continuous input of solar energy is a continuous driving force that stirs the atmosphere.

In the real world, several factors complicate the flow of air:

1. *Coriolis forces* are apparent forces due to the earth's rotation. The equator of the earth is moving eastward at about 464 m/sec (1000 mph). Therefore, air masses moving away from the slow-moving poles appear to lag behind and shift westward. Coriolis forces also produce cyclonic rotation of air masses such as hurricanes when air tries to rush radially into a low-pressure region.

2. Land masses and their associated relief affect airflow patterns.

3. Fluid-mechanical systems, depending on their dimensions and other properties, tend to set up turbulence and cellular patterns. For example, if you try to push water around in a bathtub with your hand, much of the motion you impart to the water ends up in turbulent eddies instead of smooth flow of the water. It would be extremely difficult to predict the details of the motion of the water. In the same way, we can't expect any simple model of a global atmosphere to predict the details of atmospheric motions visible on a planet from day to day. (Recall the discussion of turbulence in Chapter 3.)

ATMOSPHERIC LEVELS AND UPPER ATMOSPHERES

The principles discussed above could be applied to any part of any planetary atmosphere. However, certain additional phenomena enter the picture, especially in the upper atmospheres of the planets, because the upper atmospheres interact directly with the cosmic environment (e.g., the solar wind). The terminology used to discuss the general structure of planetary atmospheres at different levels derives from the terrestrial atmosphere and is based primarily on temperature effects. Let us begin with the lowest levels.

Troposphere, Tropopause, and Stratosphere

In a general planetary atmosphere (but particularly in the case of the earth), one might expect the lapse rate to be greatest near the ground, because infrared absorbers such as water vapor are concentrated here and also because this is where the atmosphere can be heated by contact with the ground, often producing superadiabatic lapse rates. At some higher altitude, the air is thinner and more transparent to radiation. Little energy is absorbed. At such an altitude, therefore, there would be a much lower lapse rate, as was the case in our nonabsorbing isothermal atmosphere.

This intuitive reasoning is correct in that the planetary atmospheres generally do have low-altitude regions of high lapse rates, turbulent motion, and cloud-forming activity, and higher regions with near-uniform temperature, smooth air flow, and lack of turbulence.

The region in which the temperature gradient approaches zero is known as the *tropopause,* and it is quite abrupt in the terrestrial atmosphere. It lies at a height of about 18 km (59,000 ft) at the equator and as low as 7 km (23,000 ft) at higher latitudes. The region below the tropopause is the *troposphere* (from a Greek root referring to change or mixing). It is characterized by greater lapse rates, more turbulent motions, and chemical mixing. The region above is the *stratosphere,* characterized by a more nearly isothermal temperature profile and smoother motions. Winds near the tropopause are often laminar, or uniform, and of high speed. These winds are known on earth as the *jet stream.* Air travelers will recall that once a jet airliner gets well above the turbulent cloud layer, the flight becomes much smoother. This is because the aircraft has entered the lower stratosphere.

Temperature Irregularities and the Mesosphere

In general, the region from the tropopause to the higher realm of increasing temperature is called the *mesosphere.* There is some confusion between this term and the term "stratosphere" when applied in general planetology, because the atmospheres of different planets may have different patterns of temperature irregularities. Chamberlain (1962) has proposed that the term "stratosphere" be abolished altogether in planetological use, although here we have followed the terrestrial use: The stratosphere is the nearly isothermal part of the mesosphere bordering and immediately above the troposphere, and it may in principle constitute the entire mesosphere (see Figure 12-2).

The best example of complex structure in a mesosphere is the *terrestrial ozone layer.* As the solar ultraviolet comes down through the atmosphere, it is absorbed by O_2 molecules, dissociating them into pairs of single oxygen atoms. The oxygen atoms may combine with other O_2 molecules to form O_3 molecules, ozone. The ozone itself absorbs ultraviolet sunlight, especially in certain color regions known as the *Hartley–Huggins bands.* If this high-energy ultraviolet radiation were not stopped, it would be lethal to life on earth as we know it. In the region far enough down for O_2 to be sufficiently dense, these energy absorptions cause significant heating of the air. The heating effect is maximized at about 50 km height, where there is a temperature maximum about 100°K greater than the tropopause temperature.

Depending on the compositions and radiative properties of other planetary atmospheres, there may be many variations in mesosphere structure. To date, observers have only begun to measure the fine structure in atmospheres of the other planets.

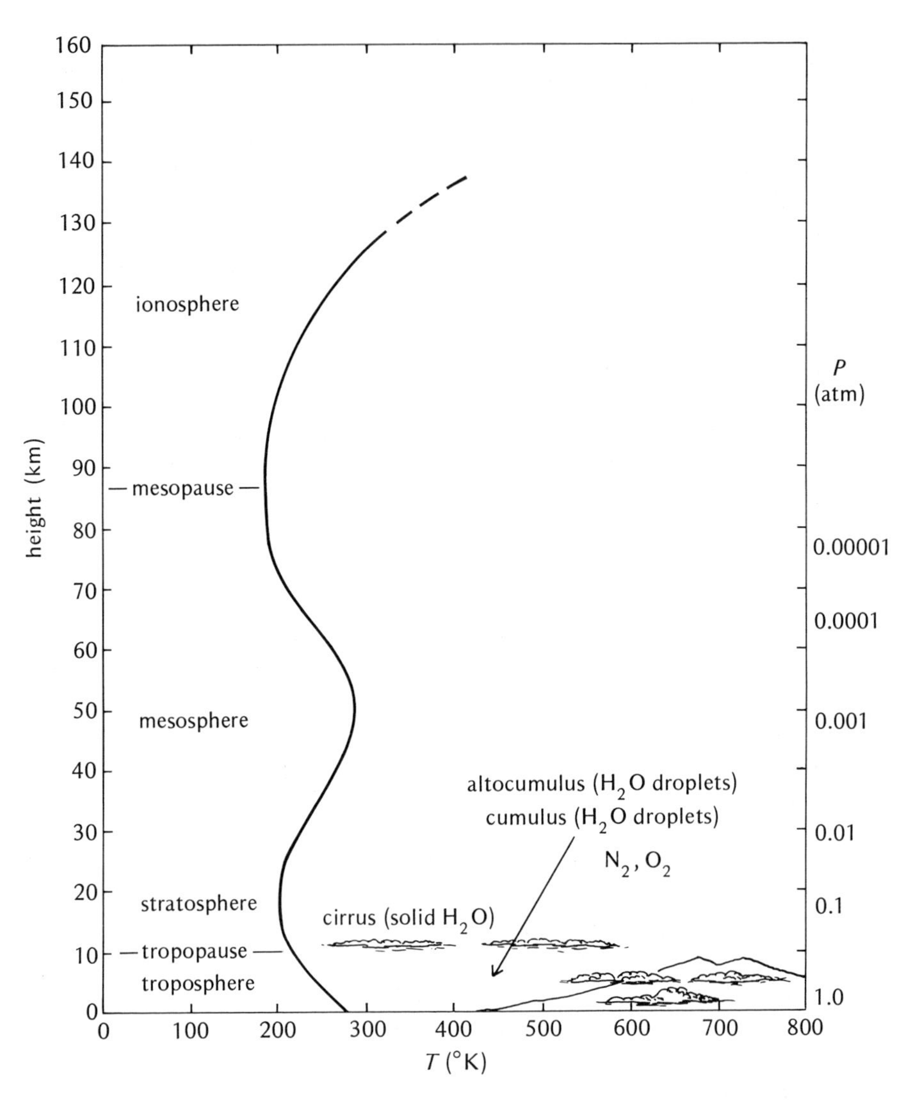

Figure 12-2

Structure of the earth's atmosphere, based on balloon and rocket observations. Typical cloud heights and high mountains are illustrated in the lower right.

Thermospheres, Ionospheres, and Exospheres

In their outer reaches, the planetary atmospheres grade into the interplanetary gas with its high-energy particles. The *thermosphere* is defined as the outer region where the temperature increases with height. An important characteristic of the thermosphere is that in the rarefied gas, molecules can move substantial distances before colliding with other molecules. Thus the mean free paths of molecules become much longer than in the denser lower atmosphere and separation of gases by diffusion becomes important, whereas in the lower atmospheres, high density and turbulent mass motions ensure mixing. Molecules of the lighter gases, especially hydrogen and helium, have higher velocities and thus can rise to great heights on near-ballistic trajectories. This produces a degree of stratification; the lightest gases concentrate in the highest layers.

The planets are cold objects moving through the hot interplanetary gas. The solar wind sweeps out past the planets with average velocities of 600 km/sec near the earth. As this corresponds to a kinetic temperature of millions of degrees,* it is not surprising that there is some heating of atmospheric boundaries. The presence or absence of a planetary magnetic field, which deflects this onrush of ionized gas, thus affects thermosphere temperatures.

Rocket and satellite observations have shown that the temperature of the terrestrial thermosphere increases from about 180°K at 80 km to a high and variable value, ranging from 700 to 1800°K above 200 km, depending on solar activity (i.e., the sunspot cycle), the time of day, and other factors. The primary cause of such high temperatures is absorption of solar ultraviolet radiation at altitudes between 80 and 200 km. Additional heating may come from the penetration of high-energy particles from the solar wind through the planetary magnetic field, to interact with the high atmosphere.

The ionosphere is a portion of the thermosphere in which charged particles, or ions, are abundant. These ions are negatively charged electrons and positively charged atoms or molecules from which the electrons have been knocked out by photons of solar ultraviolet. These photons may impart so much energy that the electrons are knocked out with much higher velocities than the local equilibrium thermal velocities characterizing the neutral molecules. Indeed, measures show that equilibrium does not exist between the radiation, the ions, and the neutral particles. Observed *electron temperatures* are often as much as twice the gas temperatures defined by the neutral particles.

Ions are so numerous that the ionosphere is an electrically conducting layer. Hence it drastically affects the passage of radio waves. Prior to 1901 it

* The reader may wonder how the planets can be embedded in a multimillion-degree gas without being vaporized. The important point is that temperature is defined in kinetic theory by the energy (or velocity) of the atoms and molecules. (Physics students should recall the equation $\frac{3}{2}kT = \frac{1}{2}mv^2$.) But for a hot gas to heat a solid mass with which it is in contact, the gas must be fairly dense. If a gas is as tenuous as the interplanetary medium, the solid mass simply will not be hit by enough atoms to absorb enough energy to maintain a high temperature. If the solid mass were somehow heated to as high a temperature as the tenuous gas, it would at once radiate at a furious rate until it cooled.

was confidently predicted that radio would be useful only over short ranges, since electromagnetic radiation travels in straight lines. The earth's ionosphere was discovered in 1901 when Marconi astonished the world by beaming radio signals across the Atlantic. Two physicists, Kennelly and Heaviside, independently but simultaneously explained Marconi's discovery by postulating a conducting, ionized layer high in the earth's atmosphere. Once called the Kennelly–Heaviside layer, the ionosphere absorbs and reflects radio waves, allowing "bounce beaming" over long distances.

The earth's ionosphere ranges from about 80 to 300 km in altitude. It contains two main maxima in electron density, the lower E-region (90 to 120 km) and the upper F-region (150 to 300 km). The latter, during the daytime, often has two maxima, F_1 and F_2. All the layers, especially the uppermost F_2-region, are highly variable and are affected by the time of day, the 11-year solar sunspot cycle, magnetic storms, etc. Electron densities in the F_2-region reach as much as 10^6 electrons/cm^3.

The outer edge of the thermosphere is the *exosphere,* a level at which a significant fraction of upward moving molecules do not hit one another but follow a long trajectory out into space, possibly escaping. The exosphere temperature is very important to the evolution of the atmosphere. If the exosphere temperature is high, exosphere molecules will have great energy and can escape into space. Certain molecules that are efficient radiators in the infrared may serve as thermostats to keep the thermosphere temperature down. "Thermostat molecules" are thus crucial to the atmosphere's history. The absence of such molecules forces the temperature up until outward radiation offsets the incoming solar radiation.

An example of the effect of "thermostat molecules" occurs in Mars's exosphere. Without a good infrared radiator its temperature would be about 2000°K. Chamberlain (1962) pointed out that CO, produced from the abundant CO_2 on the planet, would be a good radiator. Chamberlain's original calculations suggested an exosphere temperature as low as 1100°K. More recent models, based on Mariner 4 spacecraft observations, give values of 80 to 550°K, depending in part on the CO content (Monash, 1967). A theoretical model by Cloutier et al. (1969), in agreement with Mariner 4 data, gave an exosphere temperature of 490°K with an upper atmosphere scale height of 29 km.

COMPOSITION AND OBSERVATIONAL TECHNIQUES

Terrestrial Atmospheric Windows

Radiation from another planet reaches earth-based instruments only after traversing the earth's atmosphere. If the atmosphere's opacity is very high at a particular wavelength, most of the radiation at that wavelength will be absorbed. The "transparent" parts of the spectrum that lie between these high-opacity regions are called *windows*.

One of the very opaque regions of the spectrum is caused by absorption of ultraviolet radiation shortward of wavelength about 2000 Å by the ozone layer about 50 km (31 miles) above the earth's surface. This absorption prohibits ground-based observations of other planets and stars at ultraviolet wavelengths. For this reason rocket and satellite astronomy in the ultraviolet is necessary.

The atmosphere is relatively transparent in the visual wavelengths, where the sun's radiation is concentrated. This region, about 3000 Å to 1 micron (10,000 Å), is called the *visible window*. In the infrared, longward of wavelength 1 micron, the spectrum is again highly divided into windows. This is illustrated in Figure 12-3, an infrared spectrum of Mars from 1 to 2.5 microns.

Figure 12-3
"Windows" and principal absorptions in the terrestrial atmosphere.

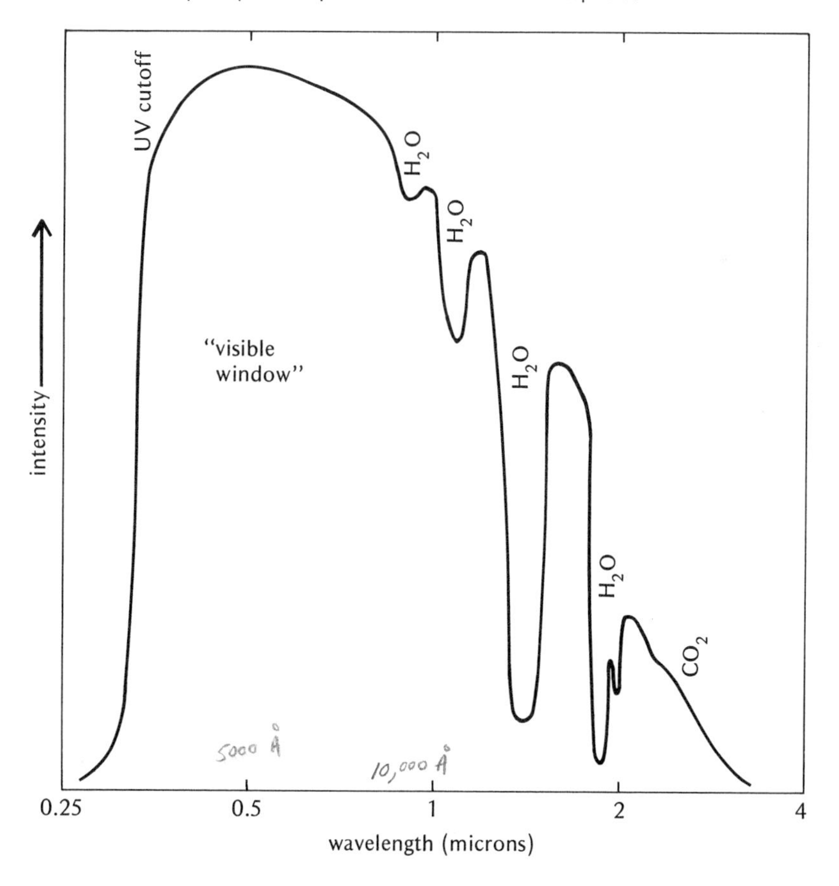

The principal terrestrial absorbing agent in the near infrared part of the spectrum is water vapor, which is concentrated in the earth's troposphere. Only from elevations above most of the water vapor, using balloons, high-altitude aircraft, rockets, satellites, or simply very high mountains, can we get good observations of extraterrestrial bodies in the infrared.

Spectral Lines and Spectroscopy

The light from a planet is reflected sunlight, modified by colors of the planet's surface materials, absorptions in its atmosphere, and, in the far infrared, by the radiation from the planet itself. The atmospheric absorptions occur in certain narrow wavelength intervals called *spectral absorption lines,* or in broader intervals called *absorption bands* (the latter in the case of absorptions by molecules instead of by atoms). By studying absorption lines, we can gain a great deal of information about a planet's atmosphere. In the immediate spectral region of an absorption line or band the background intensity is called the *continuum.* The total energy subtracted from the continuum by the line is measured by the *equivalent width* (EW), which is the width of a line with the same energy absorption (area beneath the continuum) but zero intensity and a perfectly rectangular profile. These terms are illustrated in Figure 12-4.

The equivalent width rises as the number N of absorbing atoms in the beam rises. The specific relation between these two parameters is rather complex and for each given line is presented in the *curve of growth,* in essence a plot of N versus EW. In practice, the value Nf *is used instead of* N alone.

Figure 12-4

The intensity profile of a typical spectroscopic absorption line, showing the equivalent width.

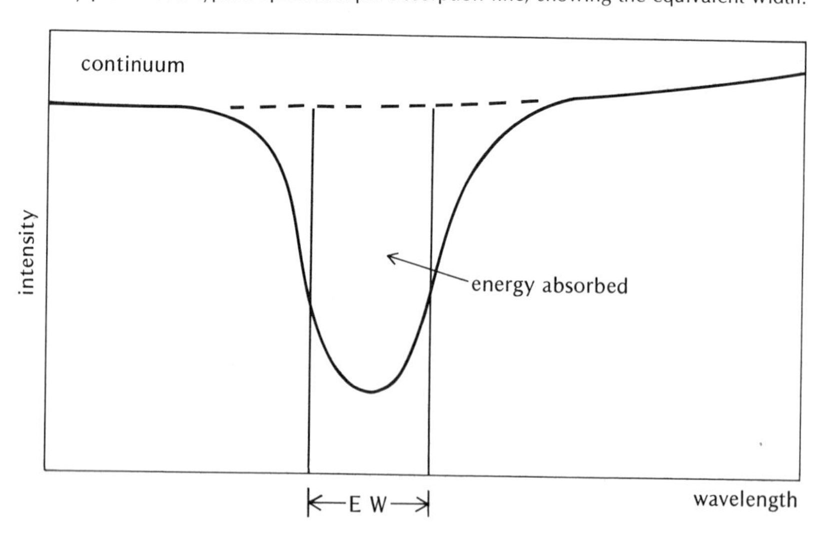

The *f factor,* as it is called, is a numerical value depending upon the particular atomic process (e.g., electronic transition) producing the given line. This can be computed, and tabular listings are available. To take a very simplified view, if we know the curve of growth for a certain spectral line of an observable constituent in a planetary atmosphere, we can take a spectrum, measure the equivalent width of the appropriate spectral line, and then read off the number of atoms in a column of the atmosphere. Thus we would have an indication of the abundance of that particular constituent. We can empirically determine the curve of growth by pumping varying amounts of gas into a tube and observing the equivalent width of the spectral line for known amounts of gas under different pressures. Thus curves of growth are found, showing a complex interdependence of equivalent width, abundance, and pressure.

A curve of growth for a typical line is illustrated in Figure 12-5. Notice

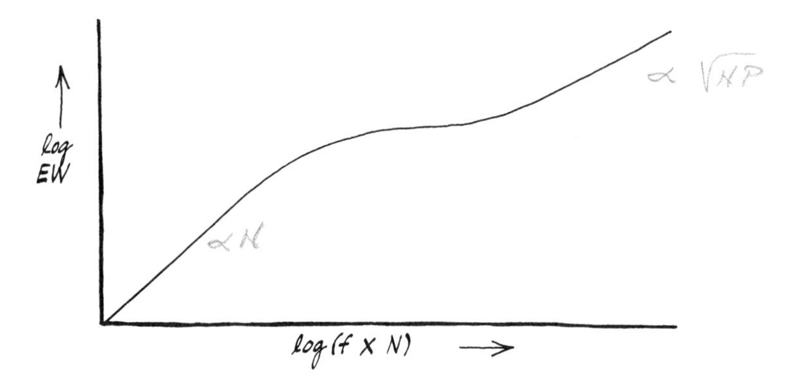

Figure 12-5

Schematic diagram of the curve of growth for a typical spectroscopic absorption line, showing the increase in equivalent width as *N*, the number of absorbing atoms, rises. The "f factor" depends on the nature of the absorption process. Different pressure environments produce different curves.

that it can be divided into three parts. For weak lines (low EW), the EW is proportional to *N*. For stronger lines, the center of the spectral line (Figure 12-4) has reached near-zero intensity and there is often a flattening of the curve of growth. Finally for the strongest lines (high EW) with many contributing atoms, the line gets very broad and the EW increases approximately as $\sqrt{NP}$, where *P* is the total pressure in the gas producing the absorption. This last situation arises as the number of collisions between atoms becomes large, at higher pressures, and is referred to as *pressure broadening*. A pressure-broadened line has a strong central core and broad "wings."

It is useful to consider an imaginary experiment in which we observe a planet at a fixed wavelength and vary the opacity at that wavelength. For high opacity—a strongly absorbing line—we would see only a small way into the atmosphere and hence get information on the outermost part of the atmosphere. Conversely, if the gas has only a weak absorption at the given wavelength, we could see down into the deeper layers of the atmosphere, where pressure broadening might be important. We would be dealing with a different part of the curve of growth and would gain information about the lower atmosphere. Thus different spectral lines with different opacities may give us data on different levels of a planet's atmosphere.

Compositions of Planetary Atmospheres

Table 12-2 summarizes some estimates and measures of the compositions of planetary atmospheres. Most of the measures are based on spectroscopic results. The types of gases are generally identified by the absorption lines found in the spectrum and the amount of gas by the curve-of-growth technique described above. In the case of Venus and the earth, direct measures have also been made *in* the atmospheres.

Students should be warned that gas compositions are given in some references as percentages by *number* or *volume,* and in others as percentages by *mass.* The latter is called the *mixing ratio* and is used in Table 12-2. In many cases the most abundant constituents (e.g., hydrogen, nitrogen, or neon) are not observable, because they lack easily identifiable absorptions. Therefore, estimates of their abundances are merely informed guesses. For observable gases, the curve-of-growth technique allows measurement of the total abundance, expressed in terms of the height of a column of the gas at 1 atmosphere pressure and standard temperature. The units of such a measure are thus meter-atmospheres, at standard temperature and pressure (STP), commonly abbreviated meter-Amagat.

OBSERVATIONS OF PLANETARY ATMOSPHERES

Mercury

At present there is no confirmed observational evidence for an atmosphere on Mercury. No dense atmosphere is expected, both because of the high temperatures (see the next section) and because of the absence of observable atmospheric effects during solar transits of Mercury in front of the sun. Yet two observational problems are of historical interest.

First and oldest are the visual observations of the dusky markings. Reports persisted, especially on the part of the well-known visual observer Antoniadi

Table 12-2
Compositions of planetary atmospheres

Initial[a] Estimated (percent by mass)		Venus Observed (percent by mass)		Earth Observed (percent by mass)		Mars Observed (m-atm STP)		Mars Estimated (percent by mass)	
H_2	63.5	CO_2	95 ± 2	N_2	75.5	CO_2	70 ± 2	CO_2	75?
He	34.9	N_2 (or NE)	3.5	O_2	23.1	N_2	$\leqslant 25$	N_2	20?
Ne	0.60	CO	0.005	Ar	1.3	Ar	$\leqslant 20$	Ar	4?
H_2O	0.34	H_2O	0.0001	H_2O	2[b]	H_2O	0.2[b]	H_2O	0.2
NH_3	0.2	HCl	0.00006	CO_2	0.05	O_2	<0.07	O_2	0.0
Ar	0.15	HF	0.000006	Ne		CO	0.1	CO	0.0
CH_4	0.1			He					
(Kuiper, 1952)		(Kuiper, 1969)				(Owen and Kuiper, 1964)			

Jupiter[c] Observed (m-atm STP)		Jupiter Estimated (percent by mass)		Saturn Observed (m-atm STP)		Titan Observed (m-atm STP)		Uranus Observed (m-atm STP)		Neptune Observed (m-atm STP)	
H_2	70,000	H_2	60	CH_4	350	CH_4	200	H_2	480,000?	CH_4	6,000
CH_4	133	He	36	H_2	190			CH_4	3,500?		
NH_3	12	Ne	2								
		H_2O	0.9								
		NH_3	0.5								
		Ar	0.3								
		CH_4	0.2								
(Owen, 1969, 1970)				(Owen, 1969)		(Kuiper, 1952)		d		d	

[a] This represents the assumed composition of primitive atmospheres with cosmic abundances.
[b] Variable or partially in the form of precipitates.
[c] Jovian planet values refer to amounts observed above the cloud tops.
[d] Various sources through 1970.

(1934), that these markings were occasionally masked by bright areas, which Antoniadi took to be veils or clouds. Cruikshank (1966) suggested that these might be strong luminescent effects, similar to effects suspected on the moon but enhanced by proximity to the sun. However, doubt has been cast on all visual observations and maps of markings by the discovery that Mercury's rotation is not synchronous (see Chapter 2). Antoniadi's interpretation of markings may have been influenced by the common belief that Mercury was synchronous. Thus there is no conclusive evidence of an atmosphere from the visual observations.

Second are the spectroscopic searches for an atmosphere. The Soviet observer Kozyrev (1964) reported spectroscopic observations of variable hydrogen emission lines in the form of small "humps" in the bottom of hydrogen absorption lines in the reflected sunlight. Kozyrev's proposal of a hydrogen atmosphere was very surprising, in view of the strong solar wind and Mercury's weak gravity, which would allow escape of a gas as light as hydrogen. Spinrad and Hodge (1965) subsequently explained Kozyrev's "emissions" as spurious features caused by the overlapping of two solar hydrogen absorptions—that in the sunlight scattered by the earth's atmosphere and that in the Doppler-shifted sunlight reflected from Mercury. Belton et al. (1967) searched for CO_2—a much more likely constituent than hydrogen—and found an upper limit of only 2 m-atm. They concluded that the pressure of any possible atmosphere (if composed mostly of CO_2) is less than 0.35 mb,* compared with about 1000 mb on the earth and that there is no observational evidence for any atmosphere. Mercury can thus be regarded as essentially airless, like the moon.

Venus

The atmosphere of Venus was discovered by a Russian scientist, Lomonosov, during a solar transit of Venus in 1761. Lomonosov observed sunlight backlighting and shining through the atmosphere. The composition of this atmosphere has been a speculative matter ever since. When the earliest observers realized that Venus was covered by clouds, they assumed that they were seeing dense water-vapor clouds like those of the earth. Science-fiction writers depicted Venus as a rainy, steamy swamp populated by dinosaurs or some fictional equivalent. However, careful observations during the 1950s and early 1960s failed to give evidence of substantial amounts of water vapor. A variety of (sometimes exotic) theoretical models of the atmosphere followed: models with planet-wide oil fields; models in which all the water was tied up in low-level clouds, ice crystals, etc.; and models with tremendous winds and blowing dust that would generate the high surface temperatures by friction. After some controversy, most investigators accepted that the clouds and general Venus meteorology are not dominated by effects of water, since the observed amount of water vapor above the clouds is very small.

On December 14, 1962, the spacecraft Mariner 2 flew less than 35,000 km from Venus. Its measurements, plus contemporary measures from earth-based observers, led to the conclusion that the surface of Venus was very hot,

* The pressure at the earth's surface is defined as 1 atmosphere = 1.0133 bars, or about 1000 millibars. A bar is 1.000×10^6 dynes/cm^2.

with temperatures at first estimated at about 400 to 700°K (Sagan, 1962). The Soviet Union began a remarkable series of Venus explorations with an automated probe, Venera 4, which parachuted into the Venus atmosphere October 18, 1967. The probe collapsed roughly 26 km above the ground, indicating extremely high pressures in the lower Venus atmosphere.

In 1967 the French spectroscopists Connes et al. made the unexpected discovery of hydrogen chloride (HCl) and hydrogen fluoride (HF) in the Venus atmosphere. The discovery of the halides together with the confirmation of the high temperature triggered a reexamination of the surface chemistry, as described in Chapter 11. Lewis's (1968) assumption that the surface rocks were in chemical equilibrium with the high-temperature atmosphere led to a model consistent with the observed halides and clouds.

Earlier models for the Venus cloud structure, such as that of Kellogg and Sagan (1961), in which an upper haze layer of C_3O_2 overlay a lower cloud deck of H_2O ice crystals, were revised. Lewis (1968) concluded that NH_4Cl crystals could form in the upper atmosphere and Kuiper (1969) proposed that high-level clouds of NH_4Cl crystals overlie a low-level yellowish haze of incompletely hydrated $FeCl_2$. Lewis (1969) subsequently proposed that the clouds are mainly mercury compounds, with the uppermost clouds being Hg_2Cl_2.

The Soviet program of Venus probes continued with Venera 5 and 6, which parachuted into the atmosphere and collapsed on May 16 and 17, 1969, and culminated with the successful landing on Venus of Venera 7 on December 15, 1970. The probe transmitted data from the surface for 23 minutes, and preliminary analysis gave a temperature of 748°K (887°F) and a pressure of 90 atm (1300 pounds per square inch)! The Venus probes also transmitted data indicating much more water in the lower atmosphere than expected by most Western scientists. Sagan pointed out that this water content matched that predicted by the greenhouse model, since Venus's CO_2 alone seems insufficient to produce the high temperatures, but others suggested that the spacecraft measures were erroneous. Thus, although the pressure–temperature structure of the Venus atmosphere is now known (Figure 12-6), the water content and cloud compositions remain puzzling.

A curious feature of Venus which has not been accounted for is the *ashen light,* which is a faint glow of the dark hemisphere of Venus. It has been detected by a number of visual observers, particularly near inferior conjunction, when the dark side of the planet is directed toward the earth. Theoreticians have been unable to account for it, in spite of the fact that it is sometimes prominent even in telescopes with apertures as small as 12 inches. It is usually described as tinted with a coppery color and is probably a high-altitude phenomenon similar to the terrestrial aurora or airglow (Levine, 1969).

Another atmospheric phenomenon visible from the earth with small telescopes during inferior conjunction is an extension of the *cusps,* or horns of Venus's crescent. When Venus passes close to a line between the earth and sun, the sun backlights a hazy layer high in the Venus atmosphere and allows

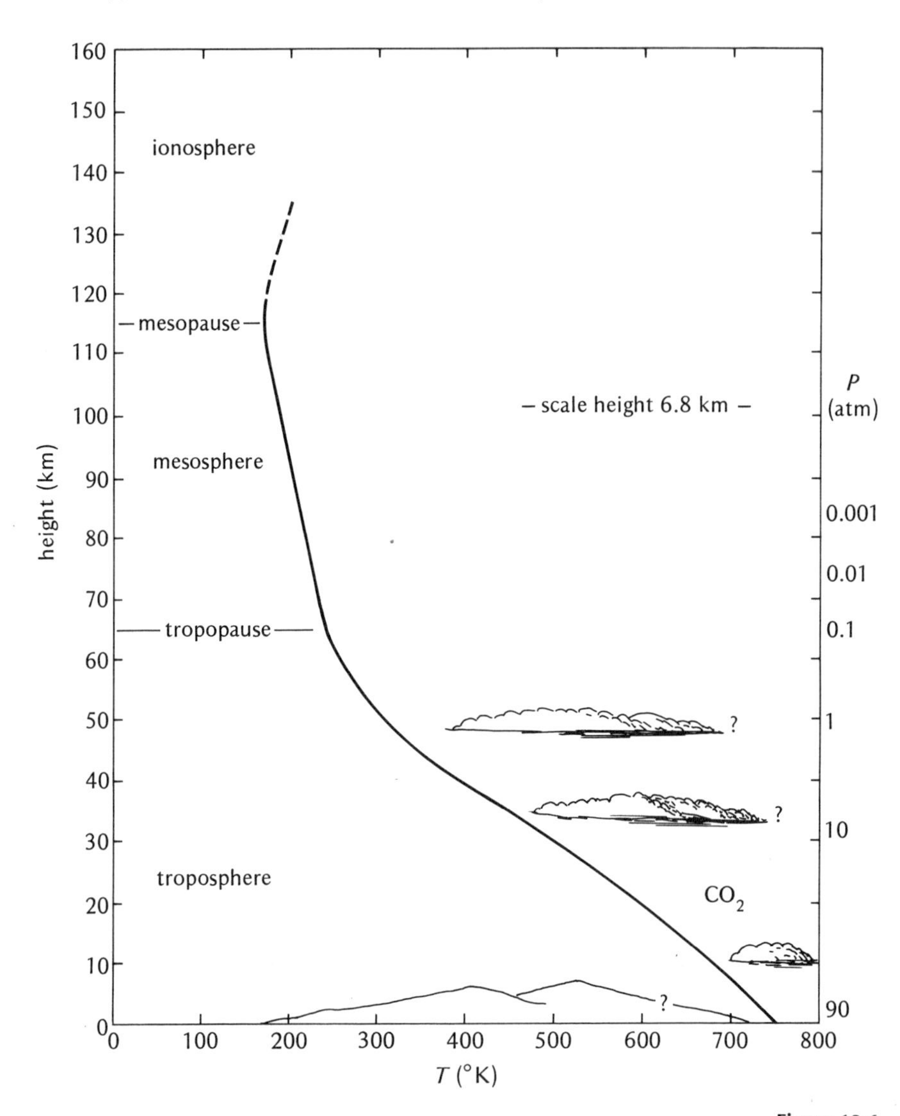

Figure 12-6

Structure of Venus's atmosphere, based on Venera spacecraft and earth-based observations.

the atmosphere to be seen all the way around the disk like a halo (Figure 12-7). It was the extreme case of backlighting—that during a solar transit—that gave the first proof that Venus has an atmosphere.

The Moon

It may seem strange to consider the moon among the planets with atmospheres. However, a number of searches for a lunar atmosphere have been made in the

Figure 12-7

Venus, showing backlighting of the atmosphere. When Venus passes between the earth and the sun, its phase narrows to a thin crescent and sunlight is scattered through the atmosphere from the far (sunlit) side, forming a halo around the disk. Brighter and dimmer parts of the atmospheric halo depend on cloud conditions on Venus at the moment when the photograph was taken. (New Mexico State University Observatory.)

past. These searches have only served to push the upper limit for lunar atmosphere density down to successively lower and lower values. A value of 1.3×10^{-7} times the earth's atmospheric density was derived in 1943. Polarimetric and photometric evidence led to a 1956 upper limit of 10^{-10}. Radio observations by the early 1960s indicated still lower values, about 10^{-13}.

Such an "atmosphere" is really only a local concentration of the interplanetary gas, with a few extra atoms derived locally, for example by bombardment of the lunar surface by high-energy solar particles or by radioactive decay. Öpik and Singer (1961) reviewed estimates of the local concentration of interplanetary gas at the lunar surface and found that such an atmosphere would be only temporary, being constantly replaced.

Michel (1964) pointed out that the total mass of gas associated with the moon at any moment is probably not more than 10^8 g, whereas the total mass expended by the rocket engines of a single Apollo landing module is of the order 5×10^6 g. Thus a number of closely spaced future expeditions or a continuously operating lunar base could create a "contaminated" lunar atmosphere of rocket exhausts. Such contaminations by rocket vehicles may hamper efforts to determine what kind of primitive atmosphere the moon may have had initially or whether any primitive biological activity may have occurred there.

Mars

Estimates of the atmospheric pressure on the Martian surface have gone dramatically downward in recent years (see review by Brandt and Spinrad, 1968). In the interval 1945 to 1955 a number of independent observers using polarimetry and photometry consistently estimated the surface pressure of Mars to be in the range 70 to 100 mb. In 1964 Kaplan, Münch, and Spinrad published their analysis of spectra of CO_2 bands in Mars's atmosphere, yielding a surface pressure of 25 ± 15 mb. This was quickly followed by new spectroscopic analyses by other observers. Owen and Kuiper (1964), who made a careful laboratory calibration of the spectra, derived a still lower value of 17 ± 3 millibars.

In the summer of 1965 the first spacecraft flyby of Mars occurred. The Mariner 4 spacecraft flew behind the edge of Mars, as seen from earth, and transmitted radio signals back to earth through Mars atmosphere. Analysis of these signals led to two values of the Mars surface pressure, 5 ± 1 and 8 ± 1 millibars—one value for each limb of Mars (Kliore et al., 1965). These, plus radar measures of Martian relief, define the range of values now accepted for the Martian surface pressure—from 5 mb at high elevations to 15 mb in the deep valleys. The experiment was performed in the following way. As the Mariner passed behind the atmosphere, the velocity of the transmitted signal was changed because the transmission medium changed from free space (with unity index of refraction) to the neutral and ionized gas of the Martian atmosphere and ionosphere (nonunity index of refraction). This circumstance was observed as a phase change in the signal received on earth. To find the phase change due to the Martian atmosphere, all other effects must be subtracted from the observed phase change. These include Doppler shifts due to the earth's motion, earth's atmosphere, spacecraft motion, etc. These totaled on the order of 10^{11} cycles, whereas the Martian atmosphere caused phase changes on the order of only 30 cycles, so the success of the experiment required measurements accurate to about 3 parts in 10^{10}!

Figure 12-8 shows the structure of the Martian atmosphere, based on spacecraft measurements and other observations.

Principal remaining problems involving the composition of the Martian atmosphere are the abundances of secondary gases, such as nitrogen, argon, and neon, which may confirm or refute hypotheses about the origin and history of the atmosphere and the planet itself.

Atmospheric activity of meteorological interest occurs on Mars. Three kinds of clouds or haze are observed: a thin blue haze, which usually obscures the surface in photographs made on blue-sensitive film, but sometimes clears; white mists and cloud layers; and yellow clouds usually interpreted as dust storms. Kuiper (1964) reported simulation of the blue haze by tiny ice particles; others have regarded it as a nonatmospheric contrast effect of the surface particles. Larger ice or CO_2 crystals would appear white and could explain the white clouds. The white clouds and bluish mist are often especially pronounced at sunrise and sunset on Mars, probably as a result of cooling during the night or as the sun sinks low in the Martian sky. This is consistent with the low, 170°K (−154°F) temperatures measured at sunrise. The yellow clouds are apparently composed of dust particles raised by Martian winds. Dollfus (1961) has summarized a number of observations of these clouds. Micron-size (10^{-4} cm) dust particles are thought to be involved. The mineral limonite, which is thought to give Mars its characteristic ochre color, is

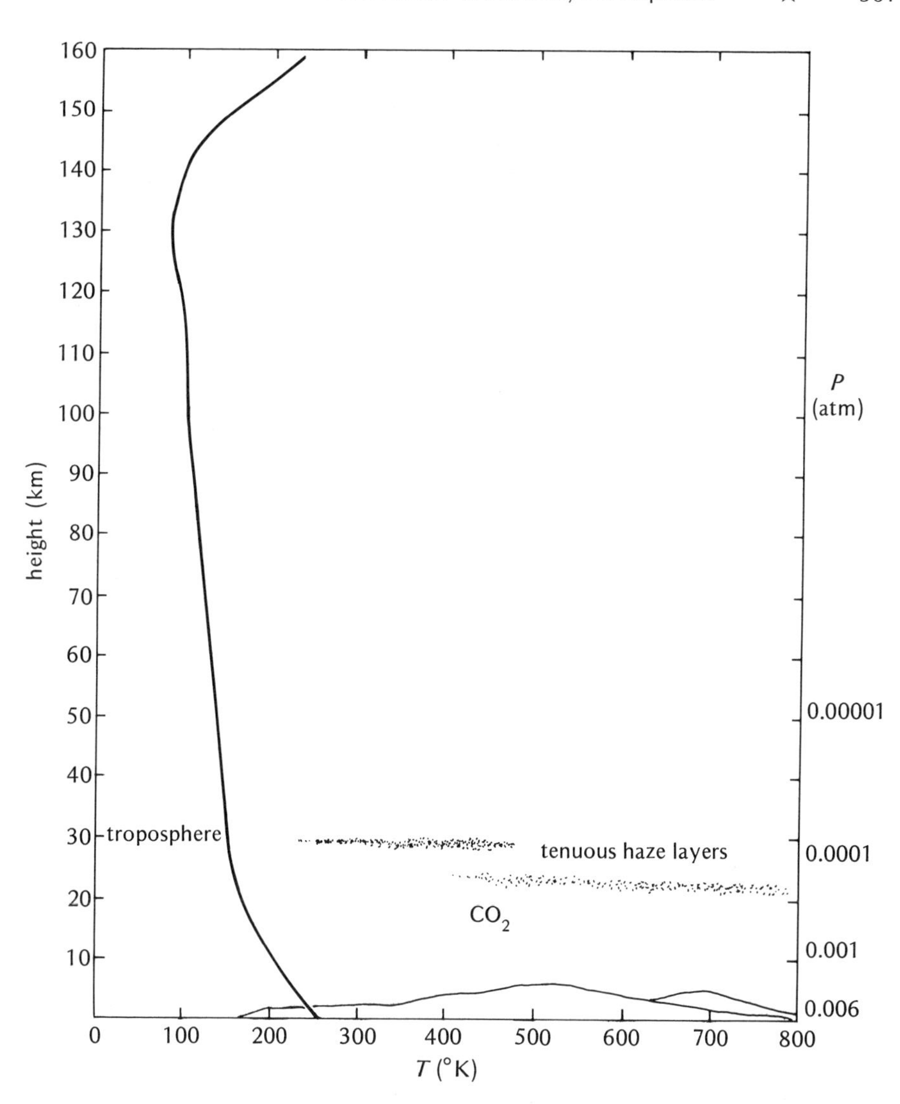

Figure 12-8

Structure of Mars's atmosphere, based on Mariner spacecraft data, earth-based observations, and theoretical results.

likely to occur as a stain on rocks and dust particles, thus accounting for the yellow color of the dust.

High-resolution photographs of Mars from Mariner spacecraft indicate that high spots, such as crater rims, are often white or brighter than their surroundings. This occurs especially at the poles and is probably a result of CO_2 frost deposits. Frost deposits along the polar-cap boundary lie on the sides

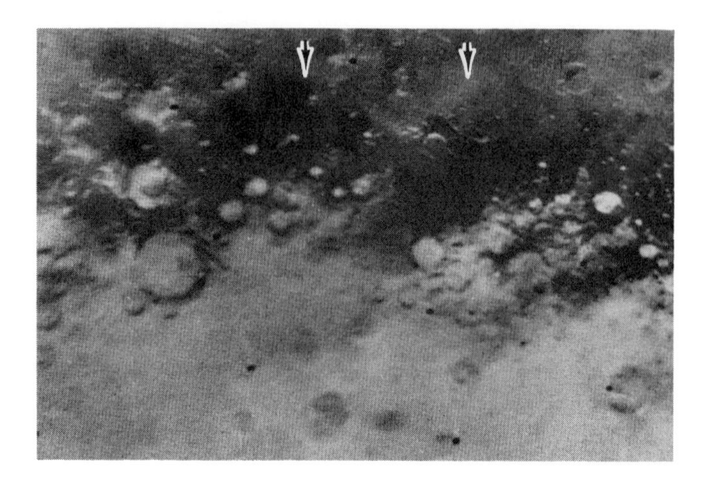

Figure 12-9
Broken frost deposits on Mars along the edge of the Martian south polar cap. Sunlight coming from the north (arrows) illuminates the south inner walls of the craters, causing the frost to melt or sublime. South inner walls of craters along the edge of the polar cap thus display dark, bare ground. Dark regions at the top are Martian deserts. (NASA.)

of crater walls facing away from the sun—where the temperature is lowest (Figure 12-9).

Jupiter and the Giant Planets

Spectroscopically, the atmosphere of Jupiter has long been known to be dominated by absorptions due to methane and ammonia, which were analyzed by Kuiper (1952). It was widely believed that a planet with Jupiter's strong gravity should have retained large amounts of hydrogen in its atmosphere, but hydrogen's absorptions are more difficult to observe than those of methane and ammonia. Eventually, molecular hydrogen was detected by means of absorptions in both the infrared and ultraviolet in a series of observations reviewed by Owen (1970). Curiously, while observations in the infrared indicated a total of 70 km-atm of hydrogen, observations in the ultraviolet indicated only 12 km-atm. This discrepancy implies that ultraviolet light does not penetrate as far down into the Jupiter atmosphere.

Similarly, detailed spectroscopic observations of different parts of Jupiter show that while we can see deep into the atmosphere in some regions, for example the polar areas, other regions have high reflecting cloud layers or mists that prevent sunlight from penetrating very far. Absorptions in the latter areas are weaker than in the former areas. This raises the intriguing question of cloud structure and formation in the atmospheres of the giant planets.

Jupiter, and to a lesser extent Saturn, are observed to have ever-changing cloud patterns. The dark *belts* and bright *zones* and a few prominent features such as Jupiter's Red Spot, are relatively stable, but many smaller features such as bright spots and wispy *festoons* can be observed to change from day to day. Such changes are visible from earth with telescopes as small as 6 inches in aperture.

The rapid changes in the cloud markings of Jupiter are the first clue to the structure of the giant planets' atmospheres. The cloud tops are apparently subjected to chaotic violent motions, suggesting turbulent activity fed by convection. If there is convection up to the region of the cloud tops, then the temperature gradient is adiabatic or slightly superadiabatic. This temperature gradient can be calculated, and is found to be in the range 2.6 to 4.0°K/km assuming that the composition is approximately that given in Table 12-2. To derive the temperature structure of Jupiter's atmosphere we can use the measured temperature of the cloud tops as a reference point, and use the adiabatic temperature gradient below this level. Above this level, the atmosphere is relatively transparent to radiation and thus is more nearly isothermal, like the terrestrial stratosphere. This is essentially the reasoning used by Kuiper in 1952 to derive an early structural model for the Jupiter and Saturn atmospheres.

Kuiper's conclusion that the Jupiter clouds are principally cirrus clouds of ammonia ice crystals has been followed by more complex models. One intriguing research question has been the cause of the rather bright colors of the clouds and the famous Red Spot. Wildt (1939) suggested that the colors were due to various solutions of metallic sodium in condensed droplets or crystals of ammonia, a proposal for which Kuiper (1952) found some support. Rice (1956, 1960) suggested that the colors could be due to charged compounds known as free radicals, frozen and stabilized in ices in the giant planets' atmospheres (unless stabilized, such radicals would react with other chemicals and lose their charge and their color properties). Examples of such radicals are amine (NH_2) and methyl (CH_3), which if joined with hydrogen would produce stable ammonia and methane, respectively. A difficulty with this hypothesis is that the colors exhibited with free radicals are mostly yellows and blues, while Jupiter is rich in reds, browns, and grays as well as yellow and blue.

Lewis and Prinn (1970) reexamined the temperature and composition data in an effort to solve the problem of cloud colors. They concluded that the colors are produced by photochemical reactions involving hydrogen sulfide (H_2S). Their model for the Jupiter cloud structure includes not only the ammonia ice-crystal clouds mentioned earlier but also clouds composed of NH_4HS or $(NH_4)_2$, or both, at a lower, warmer level. At a still lower level they propose a layer of clouds composed of ammonia-water droplets. These low-level cloud layers would account for occasional infrared observations of warmer regions seen through apparent breaks in the upper cloud decks. The ammonia

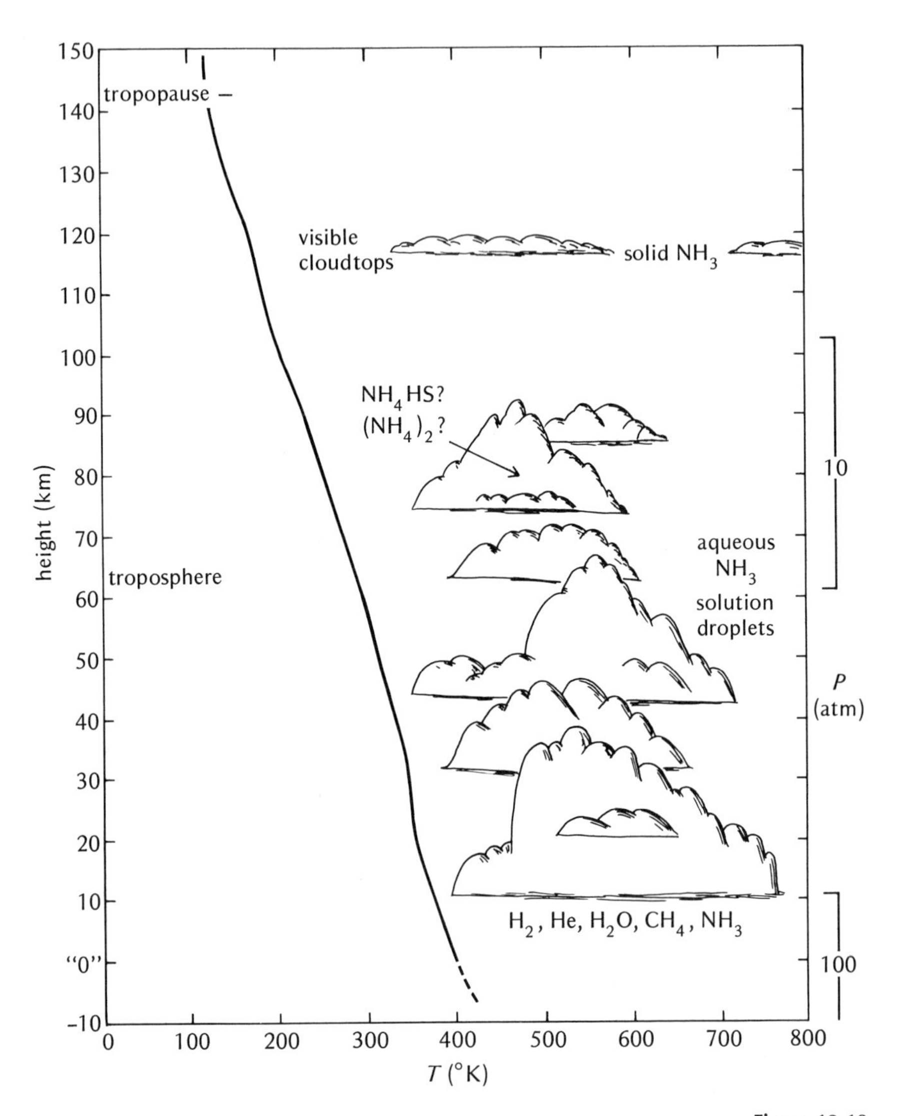

Figure 12-10

Structure of Jupiter's atmosphere, based on earth-based observations and theoretical results. (After Lewis, 1969.)

cloud deck has a temperature of 168°K; the middle deck lies about 26 km (16 mi) lower and has a temperature about 225°K; the lowest detected clouds have temperature about 309°K. This model remains theoretical since there are no spectroscopic observations of the assumed H_2S. Figure 12-10 shows an attempt to synthesize current data into an atmospheric model for Jupiter.

Satellites of Giant Planets

Kuiper (1944) discovered strong methane absorptions in the spectrum of Titan, one of the largest satellites and sixth satellite of Saturn. This was the first and only spectroscopic discovery of an atmosphere on a satellite.

Binder and Cruikshank (1964, 1966) searched for atmospheres on the larger Galilean satellites of Jupiter, using a different and ingenius technique. These satellites are eclipsed periodically by passing into Jupiter's shadow. Binder and Cruikshank's photometric observations showed that for a period of about 15 minutes after the satellite Io (JI) emerged from the shadow it was consistently brighter than at any other time. A similar but smaller effect was later found for Europa (JII), but the other satellites did not show the effect. This is possible evidence for atmospheres on these satellites. The interpretation of the observations was that during the temperature drop associated with the eclipse, methane (CH_4) snow, equivalent to a fraction of the total methane in the atmosphere, precipitated on the ground as bright snow deposits, increasing the mean albedo of the planet.

Three other observing teams have sought to confirm this effect (see *Icarus,* Vol. 14), with mixed results. Of seven eclipses observed, three showed possible brightenings, but the observers with the most sensitive equipment (Franz and Millis, 1971) observed no brightening in any of the four events they observed, throwing the Io atmosphere into doubt. Possibly the eclipse condensation of reflective materials is an intermittent phenomenon on the Galilean satellites. In view of the spectroscopic evidence for a methane-rich atmosphere on Titan and the evidence for frost on the Galilean satellites (Chapter 11), atmospheres would not be unexpected on these bodies.

ESCAPE OF PLANETARY ATMOSPHERES

The theory of gravitational escape of gases was developed by Jeans in 1916. The most widely quoted development of the theory of planetary atmospheres was given by Spitzer (1952).

Consider a molecule moving upward in an atmosphere. It has a certain probability of colliding with another molecule at a higher level. The higher the level of the upward-moving molecule, the lower the probability of a collision, as the density is less. The level at which the probability of further collisions averages 0.5 is called the *critical level.* The region above this is the exosphere, which is part of the thermosphere.

If one of the upward-moving molecules does not hit another molecule, it will follow a ballistic path—an elliptical orbit arcing above the atmosphere and eventually returning. If the molecule should happen to be one of the fastest-moving molecules and have a velocity greater than escape velocity for that particular altitude, it will follow a parabolic or hyperbolic orbit and escape from the planet (see Chapter 3). Every time this happens, the planet's atmosphere permanently loses one molecule.

Jeans, and later Spitzer, calculated the rate at which this happens. It turns out that each molecule species (H_2, N_2, O_2, CO_2, etc.) may be treated separately. The principle known as *equipartition of energy* states that on the average each kind of molecule has the same energy as any other kind. In order to have the same energy, the small ones must move fast, and the big ones must move slowly.* Therefore, on the average, the light hydrogen molecules will be moving faster than the heavier helium molecules, which move faster than the still heavier oxygen molecules, and so on.

The fastest-moving molecular species, hydrogen, will have the most molecules moving faster than escape velocity and will be the first type of gas to be depleted in the atmosphere of any planet. Because of its strong gravitational field, Jupiter has apparently retained much or most of its original hydrogen, but the earth—with its smaller gravitational attraction and lower escape velocity—has lost its hydrogen. Mars, still smaller, has lost most of its oxygen, while smaller bodies have lost even their heaviest gases. Thus the theory of thermal escape helps to explain the compositions of various planetary atmospheres.

A complication of this scheme appears due to the fact that some minor species of gases do not have a uniformly mixed distribution in the upper atmosphere but rather a distribution controlled by diffusion of molecules upward from greater concentrations at lower levels. These gases will thus have an escape rate somewhat different than that predicted by the theory of purely thermal escape.

We have already commented that the temperature of the exosphere is critical to this process. The average velocity of each molecular species depends on the temperature. The higher the temperature, the higher the molecular velocities and the quicker the escape. Figure 12-11 shows the escape time for various gases on Venus, the earth, and Mars.

★ MATHEMATICAL THEORY ★

The expression for the lifetime of a planetary atmosphere is too complex to derive here. However, a rule of thumb emerges from the fact that the escape time depends rather critically on the escape velocity of the planet, for which an expression is derived in Chapter 3. This rule of thumb can be expressed: An atmosphere will escape from a planet in a time less than the age of the solar system if

$$V \geqslant f v_\infty,$$

* You have to hit something very hard with a Volkswagen to expend the same energy and cause as much damage as a slow-moving Cadillac.

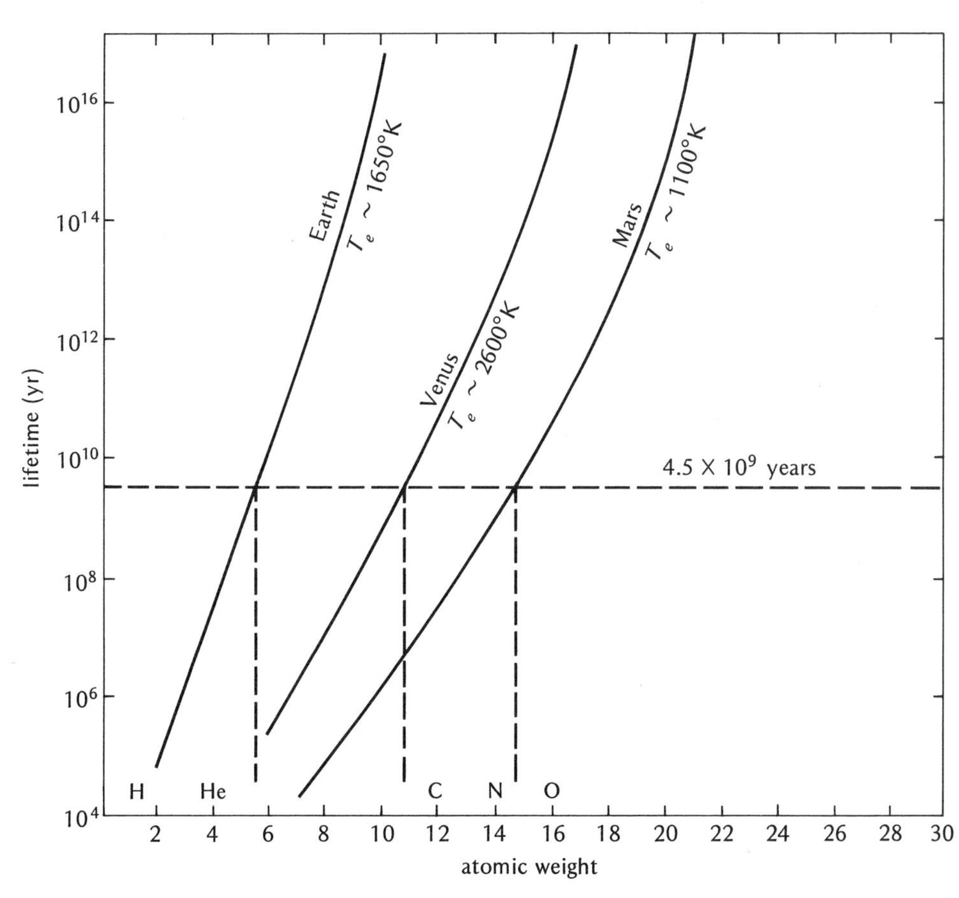

Figure 12-11

Lifetime of gases in the atmosphere of earth, Venus, and Mars, based on assumed temperatures at the exospheric level of escape (T_e). Heavier gases remain in the atmosphere longer. Mars, for example, is expected to have lost most of its gases heavier than nitrogen. (More recent measures give a lower present-day T_e for Mars.)

where V = root-mean-square molecular velocity = $\sqrt{3kT/m}$
f = a constant, about 0.25
v_∞ = escape velocity from the planet
m = molecular mass for given molecular species.

Sharonov (1958) writes a more general rule of thumb to take into account that the exosphere temperature will tend to decline roughly with the square root of the distance from the sun, to the first order (according to simple laws of radiative heating). Combining this with the equation and definitions above, Sharonov writes another rule of thumb: The ability to retain an atmosphere varies as a certain number U, which might be called the "atmosphere retentivity" and

Table 12-3
Atmosphere retentivities

Planet	U[a]	Evidence for atmosphere	Planet	U[a]	Evidence for atmosphere
Jupiter	92	Spectrum, clouds	Ganymede	4.5?	
Saturn	64	Spectrum, clouds	Io	3.7?	Eclipse photometry[b]
Neptune	55	Spectrum	Mercury	3.4?	
Uranus	45	Spectrum	Europa	3.1?	Eclipse photometry[c]
Pluto	13		Moon	2.4?	No atmosphere
Earth	12	Breathing	Callisto	1.3?	
Venus	9.6	Spectrum, clouds	Dione	1.0?	
Triton	7.3?		Rhea	0.98?	
Mars	5.6	Spectrum, clouds	Tethys	0.7?	
Titan	4.8	Spectrum[d]	Enceladus	0.3?	

[a] Atmosphere retentivity, defined by Sharanov (1958), revised from Sharanov's calculations using new planetary data given in the Appendix.
[b] Binder and Cruikshank (1964).
[c] Binder and Cruikshank (1966).
[d] Kuiper (1944).

is defined as

$$U = v_\infty a^{1/4},$$

where a is the distance of the planet from the sun. Values of U for planets and selected satellites are given in Table 12-3.

In summary, all atmospheres in time change their composition from a primitive hydrogen-rich mixture to a mixture of heavy gases, and ultimately will tend to dwindle to zero density. However, the heaviest gases require more than the age of the solar system to escape from the larger planets. The sun, which has an expected lifetime of about 5 billion more years, would "burn out" before these planets lose their atmospheres.

References

Antoniadi, E. M. (1934) *La Planete Mercure et la rotation des satellites* (Paris: Gauthier-Villars).

Belton, M., D. Hunten, and M. McElroy (1967) "A Search for an Atmosphere on Mercury," *Astrophys. J., 150,* 1111.

Berkner, L. V., and L. C. Marshall (1965) "On the Origin and Rise of Oxygen Concentration in the Earth's Atmosphere," *J. Atmospheric Sci., 22,* 225.

Binder, A. B., and D. P. Cruikshank (1964) "Evidence for an Atmosphere on Io," *Icarus, 3,* 299.
______ and D. P. Cruikshank (1966) "Photometric Search for Atmospheres on Europa and Ganymede," *Icarus, 5, 7.*
Brandt, J. C., and H. Spinrad (1968) "The Lower Atmosphere of Mars—A Progress Report," in *The Atmospheres of Venus and Mars,* J. Brandt and M. McElroy, eds. (New York: Gordon & Breach, Science Publishers, Inc.).
Brown, H. (1948) "Rare Gases and the Formation of the Earth's Atmosphere," in *The Atmospheres of the Earth and Planets,* G. P. Kuiper, ed. (Chicago: University of Chicago Press).
Cameron, A. G. W. (1963) "The Origin of the Atmospheres of Venus and the Earth," *Icarus, 2,* 249.
Chamberlain, J. W. (1962) "Upper Atmospheres of the Planets," *Astrophys. J., 136,* 382.
Cloutier, P., M. McElroy, and F. Michel (1969) "Modification of the Martian Ionosphere by the Solar Wind," *J. Geophys. Res., 74,* 6215.
Connes, P., J. Connes, W. Benedict, and L. Kaplan (1967) "Traces of HCl and HF in the Atmosphere of Venus," *Astrophys. J., 147,* 1230.
Cruikshank, D. P. (1966) "Possible Luminescence Effects on Mercury," *Nature, 209,* 701.
Dollfus, A. (1961) "Polarization Studies of Planets," and "Visual and Photographic Studies of Planets at the Pic du Midi," in *Planets and Satellites,* G. P. Kuiper and B. M. Middlehurst, eds. (Chicago: University of Chicago Press).
Franz, O., and R. Millis (1971) "A Search for an Anomalous Brightening of Io after Eclipse," *Icarus, 14,* 13.
Holland, H. D. (1962) "Model for the Evolution of the Earth's Atmosphere," in *Petrologic Studies* (New York: Geological Society of America).
Kaplan, L., G. Münch, and H. Spinrad (1964) "Analysis of the Spectrum of Mars," *Astrophys. J., 139,* 1.
Kellogg, N., and C. Sagan (1961) "The Atmospheres of Mars and Venus," *Nat'l Acad. Sci.-Nat'l Res. Council, Publ. 944.*
Kliore, A., et al. (1965) "Occultation Experiment: Results of the First Direct Measurement of Mars Atmosphere and Ionosphere," *Science, 149,* 1243.
Kozyrev, N. A. (1964) "The Atmosphere of Mercury," *Sky and Telescope, 27,* 339.
Kuiper, G. P. (1944) "Titan: A Satellite with an Atmosphere," *Astrophys. J., 100,* 378.
______ (1952) "Planetary Atmospheres and Their Origin," in *The Atmospheres of the Earth and Planets,* G. P. Kuiper, ed. (Chicago: University of Chicago Press).
______ (1964) "Infrared Spectra of Stars and Planets. IV: The Spectrum of Mars, 1–2.5 Microns, and the Structure of Its Atmosphere," *Comm. Lunar Planet. Lab., 2,* 79.
______ (1969) "Identification of the Venus Cloud Layer," *Comm. Lunar Planet. Lab.,* 6, 229.
Levine, J. S. (1969) "The Ashen Light: An Auroral Phenomenon on Venus," *Planet. Space Sci., 17,* 1081.
Lewis, J. S. (1968) "An Estimate of the Surface Conditions of Venus," *Icarus, 8,* 434.
______ (1969) "Geochemistry of the Volatile Elements on Venus," *Icarus, 11,* 367.
______ and R. G. Prinn (1970) "Jupiter's Clouds: Structure and Composition," *Science, 169,* 472.
Michel, F. C. (1964) "Interaction between the Solar Wind and the Lunar Atmosphere," *Planet. Space Sci., 12,* 1075.

Monash, E. (1967) "Mars: Upper Atmosphere," in *Mars Scientific Model, JPL Document 606-1* (Pasadena: Jet Propulsion Laboratory).

Öpik, E. J., and S. F. Singer (1961) "Density of the Lunar Atmosphere," *Science, 133,* 1419.

Owen, T. C. (1969) "The Spectra of Jupiter and Saturn in the Photographic Infrared," *Icarus, 10,* 355.

______ (1970) "The Atmosphere of Jupiter," *Science, 167,* 1675.

______ and G. P. Kuiper (1964) "A Determination of the Composition and Surface Pressure of the Martian Atmosphere," *Comm. Lunar Planet. Lab., 2,* 113.

Rasool, S. I., and C. de Bergh (1970) "The Runaway Greenhouse and the Accumulation of CO_2 in the Venus Atmosphere," *Nature, 226,* 1037.

Rice, F. O. (1956) "Colors on Jupiter," *J. Chem. Phys., 24,* 1259.

______ and D. P. Cosgrave (1960) "Some Experiments on Rice's Blue Material: Colors on Jupiter," *Nature, 188,* 1023.

Rubey, W. W. (1951) "Geologic History of Sea Water," *Bull. Geol. Soc. Am., 62,* 1111, reprinted in *The Origin and Evolution of Atmospheres and Oceans,* P. Brancazio and A. Cameron, eds. (New York: John Wiley & Sons, Inc., 1964).

Sagan, C. (1962) "Structure of the Lower Atmosphere of Venus," *Icarus, 1,* 151.

Sharanov, V. V. (1958) *The Nature of the Planets,* transl. Israel Program for Scientific Translations (Washington, D.C.: Department of Commerce, Office of Technical Services).

Spinrad, H., and P. W. Hodge (1965) "An Explanation of Kozyrev's Hydrogen Emission Lines in the Spectrum of Mercury," *Icarus, 4,* 105.

Spitzer, L. (1952) "The Terrestrial Atmosphere above 300 Kilometers," in *The Atmospheres of the Earth and Planets,* G. P. Kuiper, ed. (Chicago: University of Chicago Press).

Suess, H. E. (1949) "Die Haufigkeit der Edelgase auf der Erde und im Kosmos," *J. Geol., 57,* 600.

Wildt, R. (1939) "On the Chemical Nature of the Colouration in Jupiter's Cloud Forms," *Monthly Notices Roy. Astron. Soc., 99,* 616.

LIFE: ITS HISTORY AND OCCURRENCE

AMONG THE QUESTIONS faced by planetologists, perhaps the most profound—the question with the most implications for mankind—is the question of whether other planets in the universe harbor intelligent life. In this chapter we shall consider this question in a fivefold way. First, we note that a myriad of planets may exist outside the solar system. Second, we discuss what we mean by "life" and what conditions life would require in order to exist on such planets. Third, we review the long process of development that led to life on earth, as best we can understand it; this review draws on material in many previous chapters. Fourth, we consider the effects of the planet's own evolution upon its life forms. Finally, relying on all this information and what is admittedly some additional speculation, we attempt to estimate the probability that intelligent life exists on other planets, and the probability of our making contact with such life, or it making contact with us.

ABUNDANCE OF PLANETS OUTSIDE THE SOLAR SYSTEM

Sources of Evidence

In view of the hundred billion stars in our galaxy alone, we cannot afford to limit our search for potential life to the solar system. Rather, we must come to grips with possible abundant opportunities for growth of life on planets around other stars. For sources of evidence about extrasolar planetary systems, we must rely on the kinds of information discussed in Chapter 5.

If the origin and frequency of binary and multiple stars and black dwarfs is uncertain, the abundance of planetary systems around other stars is almost sheer speculation. Nonetheless, astrophysical theory and observations agree that the development of small secondary bodies around stars must be quite common. Sources of information which have been invoked in the literature include angular momenta, astrometric observations, and stellar mass distributions.

Angular Momentum

One bit of evidence bearing on the abundance of planets is that massive stars spin faster, as a group, than smaller stars like the sun, with a discontinuity in angular momentum at about 1.5 solar masses. Relative to these massive stars, solar-type stars have lost some of their angular momentum. The missing angular momentum appears to have been transferred from the sun to the planetary material by means of magnetic interactions between the early sun and the early nebula around it. Therefore, the observation that most solar-type stars have also lost their angular momentum has prompted a speculation that they all have planetary systems (Huang, 1965).

Astrometric Observations of Small Companions

Over the years a number of unseen small companions have been discovered to be orbiting around nearby stars. In 1963, Peter van de Kamp of Sproul Observatory announced the discovery of the smallest yet. It is orbiting around a nearby star known as Barnard's star, and was initially calculated to be only 50 percent more massive than Jupiter. This is within the mass range of planets.

This single discovery establishes that there are planet-sized bodies outside our solar system. However, as we have already noted at the end of Chapter 5, orbits may be as important as size in determining whether small bodies are "true" planets.

A few astrometric binary systems seem to have circular orbits and thus fulfill *all* the requirements to be classed as true planets, but these orbits are not in all cases reliable. Although van de Kamp's original (1963) data on Barnard's star suggested an elliptical orbit for the "planet," he later revised his calculations (1969) and found that *two* planets in circular orbits gave a better explanation of the observations. The two "planets" are estimated to have masses only 1.1 and 0.8 times Jupiter's mass. This, of course, makes the Barnard's star system even more similar to the solar system, but at the same time it serves as an example of the uncertainty of all the ellipticity data. The astrometric data can be interpreted in terms of one body in elliptical orbit or more than one body in less eccentric orbits. The very low mass companion of the star 61 Cygni has also been reported to have a circular orbit, although the data are subject to uncertainty.

To summarize, the observational evidence, though marginal, suggests that we may have detected at least one extrasolar planetary system.

Distribution of Stellar Masses

Another way of estimating the abundance of extrasolar planets has been attempted. Brown (1964) and O'Leary (1966) and others have taken the frequency of stars of various masses and extrapolated from observable stars (16 solar masses, 8, 4, 2, 1, 0.5, . . .) down to planetary masses (0.001 . . .) assuming a continuous stellar mass function. Since the stellar mass function they used shows more stars at smaller masses, they were led to the conclusion that planet-sized objects are *very* common and that virtually every star should have a planetary system. This reasoning is probably incorrect for three reasons: (1) It assumes that planets form in the same way as stars, which is at best uncertain; (2) Gaustad (1963) has found physical factors that inhibit formation of "stars" smaller than about 0.1 solar mass; and (3) detailed observations of the stellar mass function show that as we go to smaller and smaller stars, the number of stars stops increasing at about 0.2 to 0.5 solar mass. Stars of this size are the most common. Smaller stars are less abundant, so the mass function cannot be extrapolated to planetary masses (Hartmann, 1970). Thus planets cannot be arbitrarily "tacked on" to the same spectrum of objects which is defined by the known stars.

Conclusions on Occurrence of Planets

The question still remains: Are planetary systems common? The answer appears to be yes, at least outside binary and multiple star systems, for the following reasons. Planetesimal growth in cocoon nebulae appears to be a normal part of the star-forming process. Stellar-angular-momentum observations as well as solar-system observations support this. Small bodies are known to accompany many nearby stars, as shown by Table 5-1. Remembering that planets of less than Jupiter's mass are almost impossible to detect outside the solar system, even by astrometric observations, and remembering that the search for astrometric binaries has effectively only barely begun, we have promising results. Stellar systems with low-mass members are common, and some unseen companions are probably true planets. Compared to galactic distances, all these stars are right on top of us; if at least one or two other systems with planet-sized bodies exist so close by, space is perhaps rather well-populated with planets and potential homes for alien races.

A contemporary astronomer's informed but speculative guess as to the percentage of stars that have planetary bodies would be greater than 1 percent (since planet-breeding cocoon nebulae around solar-type stars are probably relatively common), but less than 50 percent (since planets probably don't appear in binary systems, multiple systems, or around very massive, short-lived stars).

HABITABLE PLANETS AND THE NATURE OF LIFE

There is no a priori guarantee that extraterrestrial planets, even if they exist, have the conditions to support life. Of the nine known planets, only the earth has advanced life forms (we believe) and probably only one other, Mars, has even low forms such as primitive plants or bacteria. Lichens are the most complex life forms usually considered for Mars.*

To be habitable, a planet must meet certain limitations on temperature, atmospheric pressure, atmospheric composition, liquid water abundance, energy flux at the surface, etc. For example, suppose we have an arbitrary star and a hypothetical planet nearby. If the star is too massive it will use up its nuclear fuel in a few million years—too short a time for life to evolve. (On the earth it took about 1 billion years for organisms to gain any foothold and about 4 billion years before any life forms were substantial enough to leave fossils.) If the star is too small, its light will be very faint and the planet would have to be in a narrowly restricted close orbit to obtain enough energy from the star to keep the surface warm.

If we rule out very small stars and stars more massive than 1.5 solar masses (lifetime less than 2 billion years), we are left with "solar-type stars." If the planet is too close to such a star it will be too hot; if it is too far it will be too cold (unless it has just the right internal heat sources or a significant greenhouse effect). If the planet is not massive enough, its gravity field will be too weak and its atmosphere will leak away into space; if it is too massive, its gravity field may be too strong and its atmosphere may be too dense. All these and other constraints must be met.

The Russian astrophysicist Ambartsumian (1965) remarked that the most important parameters necessary to make a planet habitable were (1) the energy flux falling on it from the central star; (2) the color of the star, since photochemical processes necessary for life depend on the color (i.e., wavelength) of the "sunlight"; and (3) the age of the star. Secondary properties listed by Ambartsumian were (4) inclination of the planet's equator to the plane of its orbit, determining the seasons; (5) period of rotation (length of day); (6) gravitational field; and (7) variability or multiplicity of stars in the system.

The American physicist Stephen Dole (1964) made an extensive review of the problem and estimated that stars in the mass range 0.9 to 1.0 are the most likely to have habitable planets. He predicts that about 6 percent of such stars will have planets that could be inhabited by human-like creatures.

A habitable planet is merely habitable. We have no guarantee that it is inhabited. For it to be inhabited, life must have evolved on it or come from somewhere outside.†

We are now faced with the fundamental question of the origin of life. We can get no further in speculating about the possibility of alien life unless we claim some understanding of how life originates.

First, what *is* life? Life, as we understand it, is not a status but a *process*.

* More speculative scientists in the past have considered alternatives ranging all the way to extinct Martian civilizations.

† The last possibility is tantamount to interstellar travel or transport of life and is discussed later.

It is a complex series of chemical reactions based on complex carbon-based molecules, by which matter is taken into the living system and utilized so as to contribute to the growth and reproduction of the system, with waste products being expelled. Whatever else may be involved, whatever sensitivities, perceptions, psychologies, or sensed metaphysical qualities, these chemical processes are undeniably crucial to all life that we know.

The system in which the processes occur is the cell. Known living things are composed of one or more cells.* A cell is in essence a container (cell wall) filled with an intricate array of organic and inorganic molecules (protoplasm). Coded behavior and genetic patterns are contained in very complex molecules (such as the famous DNA) located in a central body called the nucleus. The elements most prominent in the organic molecules are carbon, hydrogen, oxygen, nitrogen, and phosphorus. It is interesting to note that excepting phosphorus, these are all cosmically abundant (see Figure 11-2).

We commonly make a conceptual error in thinking of ourselves as static beings instead of dynamic systems. We casually think that we are constant entities, as if our identity is somehow related to the matter of our bodies. Our bodies are not the same ones we had seven years ago. Hardly a cell is still alive that was part of that body. As John Pfeiffer (1955) has noted, this is a far cry from the conception of life we had only a few generations back, when bodies were thought of as semipermanent machines, whose parts gradually wore out. Now we realize even our seemingly inert skeletons are living and changing, always replacing their cells. We are not constant; we are always changing. In fact, we must *keep* changing—the cells must keep processing new materials to stay alive. When the processing stops, that is what we call death.

The nature of living beings is illustrated in an analogy from the famous Russian biochemist Oparin (1962). Consider a bucket filled with water pouring in the top from a tap and flowing out at the same rate through a tap in the bottom. The water level in the bucket is constant. A casual observer would probably call it "a bucket of water." But it is not like an ordinary bucket standing full of water. The water at any instant is not the same as at any other instant. Yet the outward appearance is constant. We are like buckets with water and nutrients and air flowing through us, but with additional, much more complex attributes, such as the ability to reproduce and to be affected by random genetic changes that let us evolve from generation to generation by natural selection.

The dynamic nature of life gives us some clue to the kinds of processes in-

* This can be qualified. Viruses, for example, are simpler than cells yet have the ability to reproduce themselves using materials from host cells. There is a semantic argument over whether viruses should be considered a form of life, but probably the majority of biologists consider at least some viruses to be "living."

volved in the origin of life. We are not looking for some way to create a ready-made, living machine. Instead, we are looking for a process in which complex organic molecules can enter a cell or cell system which draws from the incoming material to create new molecules, fit them into itself, exhaust unused material, and reproduce.

Those processes are what we mean by "life." Whatever other conceptions we invent and impose on ourselves and our environment—civilization, religion, technology, art, war, romance, interstellar communication—it is those processes that define us, just as they define the spiders, moss, sea urchins, elephants, moths, amebas, redwoods, and all the other incredibly varied living creatures around us. To judge whether life may have occurred on other planets—whether in Dole's words (1964) the other habitable planets are already "taken"—we must determine how those processes began on earth.

THE ORIGIN OF LIFE ON EARTH

Four and a half billion years ago, just after the earth formed, the land was different than it is today. The surface was rugged, scarred with impact craters like those of the moon, and crossed by great mountains and cliffs formed in response to gravitational adjustments in the earth's crust. It was lifeless. The atmosphere was composed mostly of hydrogen-rich compounds common in the early solar system, for hydrogen is the most abundant material in the cosmic gas from which the solar system formed. Winds rich in ammonia (NH_3), methane (CH_4), and some water vapor (H_2O) blew across the land.

Other gases were being added to the atmosphere during volcanic outbursts and slow fumarole activity. The oceans at first were much smaller since there was not much water; the oxygen content was low and hence there was little H_2O by present-day standards.

An entire day and night lasted only about 5 hours, for the earth turned much faster than it does today.*

Modern man could not have survived in the poisonous atmosphere that characterized the early days of his own planet, yet we shall see that other forms of life not only survived but prospered.

As sketched in Chapter 12, the earth soon began outgassing. As the interior heated, the earth began to form an iron core and the lighter silicates formed a crust on the surface. Volatile materials such as water in the form of steam, hydrogen sulfide, and other gases escaped in greater quantity into the atmosphere. Masses of the crust became unstable as molten areas formed below, and fracturing and folding produced new mountain belts and valleys. The most important consequence of the rearrangement of the earth's interior was the release of water vapor.

* This is a consequence of tidal effects; see Chapter 3.

Water was crucial in the development of life. This is obvious in several ways. In the first place, just looking at ourselves we discover that most of our body weight is made up of water. Second, we recall that body fluids like tears have a salinity similar to that of the oceans. The oceans are our heritage. As embryos, we are at first more like fish than mammals; we have gills. Third, we notice that organisms deprived of water quickly die; dead organisms are shriveled and dried up. Fourth, taking a more theoretical view, we realize that water provides a medium for suspended organic molecules—a logical place for life to begin.

Let us move forward in time 1 billion years—to 3 billion years ago. The atmosphere was probably still composed largely of hydrogen compounds, ammonia, methane, and increasing amounts of water. The temperature within the earth was still increasing and volcanic and mountain-building activity was common.* The most noticeable difference between a randomly selected spot of 3 billion years ago and a spot selected today is the prehistoric spot's sterility. The land appeared stripped of all plants and animals.

The oxygen (O_2) content of the atmosphere was low, and so no ozone (O_3) had been able to accumulate in the upper atmosphere. With no protective ozone layer to absorb the energetic rays of the sun, the ultraviolet photons penetrated in great numbers to the ground, where they caused photochemical reactions. At a typical spot, the ultraviolet rays probably constituted the major source of energy for such reactions.

Three billion years ago oceans were developing from the water outgassed from the earth's interior. Dotted along unrecognizable coastlines were bays and tidewater pools where water circulation was restricted. In the air over these silent pools, energetic ultraviolet photons from the sun penetrated and interacted with the gas, rearranging atoms from simple molecules into new, complex molecules. The most common of the air molecules, as we have seen in Chapter 12, were hydrogen, ammonia, methane, water, and perhaps some

* But not as common as is usually believed. Because of the old idea of the incandescent young earth, Hollywood and Madison Avenue (those most-effective purveyors of our ideas about ourselves) tend to portray prehistoric settings as violently volcanic, with eruptions popping off at the drop of a hat, lava flows menacing every acre, and buxom starlets fleeing dinosaurs as earthquakes open fissures on all sides. This is nonsense. Volcanism and earthquakes are sporadic, not commonplace, even in places like San Francisco, which straddles a major fault in the earth's crust. A random day in the prehistoric past perhaps was not much more disturbed by eruptions than a modern day. The mixing of dinosaurs and humans is another error, since dinosaurs died out roughly 60 million years before man came along. The unattractive Australopithecine apemen of the unusually accurate film "2001" roamed only a couple of million years ago. Any drama with recognizably human characters must be set in the last few hundred thousand years.

nitrogen (respectively H_2, NH_3, CH_4, H_2O, N_2). Notice what atoms were involved: hydrogen, nitrogen, carbon, and oxygen—the elements of life.

In other words, the atoms available on the ancient earth and exposed to photochemical reactions were the ones required to build the complex organic molecules that begin life.

The idea that naturally occurring reactions in the early atmosphere could produce the building blocks of life is not idle speculation. After Nobel chemistry laureate H. C. Urey (1952) had shown that the primitive earth had a methane- and ammonia-rich environment, Urey's collaborator, S. L. Miller (1955), took a gaseous mixture of methane, ammonia, water vapor, and hydrogen in the presence of a pool of liquid water and passed electric sparks through it. The sparks represented lightning, which was suspected to be a major energy source. After the experiment had run for several days, the pool of water began to darken. Miller analyzed the liquid and discovered that the water now contained a solution of *amino acids,* the class of molecules that join together to form *proteins,* the huge molecules that occur in cells. Subsequent experiments confirmed and expanded Miller's work. Investigators have shown that many kinds of energy sources in the early atmosphere would produce amino acids: ultraviolet light, volcanic activity, even meteorite impacts. The Miller experiment proved that ordinary processes occurring on the ancient earth created pools containing organic molecules. Recently, amino acids have been found even in meteorites (Chapter 9).

The next step in the development of life is admitted by all researchers to be much less certain. One phenomenon invoked by some researchers is the existence of systems called *coacervates.* In the 1930s a Dutch chemist, H. G. Bungenberg de Jong, found that if he took a mixture of proteins in solution with other complex molecules—solutions similar to those in primeval pools—the proteins and other complex molecules would spontaneously accumulate into cell-sized clusters. The remaining fluid was then almost entirely free of the complex organic molecules. The suspended clusters are called *coacervates.* This surprising process occurred without external assistance; it is apparently triggered by the proteins. The coacervates were in the form of microscopic droplets. The interior of the coacervated droplets, although not as highly organized as the inside of a cell, were more structured than a normal fluid; for example, the particles inside the coacervates were sometimes arrranged in parallel orientations.

A. I. Oparin (1962), the Russian biochemist, and many of his American colleagues have hypothesized as a result of further experiments that coacervate droplets formed in primeval pools and began reacting with the fluids in these pools, accumulating more molecules and spontaneously growing more complex. Eventually, they suggest, the coacervates evolved into biochemical systems capable of reproducing. The American exobiologist Carl Sagan and others have objected that this theory neither explains the key problem of how hereditary properties developed and transferred nor does it specify when the nonliving chemical systems came to be living (Shklovskii and Sagan, 1966, p. 241). Although coacervates have been severely criticized as possible forerunners of cells, the fact that almost cell-like coacervates can spontaneously form is a vivid demonstration of the remarkable properties of complex organic molecules and an indication that life could begin through ordinary chemical reactions.

This view of life's gradual origin contrasts sharply with the assumptions made in past centuries, when it was thought that highly evolved life forms spontaneously arose from existing raw materials, such as garbage. Sagan (1970) quotes a seventeenth-century writer:

> (One cannot) doubt whether . . . butterflies, locusts, shellfish, snails, eels, and such life be procreated of putrefied matter, which is to receive the form of that creature to which it is by formative power disposed. To question this is to question reason, sense, and experience.

Observers erroneously concluded that vermin were produced by the "timber, dung, and putrefied matter" out of which they were seen to crawl. Today we are assured that new forms of life are no longer arising on the earth for two reasons: (1) the necessary chemically reducing environments are rare; and (2) newly assembled living organic material would at once be consumed by the vast numbers of more highly evolved life forms.

While life was beginning in scattered pools, the earth itself was not dormant. The water that was being expelled in volcanic activity not only served as a broth for life but also helped to create an oxygen-rich atmosphere. The hydrogen supply was constantly being depleted by thermal loss (see Chapter 12). As H_2O molecules entered the atmosphere from scattered volcanoes and fumaroles, some of them suffered collisions with energetic solar photons which split the bonds between the oxygen and hydrogen atoms. The hydrogen atoms could escape due to their low mass and high velocity, so there was a steady buildup of oxygen in the remaining atmosphere (Kuiper, 1952).

The eventual result was a transition from a hydrogen-rich atmosphere to an oxygen-rich atmosphere. Living things could hardly be unaffected. Whereas the earlier forms of life subsisted without oxygen, life now had to adjust to oxygen. The atmospheric and biological changes were not sudden. First, the hydrogen supply dwindled until at some intermediate stage the atmosphere may have been composed mostly of nitrogen and sulfur dioxide from the volcanic activity (Holland, 1962). As more water molecules were dissociated, the oxygen content increased and living matter became more and more complex. Biochemical processes such as fermentation and photosynthesis began. Simple plant life was probably born before the oxygen began to dominate; it is common knowledge that green plants consume water and carbon dioxide (CO_2) and produce oxygen (O_2). Alternatively, advanced animal life consumes oxygen and could not begin until the air was oxygen-rich. The slow photochemical reactions that had been adding O_2 to the air were suddenly overwhelmed by the photosynthesis process of the plants, and the oxygen content quickly increased (Kuiper, 1952). It is thought that this dramatic change was accompanied by a sudden increase in the "rate" of evolution of life (Berkner and Marshall, 1965).

Eventually, so many oxygen atoms were present that combination into O_3 ozone molecules occurred in the upper atmosphere, producing the well-known ozone layer. The importance of the ozone layer is that it absorbs ultraviolet light, preventing the energetic ultraviolet photons from getting to the ground. Thus, while ultraviolet was initially important in producing a variety of organic molecules, its screening by ozone made the ground a safer place for stable organisms by protecting them from damaging photochemical reactions.*

Let us move forward another billion years in time, to 2 billion years ago. In spite of the presence of now firmly established life, the earth still did not look much different. The land was still barren. Some of it must have looked like today's deserts; but some was moist and washed by frequent rains, and yet instead of luxurious forests and grasslands there was only bare dirt, eroded gullies, and grand canyons. Brown vistas stretched to the sea coasts. Most life was still in the oceans, and only tiny insubstantial life forms at that. Even in the oceans there was not yet anything solid enough to leave fossil remains. Single-celled forms (such as bacteria) and the blue-green algae are probably today's closest representatives of the life of that period.

In the interval from 2 to 1 billion years ago more advanced animal forms evolved. They utilized the plants as food, while the less-advanced plants could only manufacture their food from the chemicals around them. The animals also utilized the oxygen in the air and water to process the plant food. Sponges and jellyfish-like animals probably evolved in the sea during this interval.

Life forms were still very simple 1 billion years ago. Multicelled animals were only sack-like structures. A cellular construction surrounded a body cavity through which fluids containing food particles could flow and be chemically processed. Still there were scarcely any creatures in the sea that were substantial enough to leave fossils; on the land, evidences of life were scanty.

It was only $\frac{1}{2}$ billion years ago (500,000,000 B.C.) that life forms developed and became widespread enough and substantial enough to leave fossil imprints that geologists can use to trace the physical evidence of evolution. The first of the geologists' fossil-defined periods, the Cambrian, starts at this time. Although we can only surmise the story of pre-Cambrian evolution through theory and laboratory experiments, we can hold Cambrian and later fossils in our hands. To put it another way, the geologists' *stratigraphic record* began only half a billion years ago.

* The little bit of ultraviolet light that leaks through today is responsible for suntans and sunburns, testimony of its potency.

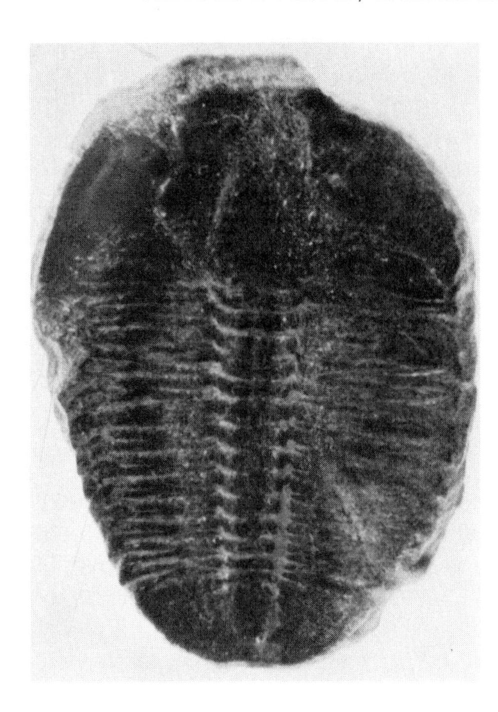

Figure 13-1
Fossilized trilobite. About 10,000 species of these extinct marine arthropods have been recognized and they are a principal index fossil for the Cambrian period, about 500,000,000 years ago. Specimen length, 3 cm.

The Cambrian period is identified by abundant fossils of trilobites, one of the first hard-bodied sea creatures (Figure 13-1). Typically a few centimeters long, these hardy animals inhabited sea bottoms throughout the world. It is easy to harbor a misconception about the Cambrian's place in time. Most books on evolution and geology, since they are concerned with physical evidence, are devoted almost entirely to descriptions of *post*-Cambrian time. Many readers thus have the false impression that nothing happened in the pre-Cambrian and that the trilobites were an "early" stage of life. We must not forget that the last $\frac{1}{2}$ billion years, and all our fossil records, and all our detailed geological knowledge represent only about 11 percent of the history of our planet! To devote a proportionate amount of space to that period, we had better finish quickly by saying that fishes and other vertebrates, dinosaurs and birds and so on, evolved, and eventually little furry creatures called mammals appeared, and recently a creature called man came along. He has a useful brain that forced an acceleration of evolution, and we are not sure what is going to happen to him.

THE EFFECT OF PLANETARY EVOLUTION ON BIOLOGICAL EVOLUTION

It used to be supposed that the earth formed only a passive backdrop for all this biological activity. Although it has long been known that some regions of present-day dry land were once under the sea, the earth's environment was regarded as being not much different on the average from what it is today. For this reason, modern geology was founded on the *principle of uniformitarianism:* The processes that shaped the landscapes of the past are the same ones we see working around us today, differing only in degree.

Using the principle of uniformitarianism, geologists were able to account for the features that they observed in the fossil record of the last 500,000,000 years. This record is known as the *geologic time scale* (Table 13-1) and is divided into a succession of *periods* which show a steady progression in biological evolution and also the vicissitudes of geologic evolution, with some periods warm and some cool, some moist and some dry, within a given region of the earth. The fact that all periods reflected in some degree geological processes known today furthered confidence in the principle of uniformitarianism.

The principle of uniformitarianism contrasted with an earlier view, known as *catastrophism,* which supposed that the ocean basins, folded rocks, and other geologic features formed in prehistoric catastrophies. The early favor in which this theory was held can probably be traced to man's predilection for fabulous interpretations of the earth's history, as is demonstrated by widespread legends of universal floods, destroyed cities, and other catastrophic events. Catastrophism, in its original form, was finally discarded. As geologic

Table 13-1
Geologic time scale

Era	Date (millions of years ago)	Period	Characteristic life forms
	0		
		Quarternary	Man
Cenozoic	2		
		Tertiary	Mammals
	70		
		Cretaceous	
	130		
Mesozoic		Jurassic	Dinosaurs
	180		
		Triassic	Reptiles
	225		
		Permian	Conifers
	260		
		Pennsylvanian	Ferns
	300		
		Mississippian	
Paleozoic	340		
		Devonian	Fishes
	405		
		Silurian	Early land plants
	435		
		Ordovician	
	480		
		Cambrian	Trilobites
	560		
Proterozoic		Pre-Cambrian	Small, soft life forms
	4500		

research strengthened the uniformitarian point of view, there was a tendency to overreact and doubt the significance of any extraordinary processes.

Nonetheless, in the last decade, geologists have been forced to accept the notion that the earth's history is full of unusual events not easily imagined by studying the contemporary world. Not all such events were sudden, but they transpired in intervals quite short compared to the earth's *total* history.

Examples of "extended catastrophes" that may have affected biological evolution are the changes in the atmospheric composition, continental breakup, continental drift, reversals of the magnetic field, and possible changes in solar radiation. Local catastrophes, such as impacts of giant meteorites, may have also affected restricted areas.

An example of the effect of these extended catastrophes on life is the influence of continental drift (Kurten, 1969). It was only some 100,000,000 years ago that two primeval supercontinents began to be rifted apart, possibly driven by convection currents in the mantle. According to recent evidence, the southern continent split first, breaking into Australia, Antarctica, Africa, and South America. The splitting of North America and Eurasia soon followed. Slowly the continents drifted apart, taking once-tropical regions toward the poles.

It may have been this extraordinary geological event and the resultant change in world climates that caused the rapid extinction of many of the great reptile species and allowed the new mammals to compete more successfully.

Each isolated subcontinent, especially Australia, Africa, and Eurasia, constituted an isolated environment. Land bridges between them were narrow or nonexistent, and so the new mammals in each area evolved separately from those in other areas. This explains a fact so commonplace that we usually forget how startling it is: The animals of the southern continents are highly varied and seem "exotic" to northerners, whereas the animals of North America and Eurasia are more nearly uniform. The explanation, according to Kurten's interpretation, is that Australian, South American, and African evolution have run independently for the last 65,000,000 years because of continental drift, while North America and Europe and Asia were more closely and more recently connected by Atlantic and Pacific land bridges. Certain mammals, such as the wolf, for example, were thus familiar throughout Western civilization in the Northern Hemisphere at the time of the voyages of discovery; but the discovery of such creatures as the duck-billed platypus in Australia seemed astounding.

So the geological evolution of the planet itself may help determine the kind of life that evolves. Perhaps it is no coincidence that the time scale of geologic revolution is comparable to that of biological revolution. Life's response to constantly and sometimes "suddenly"* changing environments is to change itself.

* On a time scale of 1 million years.

In Africa only about 3,000,000 years ago a group of mammals related to apes developed brains bigger and more specialized than those of their fellow creatures. They learned how to utilize natural materials of many kinds and how to make tools. In only 1 million years they burst forth as tool-making, family-linked, environment-altering humans. Ten thousand years ago they began to live together in clusters of houses—farming communities—and they built societies based on cooperative ventures utilizing machines to produce food, tools, and other items prized for aesthetic reasons. Humans are now learning to alter and control their environment on a large scale and in a few decades may be able to tamper with planets as a whole. It appears possible that if they do not spread themselves over several planets soon, they will exterminate themselves either by bickering among their various factions or by a simple mistake in one of their attempts to control the environment (i.e., by altering the planetary atmosphere or temperature). Other intelligent species might then rise to replace humans in another million years, unless the humans make such a disastrous mistake that their planet is rendered uninhabitable for all forms of life.

The purpose of emphasizing planetary evolution's controlling effect on biological evolution is to point out that terrestrial life forms and behavior patterns may not have any cosmic or universal validity. Feathers and fur and herd instincts and sex and seeds and what we recognize as intelligence may be products of the earth specifically and not biochemical evolution in general throughout the universe.

There are three more interesting consequences of life's planetary heritage. The first is the realization by many people of our dependence on the earth, which in turn has led to the current ecological revolution. The photographs and verbal descriptions by astronauts of the earth floating in space heightened the awareness of many that we are all passengers on what has been called the spaceship earth. We are beginning to realize that although we can use technology to make our existence more pleasant and even to transport ourselves to other worlds, we cannot employ it mindlessly on an ever-expanding scale without the dangerous consequence of altering the planet and hence evolution and hence our species. As René Dubos (1968) noted in his Pulitzer Prize-winning book on man's condition, *So Human an Animal,* man is umbilical to earth. This has been used as an argument that man will never be able to colonize other planets since the earth is the only planet for which man is adapted. There is no evidence that this is any more valid than arguing that man could not move from Asia to North America because of the new predators, diseases, and climate he would encounter. But it does mean that man will have to learn better how to live with his technology, for he will have to take his technology with him. The places in the universe must be very rare where we can walk away from our machines in our bare skin, breathe in the atmosphere, let the light of the nearest star fall on us, and find water to drink on the surface of the ground.

A second consequence of planetary evolution is that it demonstrates the great adaptability of life. Whoever has difficulty imagining the possibility of life on *other* worlds should ponder that the earth on which life initially evolved was an alien planet that would be hostile and poisonous to human beings. Life on earth developed from a hydrogen-rich atmosphere and then survived the radical transformation to an oxygen-rich atmosphere. Man himself has survived ice ages and thrives in regions from equatorial jungles to arctic plains and 5000-m (16,000-ft) Andean summits. Terrestrial life in its extremes survives much greater ranges of environment than man. Although man cannot survive unless his body temperature is within the range 303 to 313°K (a 3 percent variation), microflora are known to live in Antarctic ponds that remain liquid at 228°K (−49°F) due to dissolved calcium salts, and bacteria are known in Yellowstone hotsprings at temperatures of 363°K (194°F) (Sagan, 1970). These temperature extremes correspond to a 46 percent variation. Bacteria exist at altitudes where the atmospheric pressure is only about 0.2 atm, while more advanced organisms are known at ocean depths with pressures of hundreds of atmospheres (Sagan, 1970). These extremes range over a factor of a thousand in pressure. In Darwinian terms, life manages to adapt by natural selection to a tremendous variety of environments.

The third consequence of the planet's effect on biology is that in our search for life outside the earth we may be presumptuous to imagine alien intelligence with behavior patterns remotely similar to ours. If mushrooms and mammals and corals all evolved on one planet, how much greater may be the differences between life forms from different planets? This is a point we shall consider in more detail below.

EXTRATERRESTRIAL LIFE IN THE SOLAR SYSTEM?

Origin and Evidence of Early Life in Meteorites?

Earlier we sketched a sequence of events that probably occurred on the early earth and led to the evolution of life. The Miller experiment, as originally conceived, showed that amino acids could have originated in the early terrestrial environment. Meteoritic studies, however, have produced evidence that this very early step in the evolution of life may actually have occurred in a preterrestrial environment, in meteorite parent bodies. If this conclusion is correct, we may be led to the view that life has a universal origin in space among countless planetesimals as well as planets, and that organic materials or even living systems are delivered to planetary surfaces by meteorite impacts, after which evolution takes the course described earlier.

The evidence comes principally from carbonaceous chondrite meteorites (see Chapter 9). The first one recorded fell on March 15, 1806, in Alais, France. In 1834 the chemist Berzelius examined it and noted that its carbonaceous compounds resembled the products of decomposition of biological material. He was first to ask, "Does this possibly give a hint concerning the presence of organic structures in other planetary bodies?" (quoted by Urey, 1966).

For the next hundred years various researchers studied carbonaceous chondrites and confirmed that the organic materials resembled biogenic matter. But whether such materials were truly the product of life and, if so, whether they were native to the meteorite or terrestrial contaminants—these were controversial questions.

In 1961 Nagy and his co-workers began a series of reports in which they concluded that the carbonaceous chondrites contained native evidence of biogenic processes (see the review of these studies by Urey, 1966). Their papers indicated that the meteoritic organic material was similar to that found in terrestrial fossils and that it contained microscopic, highly organized particles that appeared superficially to be fossil microorganisms. These particles were of the order 0.01 mm in diameter and were called "organized elements" by the investigators. This work was widely criticized. The organic material did not contain certain optical properties that characterize terrestrial biogenic material. F. W. Fitch and E. Anders showed that some of the "organized elements" were probably tiny pollen grains that had contaminated the meteorite samples. Others of the "organized elements" could be explained as nonbiogenic complex mineral grains.

The presence of biogenic material in meteorites has not yet been universally accepted. Because it has such profound implications, scientists demand more definitive evidence.

One of the most difficult problems is that terrestrial biological material may contaminate the meteorite before tests can be made. For this reason, new falls of carbonaceous chondrites are the most prized specimens. A carbonaceous chondrite that fell at Murchison, Australia, in 1969 has given the most compelling evidence so far that the early steps in the evolution of life occurred in the ancient meteorites. Kvenvolden et al. (1970) found amino acids in this meteorite and then proceeded to show that the molecules were structurally different from those found in terrestrial biological matter. This determination can be made through studies of *optical rotation,* a property by which the plane of polarization in a polarized, transmitted beam of light is altered. In terrestrial amino acids, the molecular structure has a consistent symmetry, or "right-handedness," which rotates the beam only in one direction. The amino acids in the Murchison meteorite, however, were found to be evenly divided between "right-handed" and "left-handed" structures, a situation unknown in terrestrial biogenic products. This discovery has been widely heralded as the first concrete evidence of biogenic material outside the earth.

Life on Other Planets in the Solar System

The possibility that the first steps in the evolution of life—the production of amino acids—occurred in extraterrestrial planetary or preplanetary bodies does not establish the existence of substantial life forms on other planets. However, some knowledge of the first environments of biochemical activity and of the early stages of evolution of terrestrial life makes it easier to predict the extent of biological activity elsewhere in the solar system. First, we can rule out anything beyond complex organic molecules on planets without atmospheres, because such planets cannot sustain the liquid surface water needed to support the advance of life. Mercury, for example, has probably always been too hot to contain liquid water near the surface. The lunar rocks are very deficient in water. Although some meteorites give evidence of small amounts of internal liquid water at some stage in their history, the evidence of meteorites such as the Murchison meteorite shows that biochemical evolution did not get far in the planetesimals. An interesting question is thereby raised, however, about conditions on the satellites of the giant planets, at least one of which contains a CH_4-rich atmosphere and probably abundant ices. Speculation suggests that biochemical evolution on them could have proceeded further than in the meteorite parent bodies *if* internal heating or other heat sources were significant in the past. The low temperatures that apply now virtually exclude the possibility of life at present on these bodies.

Venus appears to be deficient in water and has such high surface temperatures that liquid water and life forms are ruled out.

Mars is small and has a weak gravitational field. Thus it lost its atmosphere and its volatiles more rapidly than the earth. The atmosphere is mostly CO_2, but there is probably some oxygen and a small amount of water. Recalling the tremendous adaptability of life on earth, from ocean depths to polar wastes, biologists are reluctant to dismiss the possibility that the *seasonally variable* dark markings of Mars may be indications of simple life forms (Horowitz, 1966). A leading alternative explanation of the Martian dark markings, however, is that they are patterns created by blowing dust, which is known to be present. The past Martian environment may have been more hospitable to life. More data on Mars' surface are obviously needed, and the failure of the Soviet Mars-landers in 1971 puts these data some years in the future.

The Jovian planets have atmospheres poisonous to present-day terrestrial life forms and hence have frequently been dismissed as abodes of advanced life. Nonetheless, it should be noted that these atmospheres are similar in composition to the primeval earth atmosphere; specifically they are rich in C, H, O, and N. At certain levels in these atmospheres, the temperature would be appropriate and there would probably be energy sources sufficient for the formation of organic molecules, as has been pointed out by Sagan (1970). The evolutionary stage that could be reached by airborne organic systems is uncertain, but Sagan and Khare (1971) have suggested that certain organic compounds in Jupiter's atmosphere could be responsible for the observed colors there.

It is thus generally believed that the earth is the only planet in the solar system with life forms close to the complexity of man.

ALIEN LIFE AMONG THE STARS?

Earlier we referred to highly speculative estimates that 1 to 50 percent of all stars may have planets and that perhaps 6 percent of all stars have habitable planets. Our review of the origin of life on earth, plus our knowledge of the tremendous adaptability of life, suggests that in most cases where conditions are suitable, biochemical evolution will begin.

The fraction of all stars having planets on which life has evolved may thus be on the order of 5 percent. This estimate probably reflects current thinking among astronomers and biologists, although estimates a decade or so ago varied widely. Britain's Astronomer Royal, Sir Harold Spencer Jones (1940), mentioned a figure of one star in 1000 or one in 1 million (0.1 to 0.0001 percent). Harvard astronomer Harlow Shapley (1958) also estimated 0.0001 percent. Dole (1964), concluding that life would evolve wherever temperature, pressure, and other conditions were suitable, estimated that 5 percent of all stars harbored life. Shklovskii and Sagan, the Russian and American astronomers who collaborated on the 1966 book *Intelligent Life in the Universe,* estimated that the majority of stars probably have planets, and that most planets probably have life; they estimated that life-bearing planets are attendant to at least 10 percent of all stars.

Such figures are highly speculative. Possibly our estimate of 5 percent may be within an order of magnitude of the truth. Our figures so far have referred only to the fraction of planets that have life, not the fraction that have intelligent life or structured societies. "Life" means evolved organisms of some sort, whether bacteria or grass, mollusks or dinosaurs, man or something beyond man and unrecognizable. Let us now consider the possibility of alien intelligent life. "Intelligent life" means life forms that have learned to fabricate tools and communicate. This is admittedly a vague definition, but in our case the interval from the first use of tools to the first departure into space was less than 5,000,000 years, or about 0.1 percent of the history of our planet. So our definition of "intelligent life" distinguishes rather sharply, at least in time sense, between nonintelligent and intelligent life.

The famed nuclear physicist, Enrico Fermi, reportedly interrupted a dinner party in 1943 to ask, "Where are they?" He was referring to the puzzling absence of visitations by aliens. If 5 percent of stars have life, that would make 5 billion inhabited planets in our galaxy. That would put the nearest planet with life only 10 to 20 light-years away—only four to five times as far as the nearest star. If there are planets with life that close by, we might expect some of them to have intelligent life and civilizations that would have attempted interstellar communication or travel.

In evaluating such an assertion, the real problems are whether intelligence is a normal product of biochemical evolution and whether it commonly leads to interstellar travel. Judgments on this point amount to sheer speculation, because we have only one case in point—the human race. One case is not enough for good statistics.

It is possible that biological evolution does not generally produce creatures who have a desire to build cities or travel through space. It is not obvious that human societies necessarily evolve toward a predestined sort of technocracy. Although man is universally a tool-using animal, it is not certain that even he universally has the drive or inclination to produce a technologically oriented culture. Are humans more necessarily fated to be explorers, bridge-builders, and businessmen than to be artists, athletes, or daydreamers? Is the stereotyped aggressive Westerner more representative of the essence of man than the stereotyped contemplative Easterner? Our experience suggests that our aggressive "Western" sort of technocracy may be just one type of *cultural* activity rather than a universally achieved stage of *biological* evolution.*

If humanity was never predestined to develop a technological civilization, how much less certain is the course of social development on other worlds! It is absurdly ethnocentric of us to suppose that beings on other planets would resemble us physically, psychologically, or socially. Consider the variety of highly evolved life forms on our planet alone. Ants live in ordered societies that do not regard individual survival as important. Dolphins communicate and have brains apparently almost comparable to ours but they have no manipulative organs and hence no technology. In the absence of man, would insects evolve into "intelligent" societies? We could scarcely expect aliens to be motivated by emotions that would mean much to us. We could scarcely expect, as always happens in grade-C science fiction movies, that man-like aliens will walk out of saucers and invite us to join their democratically constituted United Planets, a galactic organization structured by documents that are curiously reminiscent of the United States Constitution.

Thus one possible answer to Fermi's question—if our limited experience can teach us anything about alien life—is that aliens may be nearly incomprehensible to us at first contact. Another possible answer to Fermi's question is that technological cultures arise but destroy themselves almost as quickly as they form. One supposes that there may be a sort of hurdle for civilizations to jump—certainly there is for man: Either he must learn to control his aggressive instincts or he will wipe himself out within the next few thousand years. Shklovskii and Sagan (1966) explicitly propose two kinds of advanced

* It has been historically difficult for man to distinguish his man-made cultural patterns from what he assumes to be externally, biologically—or morally—imposed patterns.

civilizations: the ones that destroy themselves after about a hundred or a thousand years and the ones that pass the hurdle (perhaps by colonizing the universe from planet to planet before they can wipe themselves out?), continue to evolve, and last perhaps millions of years.

A third possible answer to Fermi's question is that aliens have already visited the earth but that we have not recognized them. This possibility seems unlikely for the following reasons. Although thousands of reports of "unidentified flying objects" in the last two decades have been alleged to be evidence of alien visitors by certain writers of popular books, a two-year scientific study has shown that there is no good evidence to support this (Condon, 1969). In the first place, the vast majority of the reports can be accounted for as mistakes or hoaxes, and in the second place, the few remaining reports are not very different from the kinds of unexplained observations we would expect simply as a result of the variety of natural phenomena. Furthermore, if we are generous and suppose that a *real* alien visit would be documented if it had occurred any time in the last thousand years, we must note that 1 thousand years is only 2×10^{-7} (0.00002 percent) of the history of the earth. In other words, the hypothetical aliens would have to pick a very narrow "time slot," and to be realistic it is unlikely that we in the twentieth century would be able to distinguish a real, historical visit out of all the ancient legends and modern UFO reports (Shklovskii and Sagan, 1966; Condon, 1969). Shklovskii and Sagan (1966) have surveyed the archeological and mythological literature to search for possible definitive evidence for visits by aliens or alien spaceships, but they could find no compelling evidence.

A fourth possible explanation for the lack of detectable alien visits is that the frequency of alien races may be much lower than we have estimated. Thus the stars harboring life would on the average be much farther away from us. If only one in 1 million stars had intelligent life, the nearest such stars would probably be several hundred light-years away. If the contemporary theory of relativity is correct, nothing can travel faster than light, and hence such aliens could not reach us without journeys of more than several hundred years. Such journeys may discourage interstellar travel, but they are not impossible; science-fiction writers long ago proposed giant starships on which many generations of passengers live and die before the destination is reached.

A fifth and probably the most significant explanation for the lack of obvious alien visits is that alien races may be separated from us farther *in time than in distance*. Biological evolution is so persistently experimental that even if another planet started evolving at exactly the same time as ours, and even if its biochemistry produced creatures like us, it is unlikely that those creatures would today be in a phase of evolution similar to ours. If the evolutionary "clocks" on our planet and their planet got only 0.02 percent out of synchronization, they would be 1 million years either ahead of us or behind us. In other words, even in the unlikely event that other planets produce civilizations recognizable to us, we would have to contact one of those civilizations in a very narrow time interval in order to see any recognizable common interests.

Evolution may pass through only a brief *explorative interval* in which societies on one planet would care to reach other planets; beyond that stage communication or space exploration might be no more attractive than a national program on our part to communicate with chimpanzees, ants, or dolphins. To be sure, a few of our scholars attempt this, but they "contact" an infinitesimal fraction of these lower creatures. How many ant hills or dolphin schools have we humans visited? By the same token our solar system might be ignored by advanced aliens. Aliens 1 million years ahead of us might be no more interested in us than we are in ants.

How long might an explorative interval last? Man has used tools for about 2 million years, and it appears safe to assume that he will have progressed far beyond recognizable technology in another million years, if he survives. Our explorative interval might be a few million years, then, or less than 0.1 percent of the history of the planet. Admittedly this is only an estimate of the situation, but it suggests that only one out of 1000 inhabited planets would be expected to have creatures with even remotely comprehensible technology or more than passing interest in the earth.

On this speculative basis we could, using our prediction of life in 5 percent of star systems, predict that of all the stars, only 0.005 percent, or 5 stars in 100,000, could harbor a civilization potentially interested in visiting us. If we arbitrarily suppose that out of 100 planets with life, only one experiences animal life like ours, the figure would be reduced to 0.00005 percent.

Other current estimates of this figure vary. Shklovskii and Sagan (1966) used somewhat different reasoning in their calculation but derived a similar figure of 0.001 percent. The Soviet scientist Tovmasyan (1965) lists estimates of less than 0.0001 percent. The Irish astronomer E. J. Öpik (1967) derived a much smaller figure, about 0.0000003 percent. His figure was lower primarily because of his belief that only a tiny fraction of stars would have an earth-sized planet in a temperate, stable circular orbit unperturbed by more massive stars or planets in the system.

If these estimates are right, the distance to the nearest civilization at a level remotely resembling ours is probably several hundred or several thousand light years. In the Milky Way galaxy there would be on the order of 1 million civilizations with whom we could communicate, but it would take at least several hundred years to get a radio message to the nearest one or travel there at the speed of light. If this speculation is correct, we are not alone in the universe, but we are for the time being isolated.

Our period of isolation may be nearing an end. For half a century we have been broadcasting radio communications. Already our unintentional radio alert is more than 50 light-years out from the earth. Attempts by radio astronomers to pick up possible radio responses from aliens have turned up nothing. But this does not prove that aliens may not one day pick up our signals and send either radio messages or an expedition in return.

In any case, it seems likely that our first contact with aliens, if they exist, might be as incomprehensible as the dramatized contact in the closing segment of the film "2001."

SUMMARY

We have repeated many times that much of this chapter has been speculation. Indeed, the study of possible extraterrestrial life, known as *exobiology,* has been called a science without any subject matter. Exobiology has been criticized in the spirit of Mark Twain's comment, "There is something fascinating about science. One gets such wholesale returns of conjecture from such a trifling investment of fact." The only way to reduce the conjecture and increase the amount of our knowledge is to pursue research in planetary science and the many related fields: physics, chemistry, geology, meteorology, biology. In pursuing the question of alien life, even sociology and psychology become areas for fertile interdisciplinary study.

We know many of what might be called first-order facts about the universe. We may even claim to know a good deal of second-order or "second-significant-figure" information. But there may still be some revolutionary surprises waiting for us out there, and in any case we can be sure that there is a great deal more for us to learn about ourselves, our planet, and the universe.

References

Ambartsumian, V. A. (1965) Introduction in *Extraterrestrial Civilizations,* Akademii Nauk Armyanskoi S.R., translated by Israel Program for Scientific Translations.

Berkner, L. V., and L. C. Marshall (1965) "On the Origin and Rise of Oxygen Concentration in the Earth's Atmosphere," *J. Atmospheric Sci., 22,* 225.

Brown, H. (1964) "Planetary Systems Associated with Main-Sequence Stars," *Science, 145,* 1177.

Condon, E. V., director (1969) *Scientific Study of Unidentified Flying Objects* (New York: E. P. Dutton & Co., Inc.).

Dole, S. H. (1964) *Habitable Planets for Man* (Waltham, Mass.: Ginn/Blaisdell).

Dubos, René (1968) *So Human an Animal* (New York: Charles Scribner's Sons).

Gaustad, J. E. (1963) "The Opacity of Diffuse Cosmic Matter and the Early Stages of Star Formation," *Astrophys. J., 138,* 1050.

Hartmann, W. K. (1970) "Star Formation in Clusters and the Stellar Mass Function," Proc. 1969 Liege Conference on Pre-Main Sequence Stellar Evolution.

Holland, H. D. (1962) "Model for Evolution of the Earth's Atmosphere," in *Petrologic Studies* (New York: Geological Society of America).

Horowitz, N. H. (1966) "The Search for Extraterrestrial Life," *Science, 151,* 789.

Huang, S. (1965) "Rotational Behaviour of the Main-Sequence Stars and Its Plausible Consequences Concerning Formation of Planetary Systems," *Astrophys. J., 141,* 985.

Jones, H. S. (1940) *Life on Other Worlds,* p. 447 (New York: The Macmillan Company).

Kuiper, G. P. (1952) "Planetary Atmospheres and Their Origin," in *The Atmospheres of the Earth and Planets,* G. P. Kuiper, ed. (Chicago: University of Chicago Press).

Kurten, Björn (1969) "Continental Drift and Evolution," *Sci. Am., 220*(3), 54.

Kvenvolden, K., J. Lawless, K. Pering, E. Peterson, J. Floras, and T. Hagfors (1970) "Evidence for Extraterrestrial Amino Acids and Hydrocarbons in the Murchison Meteorite," *Nature, 228,* 923.

Miller, S. L. (1955) "Production of Some Organic Compounds under Possible Primitive Earth Conditions," *J. Am. Chem. Soc., 77,* 2351.

O'Leary, B. T. (1966) "On the Occurrence and Nature of Planets Outside the Solar System," *Icarus, 5,* 419.

Oparin, A. I. (1962) *Life: Its Nature, Origin, and Development* (New York: Academic Press, Inc.).

Öpik, E. J. (1967) "Life and Intelligence in the Universe: Bottomless Speculations," *Irish Astron. J., 8,* 128.

Pfeiffer, John (1955) *The Human Brain* (New York: Pyramid Publications).

Sagan, C. (1970) "Life," *Encyclopedia Britannica.*

Sagan, C., and B. N. Khare (1971) "Experimental Jovian Photo-chemistry—Initial Results," *Astrophys. J., 168,* 563.

Shapley, H. (1958) *Of Stars and Men* (Boston: Beacon Press).

Shklovskii, I. S., and C. Sagan (1966) *Intelligent Life in the Universe* (San Francisco: Holden-Day, Inc.).

Tovmasyan, G. M. (1965) Preface to *Extraterrestrial Civilizations,* Akademii Nauk Armyanskoi S.R., translated by Israel Program for Scientific Translations.

Urey, H. C. (1952) *The Planets: Their Origin and Development* (New Haven, Conn.: Yale University Press).

______ (1966) "Biological Material in Meteorites: A Review," *Science, 151,* 157.

van de Kamp, P. (1961) "Double Stars," *Publ. Astron. Soc. Pacific, 73,* 389.

______ (1963) "Astrometric Study of Barnard's Star from Plates Taken with the Sproul 24 inch Refractor," *Astron. J., 68,* 515.

______ (1969) "Alternate Dynamical Analysis of Barnard's Star," *Astron. J., 74,* 757.

PLANETARY DATA[a]

appendix

Object	Diameter (km)	Mass (g)	Mean density (g/cm³)	Visual geometric albedo	Rotation period (days)	Obliquity	Revolution period (days unless noted otherwise)	Semimajor axis (A.U. for planets; 10³ km for satellites)	Orbit inclination (with respect to ecliptic for planets; planetary equator for satellites)	Orbit eccentricity
Sun	1,391,400	1.987 (33)	1.4	—	25.4	7°.25	—	—	—	—
Mercury	4,864	3.30 (26)	5.5	0.10	58.6	<7°	0.2408 yr	0.387	7.0	0.206
Venus	12,100	4.87 (27)	5.2	0.586	243*R*	~179°	0.6152 yr	0.723	3.39	0.007
Earth	12,756	5.98 (27)	5.52	0.39	1.00	23°.5	1.000 yr	1.000	0.00	0.017
Mars	6,788	6.44 (26)	3.9	0.15	1.02	25°.0	1.881 yr	1.524	1.85	0.093
Jupiter	137,400	1.90 (30)	1.40	0.44	0.41	3°.1	11.86 yr	5.203	1.31	0.048
Saturn	115,100	5.69 (29)	0.71	0.46	0.43	26°.7	29.46 yr	9.54	2.49	0.056
Uranus	50,100	8.76 (28)	1.32	0.56	0.45*R*	97°.9	84.0 yr	19.18	0.77	0.047
Neptune	49,400	1.03 (29)	1.63	0.51	0.6	28°.8	164.8 yr	30.07	1.78	0.008
Pluto	5,800	6.6 (26)	6?	0.13	6.4	?	284.4 yr	39.44	17.17	0.249
Moon	3,476	7.35 (25)	3.34	0.115	27.3	6°.7	27.3	384	18–29°	0.055
Phobos	18 × 22	?	?	0.06	?	?	0.319	9	1°.1	0.021
Deimos	12 × 13	?	?	0.06	?	?	1.26	23	1°.6	0.003
(1) Ceres	800	1.2 (24)?	4.5?	0.1?	0.38	?	1,681	2.767 A.U.	10°.6	0.079
(2) Pallas	490	?	?	0.1?	0.4?	?	1,684	2.767 A.U.	34°.8	0.235
(3) Juno	250	?	?	0.2?	0.30	41°?[b]	1,594	2.670 A.U.	13°.0	0.256
(4) Vesta	490	2.4 (23)?	3.9?	0.25?	0.22	25°?[b]	1,325	2.361 A.U.	7°.1	0.088
JI Io	3,500	8.0 (25)	3.6	0.92	1.769	?	1.769	422	0°.03	<0.01
JII Europa	3,100	5.0 (25)	3.2	0.83	3.551	?	3.551	671	0°.5	<0.01
JIII Ganymede	5,000	1.65 (26)	2.5	0.49	7.155	?	7.155	1,070	0°.2	<0.01
JIV Callisto	4,900	1.02 (26)	1.7	0.26	16.689	?	16.689	1,883	0°.3	<0.01
JV	170?	?	?	?	?	?	0.498	181	0°.4	0.003
JVI	130?	?	?	?	?	?	250	11,470	28°	0.158

JVII	44?	?	?	?	?	?	260	11,740	26°	0.206
JVIII	12?	?	?	?	?	?	737	23,500	33°R	0.40
JIX	14?	?	?	?	?	?	758	23,700	25°R	0.27
JX	14?	?	?	?	?	?	255	11,850	28°.5	0.135
JXI	16?	?	?	?	?	?	692	22,560	16°.5R	0.207
JXII	12?	?	?	?	?	?	631	21,200	33°R	0.16
SI Mimas	900?	3.7 (22)	0.1?	0.49	?	?	0.942	186	1°.5	0.020
SII Enceladus	550	7.2 (22)	0.8?	0.54	?	?	1.370	238	0°.02	0.004
SIII Tethys	1,200	6.6 (23)	0.7	0.84	?	?	1.887	295	1°.1	0.0
SIV Dione	820	1.03 (24)	3.6	0.94	?	?	2.737	377	0°.02	0.002
SV Rhea	1,300	1.5 (24)	1.3	0.82	?	?	4.518	527	0°.3	0.001
SVI Titan	4,850	1.37 (26)	2.3	0.21	15.95	?	15.95	1,222	0.3	0.029
SVII Hyperion	350?	?	?	?	?	?	21.28	1,481	0.5	0.104
SVIII Iapetus	1,150	1.5 (24)	1.9	?	?	?	79.33	3,560	14.7	0.028
SIX Phoebe	260?	?	?	?	?	?	550.4	12,950	30R	0.163
SX Janus	370?	?	?	?	?	?	?	160?	~0	~0
UI Ariel	1,470?	?	?	?	?	?	2.520	192	0.0	0.003
UII Umbriel	960?	?	?	?	?	?	4.144	267	0.0	0.004
UIII Titania	1,760?	?	?	?	?	?	8.706	438	0.0	0.0024
UIV Oberon	1,600?	?	?	?	?	?	13.463	586	0.0	0.0007
UV Miranda	550?	?	?	?	?	?	1.414	128	0.0	<0.01
NI Triton	3,800	1.4 (26)	4.9	0.36	?	?	5.877	353	20.1R	0.0
NII Nereid	540?	?	?	?	?	?	360	5,600	27.5	0.76

[a] Number of significant figures reflects accuracy to which datum is known, although some of the figures are known to more than four significant figures. Entries are based on C. Allen (1963) *Astrophysical Quantities* (London: Athlone Press), with revisions based on new results through 1971. Diameters and masses are reviewed by A. Dollfus, ed. (1970), *Surfaces and Interiors of Planets and Satellites* (New York: Academic Press, Inc.), where observational results are critically discussed. Asteroid data include results given at the International Astronomical Union Colloquium 12 (1971). Jupiter satellite data are reviewed by T. Johnson (1971), *Icarus, 14,* 94. The most recent mass determination for Pluto is reported by P. Seidelmann, W. Klepczynski, R. Duncombe, and E. Jackson, *Astron. J., 76,* 488. The obliquity of Venus is given in Dollfus (1970, see above).

[b] Angle from equator to ecliptic.

INDEX